Ian McHarg and the Search for Ideal Order

Ian McHarg and the Search for Ideal Order looks at the well-known and studied landscape architect, Ian McHarg, in a new light. The author explores McHarg's formative years and investigates how his ideas developed in both their complexity and scale. As a precursor to McHarg's approach in his influential book *Design with Nature*, this book offers new interpretations into his search for environmental order and outlines how his struggle to understand humanity's relationship to the environment in an era of rapid social and technological change reflects an ongoing challenge that landscape design has yet to fully resolve. This book will be of great interest to academics and researchers in landscape architectural history.

Kathleen John-Alder is Associate Professor in the Department of Landscape Architecture at Rutgers University. A practicing landscape architect with degrees from Oberlin College, Rutgers University, and Yale School of Architecture, her scholarly research bridges disciplinary boundaries in order to explore the transformative role of ecology and environmentalism in the discourse of mid-twentieth century landscape design, and its impact upon contemporary practice. Kathleen has published articles in *Landscape Journal, The Journal of Planning History, JOLA, Studies in the History of Gardens and Designed Landscapes, Site/Lines,* and *Manifest.* Her work has also received design and research awards from the Van Alan Institute, the National Park Service, and the American Society of Landscape Architects.

Routledge Research in Landscape and Environmental Design
Terry Clements, Associate Professor, Virginia Tech

Routledge Research in Landscape and Environmental Design is series of academic monographs for scholars working in these disciplines and the overlaps between them. Building on Routledge's history of academic rigor and cutting-edge research, the series contributes to the rapidly expanding literature in all areas of landscape and environmental design.

Cultural Landscapes of South Asia
Studies in Heritage Conservation and Management
Kapila D. Silva and Amita Sinha

A History of Groves
Edited by Jan Woudstra and Colin Roth

Reimagining Industrial Sites
Changing Histories and Landscapes
Catherine Heatherington

Landscape Performance
Ian McHarg's Ecological Planning in The Woodlands, Texas
Bo Yang

Desert Paradises
Surveying the Landscapes of Dubai's Urban Model
Julian Bolleter

Ian McHarg and the Search for Ideal Order
Kathleen John-Alder

For more information about this series, please visit: https://www.routledge .com/

Ian McHarg and the Search for Ideal Order

Kathleen John-Alder

Routledge
Taylor & Francis Group

LONDON AND NEW YORK

First published 2020
by Routledge
2 Park Square, Milton Park, Abingdon, Oxon OX14 4RN

and by Routledge
605 Third Avenue, New York, NY 10017

First issued in paperback 2022

Routledge is an imprint of the Taylor & Francis Group, an informa business

© 2020 Kathleen John-Alder

The right of Kathleen John-Alder to be identified as author of this work has been asserted by her in accordance with sections 77 and 78 of the Copyright, Designs and Patents Act 1988.

Publisher's Note
The publisher has gone to great lengths to ensure the quality of this reprint but points out that some imperfections in the original copies may be apparent.

British Library Cataloguing-in-Publication Data
A catalogue record for this book is available from the British Library

Library of Congress Cataloging-in-Publication Data
Names: John-Alder, Kathleen, author.
Title: Ian McHarg and the search for ideal order /
Kathleen John-Alder.
Description: Milton Park, Abingdon, Oxon;
New York, NY: Routledge, 2020. |
Series: Routledge research in landscape and environmental design |
Includes bibliographical references and index.
Identifiers: LCCN 2019018453 | ISBN 9781138681729
(hbk: alk. paper) | ISBN 9781315545639 (ebook)
Subjects: LCSH: McHarg, Ian L., 1920–2001. | Landscape
architects—United States—Biography. | Landscape architecture.
Classification: LCC SB470.M33 J64 2020 |
DDC 712.092 [B]—dc23
LC record available at https://lccn.loc.gov/2019018453

ISBN: 978-1-03-240118-8 (pbk)
ISBN: 978-1-138-68172-9 (hbk)
ISBN: 978-1-315-54563-9 (ebk)

DOI: 10.4324/9781315545639

Typeset in Sabon
by codeMantra

Contents

Figures

Acknowledgements

I thank the following individuals for their support and encouragement. William Whitaker, Nancy Thorne, and Isabel Schumacker of The Architectural Archives of the University of Pennsylvania provided patient guidance. Eeva Liisa Pelkonen instilled discipline and helped me find my voice. Peter Morin explained the intricacies of food webs and energy. Janike Kampevold Larsen, Anatole Tchikine, John Beardsley, Fritz Steiner, and Laura Lawson supplied kind, but critical suggestions. Robert Geddes, William Roberts, Laurie Olin, Dennis McGlade, Carol Franklin, and Leslie Sauer graciously provided insight into McHarg's early career. Jahangir Sedaghatfar, Nader Ardalan, Robinson Fisher, Brian Spooner, Victoria Steiger, Jon Coe, Neal Belanger, Grant Jones, and Dennis Paulson unraveled the history of Pardisan. But my greatest debt is to my husband Henry John-Alder, who accepted the disruption to our lives and the extended loss of our dining room table (and living room couch) to accommodate the piles of material I accumulated as I wrote this book.

Introduction

Ian McHarg (1920–2001) was a skilled designer, prolific writer, and influential teacher. He served as chair of the Department of Landscape Architecture at the University of Pennsylvania for over thirty years and was a founding partner of the firm Wallace, McHarg, Roberts and Todd (WMRT). Throughout his career, he positioned landscape architects as accomplished artist-technicians equipped with the imagination, skill, and force of mind to craft naturally beautiful and environmentally superior blueprints for others to follow. To advance this agenda, he developed a meticulous method of investigation that required designers to deconstruct and map the landscape's physical and cultural patterns prior to formulating a response. The procedure was synthetic, systematic, experimental, and all about order. As a scholarly enterprise, it provided a basic understanding of the biophysical history of the land, which linked design to science, facts, and empirical observation. As an applied, problem-solving practice, it forced designers to select, organize, evaluate, and compare, which made the procedure malleable and somewhat ambiguous in objective.[1] McHarg argued that his design method was a tool of discovery and a platform for action that honored the landscape and shunned preconceived notions of form.[2] To justify these claims, he wove allegorical tales of the world as an animate being whose health and welfare paralleled human existence. When people mistreated the land, the stories illustrated how they also harmed themselves. His talent for analytic thinking, combined with his flair for storytelling made him uniquely positioned to advance the critical role of landscape and landscape architecture in design discourse.[3] He pursued both passions with edifying fervor and in the process became perhaps the most famous landscape architect of the 20th century (Figure I.1).

McHarg's notoriety was due in large part to his environmental manifesto *Design with Nature*.[4] He published the book in 1969, but the groundwork for the argument occurred earlier in his career. In tone, the writing captured the confidence of the era – its material prosperity, technological marvels, and optimism – and the attendant realization that the unanticipated by-products of this brash sense of power – rampant consumerism, industrial pollution, and loss of contact with the natural world –

Figure I.1 Photograph of Ian McHarg (n.d.). The Ian and Carol McHarg Collection, The Architectural Archives of the University of Pennsylvania.

jeopardized the future. Similar to the argument presented by Rachel Carson in *Silent Spring*, McHarg wrote *Design with Nature* to bear witness to the importance of natural rhythms and processes in daily life, and to explain the consequences when they were destroyed.[5] And like Carson, he called upon values deeply ingrained in the American consciousness to illustrate what was at stake before the country thoughtlessly edged into environmental catastrophe. Charles Eames, one of McHarg's professional collaborators, described the situation as follows:

> The scary fact is that many of our dreams have come true. We wanted a more efficient technology and we got pesticides in the soil. We wanted cars and television sets and appliances and each of us thought he was the only one wanting that . . . That doesn't mean the dreams were all wrong. It means that there was an error somewhere in the wish list and we have to fix it.[6]

McHarg skillfully deployed contrasting images – a tenement courtyard versus a country village, children in a playground versus a child crawling on a dirty row house stoop, aerial views of farmland versus aerial views of the city, a cluttered commercial street versus a forested hillside, a picturesque river valley versus a polluted industrial waterway – to highlight problems and suggest possible solutions. The dramatic images of the sun and the earth that grace the front and back cover of the first edition of *Design with Nature* set the stage for this visual argument. In addition to signaling the vast scale of his design project, they illustrate his deep reverence of the natural and his desire to connect to something larger and lasting. As physical and spiritual embodiments of the unity of science, art, and religion – the trinity of subjects that comprised McHarg's vision of design with nature – they speak to objective truth, imaginative unfolding, and spiritual redemption. The photograph of the sun, which came from the Hayden Planetarium at the American Museum of Natural History, represents immanence, reason, and the immutable power of natural laws that eclipsed time and history (Figure I.2). The photograph of the earth, which came from an orbiting

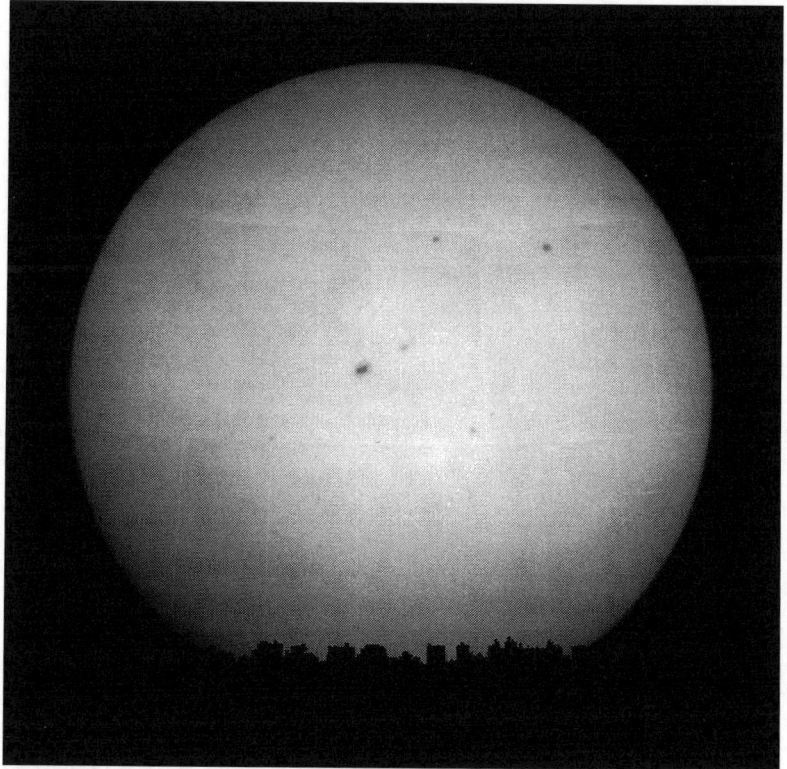

Figure I.2 Hayden Planetarium. Image # 320599, American Museum of Natural History. Reprinted by Permission.

weather satellite and also appeared on the cover of the *Whole Earth Catalogue*, represents biophysical processes, meaurement and quantification, and the understanding that even though it was impossible to fully know the natural world, the things that were known could enrich life[7] (Figure I.3). Together, they embody his hope that people would see the land and their actions towards it truthfully, and condemn the evils tacitly promoted by their inattention. The city skyline that shadows the lower edge of the sun expressed his fear that perhaps they may not.

Stories of McHarg's childhood in Glasgow, Scotland, which appear in the first chapter of *Design with Nature*, provide further insight into his hopes and fears and what he sought to achieve. At the beginning of this brief biography he stated: "I spent my childhood and adolescence squarely between two diametrically different environments, the poles of man and nature." The city offered the excitement of the railroad station, circus, theater, dances, concerts, art galleries, bagpipe parades, soccer matches, and the christening of the Queen Mary. Yet, as recounted by McHarg, these places and events were rare interludes in a city memorable for industrial

Figure I.3 November 10, 1967 satellite image of the earth from space. Courtesy of NASA.

production and soot-encrusted buildings. Emphasizing the worst, he called the city "a sandstone excrement cemented with smoke and grime. Each night its pall on the eastern horizon was lit by the flares of blast furnaces, a Turner fantasy made real."[8]

In contrast to his urban excursions, McHarg's unsupervised escapes to the countryside and the Black Woods were intimate, full of light, and teaming with life. A flashy American convertible and tales of war added exotic excitement and hints of his future:

> . . . the other path was always exhilarating and joy could be found in small events, the certainty of a trout seen in the shadow of a bridge, the salmon jumping or a stag glimpsed fleetingly, the lambing, climbing through the clouds to the sunlight above, a cap full of strawberries or blackberries, men back from the Spanish Civil War at the firepot or a lift from an American in a Packard convertible.[9]

The simple delights that he found in the countryside were both domestic and transcendent, and this, too, he wanted to share:

> The burn had familiar steppingstones, overhangs where small trout and red-breasted minnows lived, shaded by reeds, osiers and willows. Whitewashed stone farmhouses sat squarely with their outbuildings and old trees marked the ridges . . . Its gem was Peel Glen, for most of the year an unremarkable woodland, mainly beech, deep shadowed and silent, but in the Spring it was transformed. As you entered its shade there was no quick surprise – only slowly did the radiance of light from the carpet of bluebells enter and suffuse the consciousness.[10]

At the conclusion of this narrative, McHarg returned to the Black Woods and Peel Glen after facing and overcoming the challenges of war, Harvard, and tuberculosis. He expected time and experience would diminish the pleasure of his memories, but he instead discovered the woodland glade no longer existed. The ephemeral beauty of its bluebells had been replaced by much needed but uninspiring post-war housing surrounded by "sodden laundry" drying on "drunken chestnut poles."[11] What was once mysterious and magical was now mundane and ugly, and he elegiacally mourned the loss.[12]

As a coming of age tale, these stories deny a paradisiacal return to the garden. As a parable of nature and nurture, they suggest the innocent pleasures and adventurous freedoms of the countryside could counteract the enervating complexities of the city. As a form of political resistance, they call attention to the economic and social injustices he sought to reform. The recognition that the land shapes thinking and feeling is central to each of these messages. McHarg purposefully contrasted two ways of inhabiting the world – of being lost in an alienating terrain and being at home in well-known territory – to emphasize how the landscape of daily life impacts perception, thought, and action. Mirroring ideas expressed by Edith Cobb in "The Ecology of Imagination in Childhood," his adventures in the city and countryside outlined ways of seeing and knowing that oscillate between feelings of estrangement and discontinuity (what is in need of redemption and renewal) and feelings of connection and continuity (what is decent and good). Cobb claimed that the term "ecology" captured this give and take because it considered environmental interactions "not only in terms of human relations, but also in terms of man's total relations with otherness, with nature itself." She also wrote that an ecological perspective, particularly when viewed with childlike wonder, evokes playful associations that give further meaning to the phrases "the spirit of place" and "the genius loci."[13] The world offered joy and amazing possibilities, and the opening of one's life to the unexpected delights that come from its natural sights and sounds clearly inspired McHarg to write *Design with Nature*.

The decided preference for the countryside in his arguments has prompted the criticism that McHarg was anti-urban. While there is truth to this claim, it is also true, as his childhood stories indicate, that he sought an environmental synthesis – a city suffused with the light and life of the natural world and a countryside enriched by cultural and social opportunities. He would state in *Design with Nature*: "It is not a choice of either the city or countryside: both are essential, but today it is nature, beleaguered in the country, too scarce in the city which has become precious."[14] His professional projects adhered to this objective by illustrating how these landscapes and their contrasting ways of life, in consonance with his childhood memories, were inseparable. The title of the working draft of *Design with Nature*, which was *A Place for Nature in Man's World, or Man, Nature, Land and City*, signaled his allegiance to this claim.[15]

But perhaps most important, these childhood stories draw the reader into the text, and into McHarg's mind. As readers, we know very well that we are being compelled to see the world through his eyes, and through a form of clichéd storytelling. Nevertheless, the engaging way he recounts these adventures, particularly the joyous freedom of the countryside, makes it easy to suspend disbelief and willingly follow.

In 1971, *The Wall Street Journal* published a short article on McHarg that captured the essence of his personality.[16] The article reported that he loved dry martinis, good jazz, and lived in a stunning contemporary house adjacent to Fairmont Park with a backyard designed to look "exactly as nature itself might, given a century or two." It also captured the impulsive nature of his thoughts and how he raced through interviews, "his voice a blur, his hands gesturing, throwing off ideas like sparks," as he discussed how to create "a more humane, creative synthesis of man's work and nature's." Somewhat skeptically, the article noted his repeated insistence that his design method was scientific even though he had yet to prove it would "always give the same answer regardless of the planner using it."

A statement by former Secretary of the Interior Stuart Udall, who worked with McHarg during the administration of President Lyndon Johnson, testified to the importance of his work and to his tenacious efforts to ensure the natural characteristics of the land – "its network of streams, its underlying geology, its intricate life chains, its beauty" – received the consideration customarily granted to economics in development scenarios. In the same admiring spirit, the article noted that Lewis Mumford, McHarg's mentor and friend, had glowingly stated that *Design with Nature* revived "hope for the future."[17]

It is also telling that the by-line of the article asked: "Is Ian Just a Glib Showman?" In answer to this question, it was noted that his desire to put nature first had caught people's attention and he was respected for his accomplishments, but not always admired. A student interviewed for the article stated: "My first impression was that he was absolutely out of his mind," but I came to realize that "he had a much broader vision of scope than I did." Another student noted that he had "very strong ideas that he won't let go of," and a third stated, he could "talk on and on, overwhelming you with the verbal, and you don't realize that he hasn't thought it out." David Wallace, McHarg's partner at WMRT, reiterated these thoughts when he observed that Ian invited criticism by "overselling ecological determinism with his enthusiasm." Wallace did, however, observe that when he stepped away from the lecture podium he understood the necessity of compromise. The article ended on a positive note, entranced by its subject's heartfelt expression of passion for the land: "I found myself more moved by nature than anything else. My whole purpose became continuing this love affair and still making a living out of it."[18]

Design with Nature was considered a classic, and its author a minor celebrity, when *The Wall Street Journal* article was written. McHarg garnered profiles in *Life, Time, Readers Digest*, and *Smithsonian* magazines; served as host of the television program *The House We Live In*; appeared on the Mike Douglass Show; and starred in the documentary movie *Multiply and Subdue the Earth*.[19] In light of this high-profile publicity, he soon became a legend in his field, automatically referenced in discussions of environmental design and planning.

There are numerous essays and several important books on McHarg. Chief among these efforts is the work of Frederick Steiner, who assisted McHarg in the writing of his 1996 autobiography *A Quest for Life*, and the 1998 collation of papers titled *To Heal the Earth: The Selected Writings of Ian McHarg*. After McHarg's death in 2001, Steiner edited the text *The Essential Ian McHarg: Writings on Design and Nature*. In 2007, Lynn Margulis, James Corner, and Brian Hawthorne edited *Ian McHarg Conversations with Students: Dwelling in Nature*. This work consists of testimonials that explore the intellectual legacy of his work. *The Lost Tapes of Ian McHarg*, a CD that accompanies the text, allows readers to hear McHarg discuss his ideas. *Reconsidering Ian McHarg: The Future of Urban Ecology*, written in 2014 by his former student Ignacio Bunster-Ossa, examines the continuing relevance of his ideas.[20]

Building upon this rich legacy of scholarship, *Ian McHarg and the Search for Ideal Order* examines McHarg's formative years and follows his thinking as his ideas increase in scale and complexity and his design agenda moves from housing, to the city, region, and ultimately the world. The book's primary concern is McHarg's desire to understand and organize the objects, and events that comprise everyday life – his search for ideal order – and how this quest evolved in parallel with his spatial explorations and became an all-encompassing design theory that embraced natural history, social science, religion, ethics, and aesthetics. Despite his repeated claims of objectivity, this search was far from objective. He desperately wanted people to see the world as he did and to join with him to make this vision a reality.

Within this general context, *Ian McHarg and the Search for Ideal Order* foregrounds the argument presented in *Design with Nature*, and situates it within the political, social, and cultural discourse of the 1950s and 1960s. Of particular interest is the tension between McHarg's desire, as a dedicated modernist, to define and systematize a unified theory of design, and his equally strong desire, as a committed regionalist, to reveal and honor local character. To understand these related but conflicting ambitions, this book explores McHarg's broad and provocative reading of ecology and how he deployed this term to create what he claimed to be a forward-thinking and truly avant-garde unification of art, science, and society.[21] At his most visionary and optimistic moments, he even proclaimed that ecological seeing and thinking would change the world.

McHarg developed his environmental agenda when, as notably argued by the feminist theorist Donna Haraway, the traditional norms of society were beginning to fray under the onslaught of those seeking to resist patriarchy, colonialism, humanism, positivism, and essentialism.[22] As a key design figure in this transition period, McHarg actively positioned himself as a critic of the old order, but on more than one occasion opportunistically used its authoritative powers to his advantage. Moreover, following what was then common practice, he couched his ideas in terms that are not easy to accept today, including the dominance of the male voice, the condescending presumption of a design elite, and the tendency to express ideas through moral pronouncements that contain more than a smidgeon of sanctimony. Consequently, his work reflects both the strengths and the weaknesses of a watershed moment in landscape architecture. A deeper understanding of his message not only provides greater insight into his particular (and decidedly patriarchal, colonial, humanist, positivist, and essentialist) brand of ecological design, it also provides greater insight into the impact of his thinking on the theory and practice of landscape architecture and its current dominating discourses. Indeed, his insistence that natural processes and human action must co-exist in some type of cooperative transaction continues to be a critical topic in design theory and practice. The politics and language of this debate have changed in the past 50 years, but it remains as contested as it was in the mid-20th century.

Design with Nature is not a perfect book, but it asks great questions: Where should we build? How should we construct our cities? Is it possible to utilize natural resources and preserve natural processes? Can a networked system geared toward holistic resolution still honor regional differences? Is there a way past the naiveté of operative dualities? The answers that McHarg provided inform, provoke, inspire, and irritate.

What follows is my attempt to trace McHarg's vision of order, understand its structure, and assess its influence. Similar to the narrative of *Design with Nature*, the story that I weave contains intersecting histories and overlapping themes and it gains authority through the accumulation of evidence and repetition. In my years of research on McHarg, I discovered that any attempt to reduce his thoughts to a singular meaning denies the complexities, contradictions, nuances, and ironies that enrich and bedevil his thinking. Instead, I follow his lead and present his search for order on his terms – as an intellectual inquiry in which the same ideas and questions are repeated to different people and explored in different physical and social contexts. In addition to highlighting what McHarg deemed important, my intent is to illustrate how he puzzled through issues, isolated concerns, and zealously assembled and reassembled the components of his environmental manifesto until they fit his expectations and became the master narrative of *Design with Nature*.

The first section of the text discusses formative influences in McHarg's life and design training and it lays the groundwork for subsequent discussions. The second section explores the development of his theory of design and its relationship to ecology, religion, and land stewardship. This section of the text, in keeping with the kaleidoscopic patterns of McHarg's thoughts, moves between in-depth descriptions of the ideas he accumulated to create his ecological manifesto and short essays that explain how he assembled these ideas into a synthetic theory of design. The third section examines how he attempted, with varying degrees of persistence and success, to refine his theories and advance his career through teaching, scholarship, and practice. The forth section, the coda to this exploration of McHarg, examines Pardisan Park, in Tehran, Iran and positions his vision of ecological design in terms of post-colonial modernization.

The discussion of Housing and the Humane City that appears in this text is adapted from a paper published in The Journal of Planning History.

The discussion of The House We Live In that appears in this text expands an argument presented in Manifest: A Journal of American Art and Architecture, Issue No. 2- Kingdoms of God, Anthony Acciavatti, Justin Fowler, and Dan Handel eds. (New York: Manifest Journal).

The discussion of Pardisan Park that appears in this text is adapted from a paper published in Contemporary Landscapes of the Middle East, Mohammad Gharipour ed. (New York: Routledge Press).

Notes

1 For more information on the overlay method see Carl Steinitz, Paul Parker, and Lawrie Jordan, "Hand-Drawn Overlays: Their History and Prospective Uses," *Landscape Architecture* 66 (September 1976): 444–445; and Forster Ndubisi, *Ecological Planning: A Historical Approach* (Baltimore, MD: John Hopkins University Press, 2002).

2 To understand McHarg's use of Nature see Raymond Williams, *Keywords: A Vocabulary of Culture and Society* (New York: Oxford University Press, 1983), 219–224. Williams's definition incorporates the following:

    1 An inherent force, or deity, that commands natural forces and directs human beings. This definition links Nature to original innocence, a fall from grace, redemption, and rebirth.

    2 The material world, reason, observation, and the laws that govern an ideal society. This usage emphasizes reason and social development. Here Nature functions as an index, or sign, that measures progress.

    3 A concept of aesthetics in modern Western traditions where natural beauty and behavior are synonymous with the plants and creatures of the countryside, as opposed to the evils of the city.

    4 The transformation of "Nature" to "nature" that arose in conjunction with the theory of evolution. This definition links nature to the science of ecology and the study of relations between organisms, and between organisms and their environments, whether this is exemplified by inherent competition or inherent cooperation. In this definition nature, and natural laws, are considered in terms of survival and extinction rather than innocence and goodness.

3 For the purposes of this study, design is a process of production that relates an object to its purpose – function and utility – and to its ability to please independent of function – aesthetics and beauty. McHarg attempted to dissolve this distinction by arguing that form and function were complementary processes of physical and social production. See Michel Kelly, ed. *Encyclopedia of Aesthetics 2* (New York: Oxford University Press, 1998), 17–20.

4 Ian L. McHarg, *Design with Nature* (Garden City, NY: The Natural History Press, 1969). The first edition of the text was published under the auspices of The American Museum of Natural History in New York.

5 Rachel Carson, *Silent Spring* (Boston, MA: Houghton Mifflin Company, 1962).

6 Daniel Ostroff, *An Eames Anthology* (New Haven, CT: Yale University Press, 2015), xv. Eames made this statement on October 10, 1952 in a speech to the American Institute of Architects. See Manuscript Division, Library of Congress, Washington, D.C, Charles and Ray Eames Papers, "Part II: Speeches and Writings series."

7 The image of the sun is one of several from the collection of the Museum of Natural History in New York that appear in *Design with Nature*. The image of the earth is the first full-Earth picture from space and it was taken in November 1967 by a NASA (ATS) weather satellite as part of a daylong, high-resolution film shot. In addition to appearing on the cover of *The Whole Earth Catalogue*, this image of the earth appeared in ecology textbooks of the era. See Stewart Band, *The Whole Earth Catalogue* (Menlo Park, CA: Nowels Publications, Fall 1968); and Paul A. Colinvaux, *Introduction to Ecology* (New York: John Wiley & Sons, Inc., 1973), 41.

8 McHarg, *Design with Nature*, 1–2.

9 Ibid.

10 Ibid., 3.

11 McHarg revised the description of his return to Scotland in 1950 in the drafts of *Design with Nature*. Initially, he referred to the regularly spaced, four-story apartments as a "grotesque caricature of L'Unite d'Habitation." All of the descriptions emphasize the monotonous and disheveled appearance of the landscape. See The Architectural Archives of the University of Pennsylvania, Ian L. McHarg Collection, "Design with Nature Chapter Drafts and Related Research Materials." 109.II.B.1.2.1 and 109.II.B.1.3.

12 The bluebells reference McHarg's mother, who "loved the beauty of nature" and "bluebells in beechwood shadow." See Ian L. McHarg, *A Quest for Life: An Autobiography* (New York: John Wiley & Sons, Inc., 1966), 11.

13 Edith Cobb, "The Ecology of Imagination in Childhood" in *The Subversive Science: Essays Toward an Ecology of Man*, eds. Paul Shepard and Daniel McKinley (Boston, MA: Houghton Mifflin Company, 1969), 122–132. The essay originally appeared in *Daedalus* 88, no. 3 (Summer 1959): 537–548. Cobb linked imaginative play to self-knowledge and genius.

14 McHarg, *Design with Nature*, 5.

15 The Architectural Archives of the University of Pennsylvania, Ian L. McHarg Collection, "A Place for Nature in Man's World, or Man, Nature, and the City," 109.II.B.1.6. The title of this draft of *Design with Nature* is almost identical to that of an essay written by Paul Shepard and titled "The Place of Nature in Man's World." In this essay, Shepard argued that "one of the most important uses of nature in the world we [humans] dominate is simply nature being itself," and he called for the preservation of wilderness areas where people learned from nature, rather than extracted resources or heaped waste upon it. See Paul Shepard, "The Place of Nature in Man's World" *Atlantic Naturalist* (April 1958): 85–89. This particular draft of *Design with Nature* is also noteworthy because McHarg sent it to Lewis Mumford for comment. McHarg would revise the title of the *Design with Nature* several times. These revisions included *Design with Nature: A Plan for Nature in Man's World*, and *Design with Nature: A Plan for Man*. See also: The Architectural Archives of the University of Pennsylvania, Ian L. McHarg Collection, "Design with Nature: A Plan for Nature in Man's World," 109.II.B.1.9, and "Design with Nature: A Plan for Man," 109.II.B.1.10.

16 Dennis Farney, "Father Nature: How an Exuberant Scot, A Landscape Architect, Hopes to Shape the World," *The Wall Street Journal* (Monday, August 30, 1971). McHarg kept a copy of this article in his personal files. See The Architectural Archives of the University of Pennsylvania, Ian L. McHarg Collection, "Design with Nature Reviews," 109.II.B.1.14.1.

17 Ibid.

18 Ibid.

19 McHarg's autobiography contains a comprehensive list of media appearances, publications, and articles on his work and accomplishments. See McHarg, *A Quest for Life*, 400–402.

20 See Ian L. McHarg and Frederick R. Steiner, *To Heal the Earth: Selected Writings of Ian L. McHarg* (Washington, DC: Island Press, 1998); Ian L. McHarg, *The Essential Writing of Ian McHarg: Writings on Design and Nature*, ed. Frederick R. Steiner (Washington, DC.: Island Press, 2006); Lynn Margulis, James Corner and Brian Hawthorne, *Ian McHarg Conversations with Students: Dwelling in Nature* (New York: Princeton Architectural Press, 2007); Lynn Margulis, Adam MacConnell and James MacAllister, *The Lost Tapes of Ian McHarg: Collaboration with Nature* (White River Junction, VT: Chelsea Green Publishing, 2006); Ignacio F. Bunster-Ossa, *Reconsidering Ian McHarg: The Future of Urban Ecology* (Chicago, IL: American Planning Association Planners Press, 2014).

21 To explain McHarg's use of the terms ecology and environment see F. Fraser Darling, "The Unity of Ecology" in *Environmental Essays on the Planet as a Home*, eds. Paul Shepard and Daniel McKinley (Boston, MA: Houghton Mifflin Company, 1963), 207–221. The following passage, which describes both terms, appears in the Darling essay:

> Ecology deals with income and expenditures in terms of energy cycles in communities of plants and animals, deriving from sunlight, water, carbon dioxide and the phenomena of photosynthesis by which organic compounds are built. This raw definition is made more interesting by what I would emphasize as the observational study of communities of plants and animals. Here comes the possibility of a more general definition of ecology as the science of organisms in relation to their total environment, and the interrelations of organisms interspecifically and between themselves. The total environment includes all manner of physical factors such as climate, physiography, and soil, the stillness or movement of water and the salts borne in solution. The interrelations of organisms and environment are in some measure reciprocal in influence. . . But there is one outstanding difference between man and the rest of creation ecologically. He is a political animal and in our day and age it is quite unreal to ignore the political nature of man as an ecological factor.

McHarg and Darling were acquainted. A paper by McHarg appeared in a volume edited by Darling. Darling appeared in McHarg's seminar course "Man and Environment." See Ian L. McHarg, "Ecological Determinism" in *Future Environments of North America*, eds. F. Frazer Darling and John P. Milton (Garden City, NY: The Natural History Press), 526–538. See also, The Architectural Archives the University of Pennsylvania, Ian L. McHarg Collection, "Dr. F, Fraser Darling, Ecologist," 109.II.E.2.16.

22 See Donna Haraway, "A Manifesto for Cyborgs: Science, Technology and Social Feminism in the 1980s" in *Simians, Cyborgs and Women: The Reinvention of Nature*, ed. E. Donna Haraway (New York: Routledge, 1991), 74, 149–181.

Part 1
Early affinities

1 Experience and education

Whether the wish is father to the thought, or whether sentiment and idea have a common genesis, there equally arises the question – Whence comes the sentiment?

Herbert Spencer, First Principles[1]

Glasgow

Ian Lennox McHarg was born in 1920 and spent his childhood in Radnor Park, a suburban town fourteen miles from the center of Glasgow, Scotland in a house that overlooked the shipyards of the River Clyde. Behind his home, past the Great Western Highway, the hills of the Scottish Highlands loomed.[2] He was the eldest of four children in a family that aspired to bourgeoisie gentility. His father, John Lennox McHarg, contemplated a career in the Presbyterian ministry but due to the Great Depression instead weathered a number of unanticipated economic setbacks. To support his family, he worked variously as a manager for the Diesel and Edison Lamp Company, a part-time reporter for the Associated Scottish Newspapers, and a traveling salesman for a business machine and equipment company. McHarg would credit his own exuberant (and by some accounts manic) energy and interest in religion to him. His mother, Edith Bain McHarg, was a talented dress-designer and seamstress. McHarg would credit his artistic ability and love of the outdoors and gardening to her.[3]

In the 1920s and 1930s, the decades of McHarg's youth, the economy of the industrial port city of Glasgow centered on steel manufacturing, ship-building, and transatlantic trade. In *Design with Nature*, a photograph of a bleak courtyard surrounded by tenement housing represented the character of the city. A corresponding photograph of picturesque stone buildings surrounded by mountains represented the character of the countryside. As mentioned previously, he adventurously explored both terrains (Figure 1.1).

By his own admission, McHarg was a talented student. He scored in the top percentiles of the high school qualifying exam, entered the "A stream" curriculum, studied two foreign languages (one old and one new), and began to prepare for university and a career in the ministry, law, medicine, or civil

Figure 1.1 Photographs of Glasgow and the Highlands from *Design with Nature*. © 1992 by John Wiley & Sons, Inc. Reprinted by permission of John Wiley & Sons, Inc.

service. He spent Saturday mornings drawing and painting at the Glasgow Art Gallery and won many awards for his efforts. He visited the library on the way to the gallery. His path in life seemed obvious, and yet he was restless and felt confined by the conventions of middle-class life. He longed for adventure.

At age sixteen, he resolved to quit school, become a cadet officer for the Cunard Line, traverse the globe, and experience the excitement of faraway places. His father discovered his intent and soon put a stop to his plan.

Following this aborted attempt at independence, the young rebel acquiesced to his father's demand that he complete his education. A careers counsellor noted his artistic talent and love of the outdoors and suggested a five-year apprenticeship in landscape architecture that he could complete under the auspices of Donald A. Wintersgill, a Beaux Arts trained architect who specialized in the design of country estates and rock gardens. When they first met, Wintersgill impressed his protégé with a climate-sensitive reading of the landscape that included a grand scheme for a forest, lake, and small village nestled in the bracken-covered mountains of the Scottish Highlands. Wintersgill's dramatic style of dress – an Inverness cape, deer-stalker hat, bow tie, and spats – and dramatic style of presentation – a 180-degree sweep of his walking stick as he described his ideas – provided additional enticement. McHarg soon learned to draft design plans, produce watercolor renderings, calculate cost estimates, coordinate the dispatch of plants from nurseries, and supervise construction. This experienced culminated in the landscape design of the Empire Exhibition held in Glasgow, Scotland in 1938.[4] To appease his father, he attended night courses at the West of Scotland Agricultural College and the Glasgow College of Art. Perhaps best of all, Wintersgill taught his apprentice how to drive a car.

World War II and the military

In May of 1938, in a flurry of pre-war patriotism, no doubt abetted by his insatiable energy and desire for adventure, McHarg resigned his apprenticeship, quit night school, and enlisted in the British Army. Over the next seven years, he was deployed to North Africa, Italy, France, and Greece. In each field of battle, he served with distinction. The successful completion of Officer Training School led to a promotion to second lieutenant and assignment to the Second Parachute Division of the Royal Engineers. He achieved the rank of major and the command of a brigade.

One of his notable military achievements involved the production of a document titled the *Report on the Damage Caused to the Working of the Apulian Aqueduct by the German Army and on the Work Done to Repair the Aqueduct*. McHarg edited and signed the report, which contained an impressive array of photographs and drawings that recorded the destruction of war, and the subsequent repair of the siphons, pumps, and pipes by Pietro Celentani-Ungaro, the Director of the Aqueduct Corporation[5] (Figure 1.2).

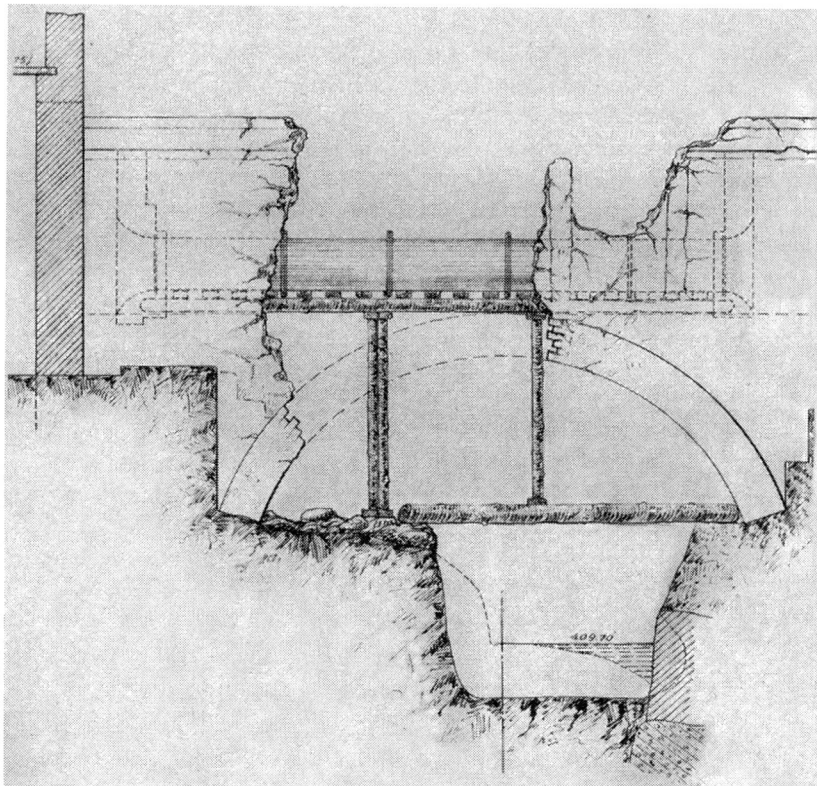

Figure 1.2 A drawing from the Apulian Aqueduct study. Ian L. McHarg Papers, The Architectural Archives of the University of Pennsylvania.

McHarg also enrolled in a correspondence course in town and country planning established for military personal by the London School for Reconstruction and Development in conjunction with the Association for Planning and Regional Reconstruction (APRR).[6] The course singled out "men in the Forces who either intend to become Town Planners, or who wish to be able to take responsible and informed part in the physical reconstruction of the country after the war."[7] The curriculum, formulated by the planner Jacquelyn Tyrwhitt, consisted of three modules – Background of Planning, Planning Factors, and Planning Practice. An acolyte of the Scottish planner Patrick Geddes and a member the Modern Architectural Research (MARS) Group – the English affiliate of the Congrès Internationaux d'Architecture Moderne (CIAM) – Tyrwhitt introduced the students to bioregionalism and architectural modernity. Following the methods developed by Geddes and MARS, course exercises required students to first collect and map a broad array of social and economic data and then correlate this information to physical geography (physiography) and human settlement patterns. In the process, they learned how to produce a balanced synthesis of multidisciplinary information. The pedagogical prototypes for this activity included the regional survey maps prepared by the APRR for the National Plan.[8] This set of black and white maps of England, Scotland, and Wales documented a wide range of information: Fog and Sunshine, Water Catchment Areas, Arable Land, Hills and Valleys (topography), Woodland, Extractive Industries, Farm Workers, Employment, Unemployment, Hazardous Employment, Trade and Transport, Local Government Boundaries, the Electric Grid, Women in Industry, Overcrowding, Tuberculosis, Doctors, and Secondary Schools. The booklet of maps came with two acetate overlays – one in blue that located chief urban areas and the other in red that depicted population distribution. When placed over the black and white maps, the overlays established a visual correlation between the statistics depicted in the maps and the loci of urbanization. McHarg completed all three modules of the correspondence course and obtained the credentials needed to enroll in the post-war completion course. To become a professional planner, he only needed to sit for the Intermediate Examination of the Town Planning Institute, the last step in this professional certification route.[9]

The correspondence course clearly presaged the central focus of McHarg's later career, but he still retained a strong interest in design. In 1945, he entered and won a competition for a World War II memorial and cemetery in Athens, Greece for British soldiers killed during the Battle of Greece. The hillside cemetery was built and it is located several kilometers southeast of the city overlooking the harbor. A memorial arch anchors the design and frames views of a stone of remembrance and a white cross. Stone plinths, engraved with the history of the Greek campaign and the names of the dead, flank the arch. Gravestones flank the plinths (Figure 1.3). Although more modest in scale and execution (but not necessarily in its derivative ambition),

Figure 1.3 World War II memorial and cemetery in Athens, Greece. Ian L. McHarg Papers, The Architectural Archives of the University of Pennsylvania.

the repertoire of elements calls to mind the iconic Thiepval Memorial to the Missing of the Somme designed by Sir Edwin Lutyens in 1932.[10]

Shortly before his military discharge, McHarg applied for, and received, an Army Release Scheme Scholarship that provided tuition and a stipend to attend the higher education institution of his choice. He used the monies to pay for study at the Graduate School of Design (GSD) of Harvard University in Cambridge, Massachusetts. No post-degree employment stipulations were attached to the funding; however, all parties to the agreement, including McHarg, tacitly assumed he would return to Scotland following graduation and contribute his expertise to post-World War II reconstruction.

When he submitted his application to the GSD, McHarg feared that he would not be admitted because he lacked the requisite high school diploma. Intent on pursuing a university degree in landscape architecture, which was exceedingly difficult to accomplish in Britain at that time due to the scarcity of institutions offering instruction, he went on the offensive.[11] He sent a telegram to Bremer Pond, the chair of the landscape architecture department, announcing his decision to study at Harvard and the date of his arrival. Left out of his confession – an omission that highlights his willingness to dramatize the truth in deference to a good story – was the fact that his apprenticeship, noteworthy achievements as a military engineer, successful completion of Tyrwhitt's highly regarded correspondence course (and her close ties with the faculty of the GSD), winning entry for the Athens cemetery competition, and full scholarship administered by the Department of Health in Scotland, indicated he was a well-qualified candidate and all but ensured his acceptance. Moreover, half of Harvard's 1946 admissions consisted of veterans, as the institution, in the words of Provost Paul Buck, sought to blend "democratic selection into aristocratic achievement."[12]

Harvard

In September 1946, one month shy of his military discharge, McHarg began his studies at the GSD with the proviso that he would enter the graduate program upon satisfactory completion of the first year of undergraduate course work. Following the pattern that characterized his military career, he excelled in academia and received three degrees – a Bachelor of Landscape Architecture (1949), a Master of Landscape Architecture (1950), and a Master of City Planning (1951) – in five years. He also served as chairman of the student council, which brought him in weekly contact with Joseph Hudnut, the dean of the GSD, and Walter Gropius, the chair of the architecture department.[13]

McHarg soon became disillusioned with the landscape architecture program. Years later, in his autobiography, he attributed his discontent to an "ossified" curriculum that failed to "engage the mind, far less challenge it."[14] When considered in relation to his military service, it is also fair to suggest that following his war experiences and mastery of the technical aspects of landscape construction, garden design and estate planning no longer captured his imagination as they had in his youth. His critique of the curriculum, perhaps intentionally, echoed the call for training informed by science, sociology, and modern lifestyles made in the late 1930s by the GSD students Garrett Eckbo, James Rose, and Dan Kiley.[15]

To rectify the situation, he decided to simultaneously pursue a degree in city planning. As he rightly surmised, a planning degree would provide the training in sociology, geography, economics, and government policy that he would need to successfully conduct post-war reconstruction when he returned to Scotland; it was also a logical decision in light of the military correspondence course that he had recently completed. The architect G. Holmes Perkins served as his adviser.[16]

In addition to the requisite landscape architecture courses in construction, planting design, and design history, McHarg studied political science, geography, and economics The courses taught by professors Arthur Maass, an expert on the Army Corp of Engineers, and Morris Lambie, an expert on municipal government, proved to be particularly influential. The course by Maass stressed the importance of water resources in regional planning in the United States. The achievements of the Tennessee Valley Authority, the Bureau of Reclamation, and the Bonneville Power Administration served as case studies. Another important legacy of this class, at least in regard to McHarg's future environmental activism, was his contact with fellow student Avis De Voto, the wife of the Pulitzer Prize winning historian Bernard De Voto. Avis provided her classmates with an article written by her husband, titled "Sacred Cows and Public Lands," that had recently appeared in *Harpers*.[17] The essay critiqued the undue influence of outside pressure groups (today's ubiquitous lobbyists) in the formulation of grazing rights policies on government land. De Voto's anecdotal style of writing and willingness to battle special interest groups resonated with McHarg who kept the essay in his personal files.

Professor Lambie's course on municipal government required students to read and write concise summaries of influential planning literature.[18] McHarg's selection of *The Culture of Cities* by Lewis Mumford from a list of choices possibly reflected the fact that it was required reading in Tyrwhitt's correspondence course, but in light of Mumford's later mentorship of McHarg, it is worth reviewing his summary of Chapters 5 and 6 of the text in greater detail. The summary noted Mumford's claim that cities faired better, survived longer, and achieved their greatest evolutionary potential when their constituent elements were "enriched by diversity, and organized to maintain a balanced but dynamic equilibrium between the expediency of human action and the geologic pace of natural change." It also included Mumford's observation that cities were plant-like organisms that required careful cultivation if they were to advance the civilizations that gave rise to them.[19] The most tangible impact of this assignment, however, was Mumford's promotion of a regional survey that consisted of "geologic structure, soil, surface relief, drainage, climate, vegetation and animal life, reformed and partly defined by the settlement of man, the nucleation of communities in villages and cities, power, climate and the movement provided by the existing technics."[20] This method of analysis, which reflected both Mumford's work with The Regional Planning Association of America and his interpretation of the bioregionalism of Patrick Geddes, reinforced the instruction McHarg received in Tyrwhitt's correspondence course, and in every practical sense established the criteria for the landscape surveys presented in *Design with Nature*.[21]

At this stage of his career, McHarg was clearly drawn to the ideas of Lewis Mumford, but he had not yet become an acolyte, and this was due in large part to Walter Gropius, the German émigré and founding Director of the Bauhaus, who led the GSD architecture department from 1938 to 1952.[22] Gropius secured his position as an architecture cause-célèbre through design success, articulate writing, and political acumen. His ascent into stardom began in 1925 with the Fagus Factory, jointly designed with Adolf Meyer, and the 1926 design of the Bauhaus at Dessau. Alfred H. Barr, the founding Director of the Museum of Modern Art (MoMA) – the foremost American institution promoting architectural modernism at that time – celebrated his arrival to the United States by staging a Bauhaus exhibit that praised both the institution and Gropius for their vital contributions to the artistic, technical, social, economic, and spiritual developments of the modern world.[23]

When Gropius assumed his position at the GSD, he sought to instill the curriculum with a Bauhaus pedagogy that emphasized experimentation, practical experience, and technical proficiency. His objective was complete building design – a quasi-mystical merger of art, science, and life in which "the immaterial space of inward vision and inspiration" is given outward form.[24] Gropius claimed his pedagogy did not promote a particular style, but instead addressed physical and psychological problems using contemporary materials and idioms. With this aim in mind, students were encouraged to abandon object-oriented design and think in terms of relationships. Form, as the physical embodiment of these thoughts, was a collaborative effort among disciplines, a utilitarian action, the natural handling of materials, a rationalized production effort, a spatial theory, and a creative experiment. The end product was functional, and yet expressive design distinguished by simplicity, ingenuity, and a strategic merger of thinking and feeling.[25]

To advance his agenda, Gropius hired two Bauhaus colleagues, the architect Marcel Breuer and the planner Joseph Wagner. He also advocated the hiring of the artist Joseph Albers to teach a six-month introductory course in basic design. Hudnut, however, thwarted this request because he believed the proposed course separated the artistic exploration of form from the functional aspects of construction, and, equally important, it rejected history.[26]

Although Gropius was unable to transform the pedagogy of the GSD to fully correspond to the Bauhaus program, when McHarg began his studies in the fall of 1946, the curriculum included four core design studios that all students were required to take – Design 1, Planning 1, Construction 1, and Architecture 3.[27] This curriculum change also fomented an active debate among the faculty and students about the meaning and practice of design: Was it primarily aesthetic or the fulfillment of functional and social need? Conversely, was it the purview of artistic genius or the product of multidisciplinary collaboration? Joseph Hudnut, as previously noted, believed these questions could only be addressed if history remained an integral component of the curriculum.

For younger members of the faculty, including Norman T. Newton, the core studio instruction was simultaneously aesthetic, social, and functional, and it promoted artistic genius and interdisciplinary collaboration. In *An Approach to Design*, a text written by Newton to explain the modernization of the GSD design curriculum under Gropius, he enthusiastically claimed these courses explored "the whole human experience, about the world we live in and about relations of humans to that world." They were also challenging and required students to make selections, gauge differences, organize variables, and assign relative values as they proceeded to develop an inclusive "attitude toward the *unified field* of design." To explain the synthetic nature of this instruction, Newton singled out the field of ecology and praised its relational thinking and refusal to reduce decisions to false dualisms. "That is to say," he wrote, "we have come to realize a strong parallel exists between the structure of design and the biological gestalt of the *organism-as-a-whole-in-an-environment-as-a-whole.*" By the end of the text, the design sequence embraced "the dynamic relations of all organisms – both fauna and flora – in natural association with each other and with the other forces of the total environment in a given area of the surface of the earth." Accordingly, Newton urged students to follow the path of ecology as far it goes:

> It will repay you in countless ways. It will give you not only a deeper understanding of human life, but a clearer vision of design as a *way of life* – a way bright with the promise of opportunity to serve your fellow men in years to come.[28]

Suffice it to say, an exploratory approach to design that challenged convention appealed to McHarg and he became a dedicated crusader for the cause. In addition to his study of Mumford, Garden Cities, and British New Towns, he investigated Le Corbusier's schemes for Paris, Frank Lloyd Wright's Broadacre City, and Ludwig Hilberseimer's plans for Chicago. His studio projects were notable for their minimalist aesthetic and sophisticated integration of topography and built form. The work that he photographed included a dramatic blue and black study of shape and color, and two study models that sensitively situate architecture in the landscape. In one model, a boardwalk and pavilions enclose a small rectangle of land. The architecture is raised on posts, and the topography and vegetation flow through the site. In the other model, a house on the shore of Loch Lomond overlooks a small cove carved into the land.

McHarg's thesis – a redevelopment and revitalization scheme for Providence, Rhode Island that he collaboratively developed with the architecture students Robert Geddes, William Conklin, and Martin Sevely – reflected the design debates swirling through Robinson Hall during the Gropius era. The project included an on-grade highway that rerouted traffic around a new urban core; three shopping courts, a department store, a new city hall, a federal building, a church, and a public library; and a pedestrian mall with gardens, trees, and water. The proposed plan carefully considered work, transportation, and recreation, but made little if any provision for housing. It was instead assumed the people who worked in the city would commute via automobile from their suburban homes (Figure 1.4). A model of the project, replete with superblocks and buildings that imitate the work of prominent architects, indicated the students were versed in the language of architectural modernity.

Figure 1.4 Ian McHarg and William Conklin at work on the Providence Rhode Island Collaborative Thesis. Ian McHarg Collection Loeb Library, The Graduate School of Design Harvard University.

In a remark that reflected their faith in the ability of design to instill cultural values, McHarg and his colleagues claimed the proposed redevelopment and revitalization scheme paid homage to the "agora, the forum, the village green," and thus combined the democratic aspirations of ancient Greece and American society. "Our interest in urban redevelopment," they idealistically continued, "stems from our belief that cities are the fruit of civilization," and they must "speak of life and its joys."[29]

A four-page article in the *Providence Sunday Journal* pragmatically noted the plan would keep businesses and shoppers in downtown Providence, and this would combat the decentralization trends engendered by the automobile and highway construction. The article also reported the following statement by Gropius: "They have worked with their brains and their hearts, but most of all they have worked as a team. The work of a team is greater than the sum of the work of the individual."[30]

Several key aspects of the Providence master plan indicate why Gropius praised the project as an exemplary model of interdisciplinary design. First, the students defined the economic and social variables that support a viable city center prior to their study of spatial organization. Second, the design presented the city center as an ensemble of buildings, streets, and parks. Third, the proposal deployed technology to re-tool the city in humanist terms. Here, the automobile – a machine that abets urban congestion and suburban decentralization – played a central, albeit ambiguous, role. Once cordoned off by a highway, it was possible to strategically repopulate the pedestrian-friendly, traffic-free urban core of Providence with new businesses, cultural institutions, and socially dynamic public spaces.[31] Fourth, and as indicated by McHarg's extensive social programming, a newly revitalized Providence would attract people through the magnetic draw of shopping, nightclubs, movie theaters, and, in a recapitulation of the activities that he enjoyed in pre-war Glasgow, the pleasures of circuses, band concerts, puppet shows, pageants, parades, holiday festivities, and political rallies.

McHarg's description of the landscape design began with a brief overview of garden history appropriated from the lectures of Bremer Pond. His summary denounced the ostentatious designs of the Renaissance aristocracy, praised the "meticulous use of water and skillful manipulation of microclimate" in the gardens of the Middle East, and approved the natural aesthetic of the English Landscape garden even though he faulted its emphasis on the pictorial and found it limiting.[32] He further noted that each of these traditions had basic strengths and weaknesses, but none of them addressed modern urban life. This condition required an "order and amenity" that could only be achieved through the unity of interior and exterior space, which he primarily defined in terms of physical controls located outside the building envelope. His subsequent recommendations included the deployment of plants and water to ameliorate the microclimate.[33] To explain his intent, he cited a passage (misquoted) from *American Building: The Forces That Shape It* written by the architect James Marston Fitch. This passage nullifies distinctions between artistry and function, while the perhaps purposeful misquote of Fitch elevates the agency and stature of landscape architecture vis-à-vis architecture:

> Landscape architecture [substituted by McHarg for landscaped areas] must be judged by actual performance as well as beauty. No more than architecture are the two contradictory; on the contrary, such an integration will yield new and higher aesthetic standards.[34]

McHarg provided no quantitative evidence that the provision of trees and gardens in the Providence design actually ameliorated the microclimate. At this stage of his career, he was content to list the functions enumerated in *American Building* – moderation of air temperature, filtering of atmospheric pollutants, and reduction of glare – and to associate these attributes with qualitative values. Like Fitch, he assumed there would be a corresponding perception of beauty correlated to physical comfort as mediated by the use of natural materials. A drawing that illustrated these thoughts

Figure 1.5 Courtyard design by Ian McHarg from the Providence Rhode Island
        Collaborative Thesis. Ian McHarg Collection Loeb Library, The
        Graduate School of Design Harvard University.

contains clear references to the work of Mies van Der Rohe and Christopher
Tunnard – two designers that he cited in the project description for their
ability to thoughtfully merge internal and external space (Figure 1.5).

Later, in his autobiography, McHarg proudly called the study, and his
presentation to members of the faculty and Gropius, the highpoint of his
GSD experience. The project was well received, and subsequently included
in the 1952 CIAM publication *The Heart of the City: Towards the Humanization of Urban Life.*[35] In addition to essays by the editors – Jacqueline
Tyrwhitt, Josep Luís Sert, and Ernesto Rogers – the publication featured
articles by the CIAM stalwarts Sigfried Giedion, Le Corbusier, and Walter
Gropius. Each of these individuals, in their various pronouncements on civic
design, promoted a well-ordered and aesthetically pleasing arrangement of
buildings, arcades, squares, and parks. Considering the publication's interest in urban revitalization that rekindled the democratic spirit of ancient
Greece, it seems inevitable the Providence project became a case study.

Pauline

The same month that he entered the GSD, McHarg met Pauline Crena de
Iongh at a social for foreign students. Pauline, who was studying English
literature and philosophy at Radcliffe, came from a prominent Dutch

family. She spoke four languages, played the cello, and was conversant in philosophy, history, art, and literature. During the German occupation of Holland, she helped the resistance, suffered through the Hunger Winter of 1944, and tragically lost her fiancé, who died in a concentration camp. To facilitate her recovery from these traumas, she was sent to the United States to study. Her father, Dr. Daniel Crena de Iongh, was in Washington, D.C. serving as the Treasurer of the World Bank and Executive Director of The International Bank for Reconstruction and Development.[36]

One year later, they married and together established a post-war identity that included the decoration of their modest Cambridge apartment using color, texture, and imagination – an endeavor they proudly documented with photographs (Figure 1.6). The design, which involved the construction of a two-person desk where they both studied, typified the following manifesto by the architect Marcel Breuer: "Let our dwelling have no particular 'style,' but only the imprint of the owner's character." Breuer further noted that any object "properly and practically designed should 'fit' in any room in which it is used as would any living object, like a flower or a human being."[37] The décor of the Cambridge apartment, accented with a carefully positioned vase of flowers, indicated the strong imprint of European architectural modernity on the owner's character.

McHarg, it seems, considered the apartment to be a living laboratory where he could, with the help of Pauline, explore the design ideas circulating among the GSD faculty and students. The finished product, echoing this inspiration, was for a certain type of person (educated), interested in a certain type of conversation (design), about a certain type of self-expression (avant-garde modernity), which as Tom Wolfe humorously noted in *From Bauhaus to Our House*, was de rigueur for every

Figure 1.6 The Cambridge apartment. Ian and Carol McHarg Collection, The Architectural Archives of the University of Pennsylvania.

aspiring architect.[38] McHarg was, in essence, emulating the members of the faculty who designed their own houses. The most famous example of these residences was, of course, the house that Gropius built in Lincoln, Massachusetts after he began teaching at Harvard. The Gropius House exemplified the modernist union of interior and exterior space as expressed by the seamless merger of architecture and landscape. A roof deck and large windows provided views of a fieldstone wall, and an oak tree shaded the building in the summer. The furniture, which included a two-person desk designed by Breuer, consisted of select pieces handcrafted at the Bauhaus. A supreme publicist and self-promoter, Gropius had the finished product photographed for dissemination in *Architectural Forum*.[39]

The prominent position of the molded plywood chairs by Charles and Ray Eames in these photographs, however, indicates that McHarg was simultaneously paying close attention to the innovations of other, non-GSD architects.[40] The Eames's design approach, although similar to the Bauhaus in its emphasis on creative problem solving, exploited the potential of mass production, the post-war building boom, and the democratic ideal that every American family had the right to live in a well-designed house furnished with products by well-known designers. As Charles Eames observed, the plywood chair, which was showcased in a 1946 exhibition at MoMA, was about providing "the largest group of people good furniture within their means, without sacrificing quality or performance."[41]

The photographs of the apartment also highlight the important role that Pauline played in McHarg's intellectual development and personal ambitions. Prior to her untimely death from cancer in 1974, her love of language, witty observations, engaging social skills, and family connections helped facilitate his career, including his professional interactions with the Dutch architects Steen Eiler Rasmussen and Aldo van Eyck, and his first published paper, "Architecture in the Netherlands."[42]

The young couple's interchange of ideas, while clearly seen in the apartment décor and two-person desk where they jointly studied, is even more apparent in a journal they co-wrote to document a road trip across the United States that occurred in the summer of 1949. Nominally Pauline's journal, the entries flow effortlessly between the two of them, making it at times all but impossible to recognize – especially when the pages are typewritten and there is no typographical distinction in handwriting – who made the observation and who wrote the entry.[43]

The road trip began with the packing of camping gear into a second-hand, gas-guzzling Studebaker Commander that McHarg purchased on the advice of a classmate, who, based upon his ability to draw vintage cars, supposedly knew what to buy. "Armed with a large can of oil and letters of introduction," they embarked on a grand tour of the United States.[44] The landscape architect Hideo Sasaki, a family friend, helped devise the itinerary. The route included the Skyline Drive, Black Mountain College, and the Tennessee Valley and Columbia River Authority projects. The young couple also visited Bryce and Grand Canyon national parks, crossed the Golden Gate Bridge, admired the redwood forests, saw Mount Rushmore and Devils Tower, and stopped at Wall Drug Store in the South Dakota Badlands.[45]

During their time in Los Angeles, McHarg traversed the city in search of exemplary models of housing. He sought, but was unable to find, the George Sturges House and Ennis House by Frank Lloyd Wright. He did, however, meet Garrett Eckbo who provided a tour of his recent residential projects. The tour included the Mar Vista Tract, a housing development with architecture by Gregory Ain, the Crestwood Hills Mutual Housing Association project, and the Goldstone, Hartman, and Sam Taylor residences. McHarg documented his impressions of these designs with quick sketches of a concrete block wall decorated with glass bottles and a grapevine panel used as a garden wall. Later, in San Francisco, he met the landscape architects Lawrence Halprin, Rob Royston, and Francis Violich, and the architect William Wurster.[46]

The description of the trip in McHarg's autobiography almost exclusively discussed the couple's attendance at the inaugural debut of the Aspen Institute and Music Festival. This cultural venue, conceived by the German-American businessman Walter Paul Paepcke with the assistance of Walter Gropius, sought to remind the world in the aftermath of World War II that Germany had made important contributions to the arts and humanist thought.[47] During the 1949 inaugural event – a bicentennial celebration of Goethe – they sat under an orange tent designed by Eero Saarinen and enjoyed the music of the cellist Gregor Piatigorsky. They also listened to talks on world order presented by the theologian and medical missionary Albert Schweitzer and the Spanish philosopher José Ortega y Gasset.[48] In contrast to McHarg's later autobiographical attempt to situate their grand tour within the highest echelons of culture, Pauline's entries noted the daily lives of their friends, their humorous attempts to use the Studebaker car seats as mattresses, and their conversations with fellow tourists, such as the one with a family from a town in Wisconsin where "they make fake Edam cheeses."[49]

Even though the young travelers were at times daunted by "the less-than-informative U.S.A. road map," they marveled at the vast scale of the country and its scenic beauty, as seen, for example, in this description of the Sangre de Christ Mountains:

> The road led along the Black Canyon of Gunnison [National Park], the river deep down, the sides steep down and the road winding and twisting unprotected along the side. Alpine meadows with thrift and Scottish Thistles. Then the view widened suddenly and the west spread out at our feet. The green distances lined by the blue mountain range. The essence of the American landscape, the vastness of it.[50]

When they ventured outside the exclusive protection provided by the ivy-covered walls of Harvard and Radcliffe, they also witnessed racial inequality and segregation:

> Before leaving the subject of St. Louis it seems appropriate to note our first acquaintance with the colored problem. During our stay in the South we noticed rather little of it, but in Missouri which is an intermediary state, the situation had suddenly come to a point on the subject of bathing pools. In this heat wave swimming is an important issue and the Mayor of St. Louis had opened the public bathing pools to the entire population. In the riots resulting from this decree the police had been called out and the white mob, armed with pieces of lead pipe killed two would-be swimmers. Little was said about this riot, and there were no large-scale investigations into these murders, but while in St. Louis the papers mentioned a similar incident in Washington, D.C.[51]

Particularly striking is the correspondence between the journal and written passages that appear twenty years later in *Design with Nature*. Note, for example, the following entry by Pauline that chronicled their route south from New York City:

> We discovered that there are no roads leading around New York. The traffic is atrocious and the N.Y. area is blighted by city development; first dumps and scattered houses, gas stations etc. and then gradually narrowing roads, slums, plants, altogether a heartbreak for the tourist . . . The city blight seemed to go far into New Jersey, past Newark, Elizabeth. Hot dog stands and gas stations dominate whatever landscape there is and the traffic is very bad.[52]

A similar description, purposefully dramatized and politicized, appears in *Design with Nature*:

> What are the visible testaments to the American mercantile creed – the hamburger stand, gas station, diner, the ubiquitous billboards, sagging wires, the parking lot, car cemetery and the most complete conjunction of land rapacity and human disillusion, the subdivision.
>
> And what of the cities? Think of the imprisoning grey areas that encircle the center. From here the sad suburb is an unrealizable dream. Call them no-place although they have many names. Race and hate, disease, poverty, rancor, and despair, urine and spit live here in the shadows. United in poverty and ugliness, their symbol is the abandoned carcasses of automobiles, broken glass, alleys of rubbish and garbage.[53]

Needless to say, the summer adventure culminated a series of formative experiences that enriched McHarg's understanding of the world and his place in it. As he traversed the country with Pauline, the landscape of the United States shaped, both positively and negatively, his perceptions and he began to think critically about the land and the way it impacted cultural norms and values, and conversely, the way cultural norms and values impacted the land. By the end of the journey the thoughts he would express in *Design with Nature* were not fully formed, but they were fully in place. McHarg would subsequently explore these ideas, with varying degrees of persistence and success, in teaching, scholarship, and practice.

Notes

1 Herbert Spencer, *First Principles* (New York: A. L. Burt Company Publishers, 1862), 11.
2 John Lennox McHarg is the name that appears on McHarg's birth certificate. He and his family used the name Ian, an Anglicized Gaelic version of John, to differentiate the son from his father. Personal conversation, William Whitaker, February 8, 2019 and April 1, 2019.
3 Unless otherwise noted, the information for this brief discussion of McHarg's life comes from his autobiography. See McHarg, *A Quest for Life*, 9, 17–19, 21–24, and 31–55.
4 In his role as apprentice, McHarg worked on the design of the of the 1938 Glasgow Empire Exhibition. Personal conversation, William Whitaker, April 1, 2019.
5 See The Architectural Archives of The University of Pennsylvania, Ian L. McHarg Collection, McNeil, R.E. and Ian McHarg, *Report on the Damage Caused to the Working of the Apulian Aqueduct by the German Army and on the Work Done to Repair the Aqueduct*, October 14, 1943, 109.1.B.1.1. The introduction to the report states: "The original report was transposed by Major McNeil R.E. and subsequently edited and collated by Lt. McHarg, who acted as Liaison Officer to the Corporation."
6 See The Architectural Archives of the University of Pennsylvania, Ian L. McHarg Collection, "Correspondence Courses for Members of H. M. Forces. Part 1: Background to Planning, Part II, Planning Factors, and Part III, Planning Practice (British War Office, 1944 and 1945)," 109.I.B.1.2. This folder contains: *Syllabus of a Course on The Background of Planning* (A.E.3 Series T.P. 1, June 1944); *Syllabus of a Course on Town Planning* (A.E.3 Series T.P. 2, n. d.); *Syllabus of a Course on Planning Practice Part I* (A.E.3 Series T.P. 3, April 1945).
7 See Ines Maria Zalduendo, "Jacqueline Tyrwhitt's Correspondence Courses: Town Planning in the Trenches" Digital Access to scholarship, https://dash.harvard.edu/handle/1/13442987, accessed March 4, 2019; Ellen Shoshkes, "Jacqueline Tyrwhitt Translates Patrick Geddes for Post World War Two Planning" *Landscape and Urban Planning* 166 (2017): 15–24. See also, Jacqueline Tyrwhitt, "Town Planning" in *Architects Yearbook 1*, eds. Trevor Dannatt and Jane Drew (London: Elek Books Limited, 1945), 11–29. In anticipation of post-war reconstruction, the correspondence course taught students how to create land and resource surveys. As noted by the architectural historian Volker Welter, the course allegedly enrolled 2,000 students, and 170 men and two women completed their studies. See Volker Welter "The Valley Region: From Figure of Thought to Figure on the Ground" in *New Geographies: Grounding Metabolism*, eds. Daniel Ibanez and Nikos Katsikis (Cambridge, MA: Harvard University Press, 2014), 78–87.

8 See Patrick Geddes, *Cities in Evolution* (London: Williams & Norgate, LTD., 1949); and Association for Planning and Regional Reconstruction, *Maps for the National Plan: A background to The Barlow Report, The Scott Report, The Beveridge Report* (London: Lund Humphries & Co Ltd, 1945). Several of the maps produced for the National Plan appear in *Cities in Evolution*.

9 See British War Office, *Syllabus of a Course on The Background of Planning*, 1.

10 McHarg, *A Quest for Life*, 56.

11 See Jan Woudstra, "The 'Sheffield Method and the First Department of Landscape Architecture in Great Britain" *Garden History* 38 no. 2 (Winter 2010): 242–266.

12 Glenn C. Altschuler and Stuart M. Blumin, *The GI Bill: The New Deal for Veterans* (New York: Oxford University Press, 2009), 114. In a 1979 address delivered to Royal Society of Art in London, McHarg thanked G. A. Jellicoe, who was in the audience, for recommending that he attend Harvard and study landscape architecture. McHarg clearly had an illustrious group of patrons that supported his application to the GSD. See The Architectural Archives of the University of Pennsylvania, Ian L. McHarg, "Garden as Metaphysical Symbol (1979)," 109.II.C.37.

13 McHarg, *A Quest for Life*, 72.

14 Ibid., 71.

15 See Anthony Alofsin, *The Struggle for Modernism: Architecture, Landscape Architecture, and City Planning at Harvard* (New York: W. W. Norton & Company, 2002).

16 McHarg, *A Quest for Life*, 71–73.

17 Bernard DeVoto, "Sacred Cows and Public Lands," *Harper's Magazine* 197 no. 1178 (July 1948): 44–55. See The Architectural Archives of the University of Pennsylvania, Ian and Carol McHarg Collection, "Harvard Courses Spring Term 1948–1949," 365.I.11.

18 The Architectural Archives of the University of Pennsylvania, Ian L. McHarg Collection, "Government 17a _ Municipal Government General Information, Recommended Reading List" and "Reading Period Assignment Government 17a-1947/48," 365.I.10; and The Architectural Archives the University of Pennsylvania, Ian L. McHarg Collection, "Notebooks from Government Course (1947)," 109.I.B.2.5.1. See also Lewis Mumford, *The Culture of Cities* (New York: Harcourt Brace & Company, 1938, 1970). McHarg's personal library contained *The Culture of Cities*.

19 Ibid., 8.

20 Ibid.

21 The Regional Planning Association of America was founded in 1923. Initial members included Clarence Stein, Lewis Mumford, Benton MacKaye, Alexander Bing, and Henry Wright. In 1925, Mumford edited *The Survey Graphic*, which contained essays that served as touchstones for McHarg. See Lewis Mumford, "The Fourth Migration," *The Survey Graphic* LIV no. 1 (May 1, 1925): 130–133; Clarence S. Stein, "Dinosaur Cities" *The Survey Graphic* LIV no. 1 (May 1, 1925): 134–138; Benton MacKaye, "The New Exploration, Charting the Industrial Wilderness" *The Survey Graphic* LIV no. 1 (May 1, 1925): 153–157.

22 Alofsin, *The Struggle for Modernism*, 138–195.

23 Herbert Bayer, Walter Gropius and Ise Gropius, *Bauhaus 1919–1928* (New York: The Museum of Modern Art, 1938), 8.

24 Ibid., 25.

25 See Walter Gropius, "Education" *Task* 1 (Harvard Graduate School of Design Magazine), 34–35.

26 Jill Pearlman, "Joseph Hudnut's Other Modernism at the Harvard Bauhaus" *The Journal of the Society of Architectural Historians* 56 no. 4 (December 1997): 452–477.

27 Alofsin, *The Struggle for Modernism*, 196–197. See also "The Graduate School of Design, Courses in Architecture, Landscape Architecture, City and Regional Planning," Archival Collections [GSD], Subseries AAOOO, Subseries AA, Academic Programs and Curriculum Development 1941–1991.

28 Norman T. Newton, *An Approach to Design* (Cambridge, MA: Addison-Wesley Press, Inc., 1951), vii–viii, 4, 86, 143–144. McHarg's personal library contained the book.

29 William Conklin, Robert Geddes, Ian L. McHarg, and Martin Sevely, "Introduction," *Collaborative Thesis Program*. (1949–1950) Special Collections Department, Francis Loeb Library, Harvard Design School. Reprints courtesy of Mary Daniels. McHarg lived in Providence for several months during the project.

30 Edward J. Milne, "Providence Tomorrow?" *Providence Sunday Journal* (June 11, 1950): 5.

31 Robert Geddes noted that each student worked independently and pursued different interests, and, in his opinion, failed to fully collaborate and create a unified design. Personal communication, Robert Geddes, August 18, 2015.

32 Ian L. McHarg, "Landscape Architecture" Appendix D, Part 1, page D1-1, *Collaborative Thesis Program*. Special Collections Department, Francis Loeb Library, Harvard Design School. See also Alofsin, *The Struggle for Modernism*, 208–211.

33 Ibid., D1–3.

34 Ibid., D1–4. See also James Marston Fitch, *American Building: The Forces that Shape it* (Boston, MA: Houghton Mifflin Co., 1948), 295.

35 J. Tyrwhitt, J. L. Sert, E. N. Rogers, *The Heart of the City: Towards the Humanization of Urban Life* (London: Lund Humphries & Co. Ltd., 1952). See also Eric Mumford, *The CIAM Discourse on Urbanism 1928–1960* (Cambridge, MA: The M.I.T. Press, 2000).

36 McHarg, *A Quest for Life*, 68. See also Robert W. Oliver and the Oral History Research Office, Columbia University, 1968, The World Bank Oral History, "Crena de Iongh, Daniel," 1961 transcripts, http://oralhistory.worldbank.org/person/crena-de-iongh-daniel; and *The New York Times* Archives, "Daniel Crena de Iongh, 82, Dies; Served World Development Bank, November 28, 1970, www.nytimes.com/1970/11/28/archives/daniel-crena-de-iongh-82-dies-served-world-development-bank.html.

37 Bayer, Gropius and Gropius, *Bauhaus 1919–1928*, 128.

38 Tom Wolfe, *From Bauhaus to Our House* (New York: Picador, Farrar, Straus and Giroux, 1981), 42.

39 George Nelson, "The Modern House in America" *The Architectural Forum* 71 no. 1 (July 1939). Featured residences in this special volume on the modern house included the homes of G. Holmes Perkins, Marcel Breuer, and Walter Gropius.

40 See John Neuhart, Marilyn Neuhart and Ray Eames, *Eames Design: The Work of the Office of Charles and Ray Eames* (New York: Harry H. Abrams, Inc. Publishers, 1989), 28–36, 52–53, 58–61. The plywood chair developed from studies the Eames conducted on the manufacture of compound curves with wood veneer. The studies began in World War II with the design of a molded plywood leg splint for use by doctors in the field of battle. The Herman Miller Company subsequently marketed the chair to "young homemakers" at an affordable price. The cost of the plywood chair in 1946 was $17.85. Personal communication, Nick Butterfield, Communications and Media Director, Global Marketing Herman Miller Company, August 18, 2016.

41 Daniel Ostroff, *An Eames Anthology*, 6, 131. See also Charles Eames, "What is a House?" *California Arts & Architecture Magazine* 61 (July 1944): 17–22.

42 Ian McHarg, "Architecture in the Netherlands" *Quarterly Journal of the Royal Incorporation of Architects in Scotland* 94 (1953): 41–46.

43 The Architectural Archives the University of Pennsylvania, Ian L. McHarg Collection, Pauline Crena de Iongh, 1949, "Pauline Crena De Iongh's Journal of Trip Across America," 109.I.B.2.9. G. Holmes Perkins supported the trip, and he composed a letter to the Department of Health of Scotland, the institution that sponsored McHarg's education to help garner funding. The letter stated the trip would provide an unmatched educational experience. See McHarg, *A Quest for Life*, 78.

44 Ibid.

45 The Architectural Archives of the University of Pennsylvania, Ian L. McHarg Collection, "Misc. Directions and Intended Stop Locations (1949)," 109.B.2.10.

46 Ibid.

47 See: The Aspen Institute, "A Brief History of the Aspen Institute," www.aspeninstitute.org/about/heritage/.

48 McHarg, *A Quest for Life*, 79.

49 Pauline Crena de Iongh, "Journal of Trip Across America," 28.

50 Ibid., 12.

51 Ibid., 10.

52 Ibid., 1.

53 McHarg, *Design with Nature*, 20.

2   Housing and humane cities

To have imagination and taste, to love the best, to be carried by the contemplation of nature to a vivid faith in the ideal, all this is more, a great deal
more, than any science can hope to be. The poets and philosophers who
express this esthetic experience and stimulate the same function in us by
their example, do a greater service to mankind and deserve higher honor
than the discoverers of historical truth. Reflection is indeed a part of life,
but the last part. Its specific value consists in the satisfaction of curiosity,
in the soothing out and explanation of things: but the pleasure we actually
get from reflection is borrowed from the experience on which we reflect.

George Santayana "The Sense of Beauty"[1]

Following the completion of his studies at Harvard, McHarg returned to
Scotland. He accepted a position in Edinburgh as a junior planning officer
in the Department of Health, the institution that provided the funds for his
education, and he began to work on his country's post-war reconstruction[2]
(Figure 2.1).

His plans were interrupted by a diagnosis of tuberculosis. For six months
he was confined to the Southfield Colony for Consumptives, a medical
facility associated with the Edinburgh School of Medicine, where he un-
derwent pneumolysis – a surgical procedure commonly performed prior to
the development of antibiotic treatments that collapsed the infected lung
to prevent the spread of tubercular tissue. After the operation he developed a persistent fever and became so weak that he was unable to sit up,
read, wash, or shave. Appalled by the sanitarium's "Victorian" medicine,
"lackluster" staff, and "general air of dirt and decay," he went on the offensive, and, consistent with his campaign to enter Harvard, he used his
military contacts to advantage. This action resulted in a transfer to the
Hotel Belvedere in the high Alps near Leysin, Switzerland. The Belvedere,
which was doubly famous as a ski resort and a spa that treated pulmonary
disease, cared for British military personnel with tuberculosis. He spent six
months in this luxurious alpine setting, and would later credit the recovery
of his health, dignity, and hopes for the future to the fresh air, sunlight,
food, and attentive staff of this mountaintop retreat.[3]

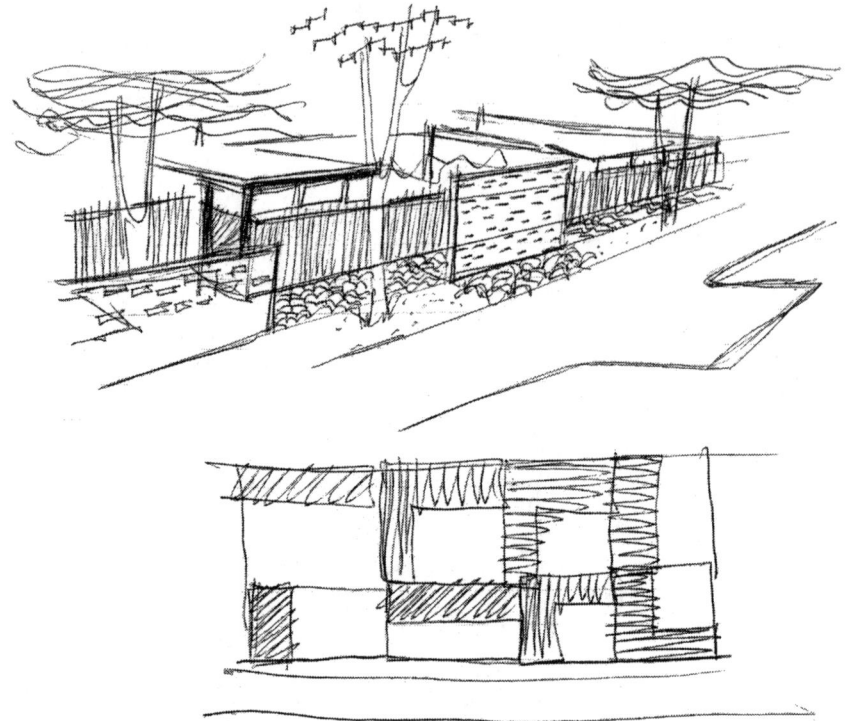

Figure 2.1 Sketch of a house with an attached courtyard by Ian McHarg. Ian
L. McHarg Papers, The Architectural Archives of the University of
Pennsylvania.

McHarg returned to work brimming with ambitious plans for post-war
reconstruction. He quickly submitted proposals for the upgrade of Scotland's power infrastructure that followed the precedents of the Tennessee
Valley Authority; sought to reform coal mining practices based on his study
of strip mining and landscape restoration operations in West Virginia; proposed an ambitious overhaul of urban transportation that emulated the
separation of traffic that Frederick Law Olmsted and Calvert Vaux proposed for Central Park; and argued for scenic roadways that matched the
topographic bravura of the Bronx River Parkway and Skyline Drive.[4] In
his spare time, he taught landscape architecture at The Glasgow School of
Architecture.[5] The course, which provided a broad overview of the profession, included lectures on history, open space and housing, and plant
and design materials. The Santayana quote that introduces this chapter appeared in his lecture notes under the heading the experience of beauty.

During the manic burst of creativity that followed his recovery, he also
developed a proposal for a New Town on a hillside site near Glasgow that

overlooked the Firth of Clyde.[6] The design, which incorporated many aspects of the Neubühl housing estate in Zurich, Switzerland, positioned tall residential towers on the hilltop and single-family housing on the slope[7] (Figure 2.2). To ensure the proposed architecture reflected the climate sensitive principles that he had earlier outlined in his GSD thesis, he collaborated with J. Blanco White – a Civil Service architect and the wife of the famed Scottish geneticist C. H. Waddington – to devise a topographically appropriate terrace scheme that embedded the single-family housing in the hillside.[8] This design feature, in conjunction with the passive solar heating provided by large south-facing windows, mediated the microclimate and reduced energy consumption. Additionally, and no doubt a hygienic battle cry against tuberculosis, the terrace housing contained roof gardens to maximize light and air. The stepped profile and exuberant profusion of sunlight and plants made this supremely functional design visually exciting and aesthetically appealing.

Impatient for reform, McHarg soon became disillusioned with the work produced by the Department of Health. Similar to his earlier student rebellion against the ossified landscape architecture curriculum at the GSD, he believed his colleagues were not taking advantage of the new ideas revolutionizing practice. This negative assessment was due in no small part to the senior career administrator's rejection of the New Town design that he had created for the Firth of Clyde. Upset with the decision,

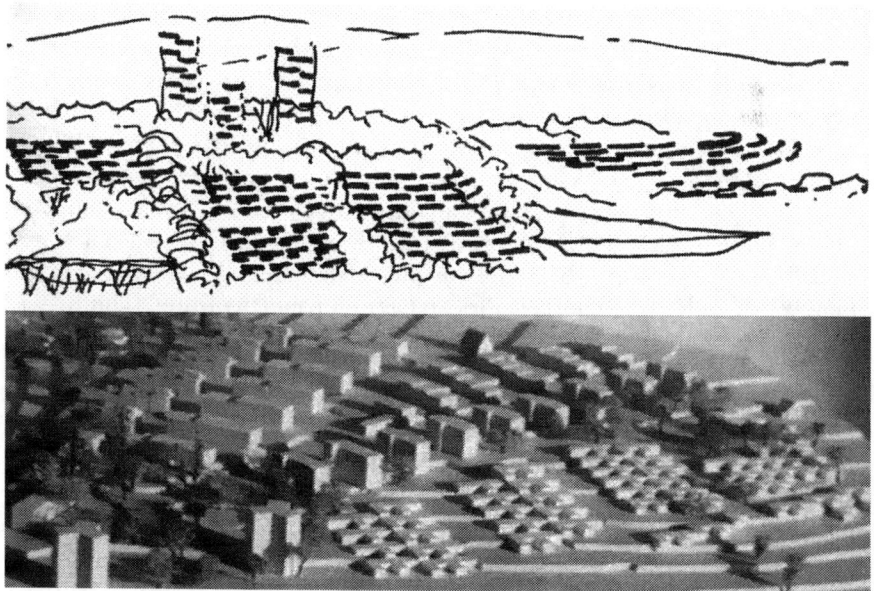

Figure 2.2 Sketch of the Firth of Clyde Housing Proposal (top) and a model of the Neubühl Housing Estate (bottom). Ian L. McHarg Papers, The Architectural Archives of the University of Pennsylvania.

and seeking to redress the situation, he delivered several public lectures that outlined the merits of his housing proposal and why its dramatic views, energy-efficient design, and hygienic embrace of light and air were superior to existing planning models. Also aware that his public critique would limit his future within the Department of Health, he wrote to his former graduate adviser G. Holmes Perkins to inquire about work.[9]

In 1954, at the bequest of Perkins, who was now the dean of the School of Fine Arts at The University of Pennsylvania and needed bright young talent to advance his academic agenda, McHarg returned to the United States to lead the landscape architecture department at Penn.[10] In addition to building a program that rivaled the GSD in the quality of its education, he continued to explore housing, and, in keeping with the requirements of academic scholarship, he published.

Over the next four years, McHarg refined the ideas he explored in his studies at Harvard and his professional work in Scotland, and he presented the work in the essays "Open Space and Housing," "The Court House Concept," and "The Humane City." Although these essays are considered of minor significance in comparison to *Design with Nature*, the vision of environmental order that they present, written by an ambitious landscape architect still in search of his true métier, warrants further examination.

1955: Open space and housing

McHarg wrote "Open Space and Housing," which appeared in *Architects' Year Book 6*, shortly after the Department of Health refused to sanction his Firth of Clyde housing scheme.[11] Composed as a retort to his former colleagues for their dismissal of his ideas, he began the essay with the observation that New Town planners misapplied Garden City density models and disregarded modern use-based programming. As a result, their work suffered from the application of "ill-used," "unused," and "useless" facts. The landscape in these projects may be expansive, he noted, but it lacked spatial and social focus as evidenced by roads devoid of people and backyard gardens composed of "a forest of clothes lines and anarchic horticulture" (Figure 2.3).

To address the spatial and social deficiencies of the New Town landscapes that he had catalogued during his brief sojourn at the Department of Health, McHarg proposed a utilitarian retrofit of the architecture that involved the addition of an outdoor courtyard. In contrast to the empty streetscapes, unsightly clotheslines, and garbage pails that he claimed disfigured the landscape of public housing estates, the "outdoor room" was ideal for conversation, dining, and the supervised play of children.[12] Ignoring the fact that this proposal did not resolve (and perhaps exacerbated) the issue of empty streetscapes that began his critique, he insisted that all New Town designs include houses with courtyards. Implied, but not explicitly stated, was the understanding that the inward focus of this intervention would strengthen the family, the neighborhood, and community.

Figure 2.3 Photographs of British New Towns from the personal files of Ian McHarg. Ian L. McHarg Papers, The Architectural Archives of the University of Pennsylvania.

Figure 2.4 Photograph of Frankendael from the personal files of Ian McHarg. Ian L. McHarg Papers, The Architectural Archives of the University of Pennsylvania.

Housing projects from Scandinavia, one of the epicenters of architectural modernity, exemplified ideal spatial organization. McHarg admired Frankendael (1939), by the Dutch architects Merkelbach, Elling, Karsten and the landscape architect Mien Ruys, for the way the buildings enclosed a central green (Figure 2.4). Instead of dark service courts and unsightly clotheslines, people who lived in this housing estate enjoyed views of, and easy access to, a lawn with trees and a small playground. The competition entry for Skånska Hustyper by the Danish architects Møgelvang and Utzøn (1953) illustrated the integration of architecture and landscape at the scale of the housing unit. More inward looking than Frankendael, in this design each room faced an interior courtyard. If family needs changed, and additional rooms were required, this could be accommodated by modifying the size and layout of the courtyard.[13]

To justify his courtyard proposal, McHarg referenced a claim made by the architect William Wurster in the essay "The Outdoors in Residential Design" that stated a private, outdoor space for children, gardening, seating, eating, and sunning reflected "how people actually want to live and carry out their family and social life."[14] A passage from the *Social Pressures of Informal Groups*, a study of human ecology in student housing provided scholarly support:

> The architect who builds a house or who designs a site plan, who decides where the roads will and will not go, and who decides in which direction the house will face and how close together they will be, also is, to a large extent deciding the pattern of social life among the people who will live in the houses.[15]

Health and hygiene played a prominent role in McHarg's argument. He discussed optimum solar orientation, noted the therapeutic qualities of sunlight via the production of vitamin D, and detailed the germicidal benefits of ultraviolet light. These concerns clearly reflected his recent battle with tuberculosis.[16] More generally, his references to sunlight, fresh air, and hygiene indicated an enthusiasm, in the true modernist sense, for progressive designs that incorporated the latest advances in science and technology. On another level, and in parallel with his factual analysis, McHarg observed that open space had to transcend utility and embrace an "authentic inner life." This "experience of nature," he wrote, "cannot be defined as a function," but must instead be considered a "profoundly important" need that only the "sunlight, sky, leaves and grass, sticks, stones and the changing seasons," can satisfy.[17] Even though the bulk of his argument expressed confidence in quantifiable evidence, this comment, like the thoughts expressed by Santayana, suggests that rational analysis, in and of itself, ignored a critical aspect of design.

The views expressed in "Open Space and Housing," leave little doubt that McHarg was excited about what he had recently learned in school. But his derivative repetition of these ideas demonstrates that he was still a young designer who had yet to put his own imprint on his ideas. His most original and subversive thought, which is not that inventive or rebellious, was his concluding recognition that the landscape united Le Corbusier's high-density "unity of nature" and Lewis Mumford's low-density "life-centered" community.[18] The desire to impose order through the mediation of difference would remain central to his work.

1957: The court house concept

In 1899, the utopian planner Ebenezer Howard considered whether the "country could be made more attractive to a workaday people than the town." Howard saw the issue as a choice between two extremes, which he diagrammed in the *Garden Cities of To-morrow* as town and country magnets (Figure 2.5). The city magnet offered business enterprise, social opportunity, places of amusement, and high wages; but these advantages produced high rent, foul air, gin palaces, and overcrowded slums. The country magnet offered low rents, the beauty of nature, forests, meadows, clean water, and bright sunlight; but these advantages produced a lack of society, no public spirit, and deserted villages. The third magnet in his diagram, his proposed Garden City, enjoyed the advantages of both. This ideal "town-country" mix provided the beauty of nature, business enterprise, social cooperation, and a house with a garden.[19]

Fifty-eight years later in "The Court House Concept," McHarg reversed Howard's query and asked if there was "an urban house which, exploiting the advantages of city life, offers in addition a residential environment at least equaling the sum of advantages and disadvantages of the modern house

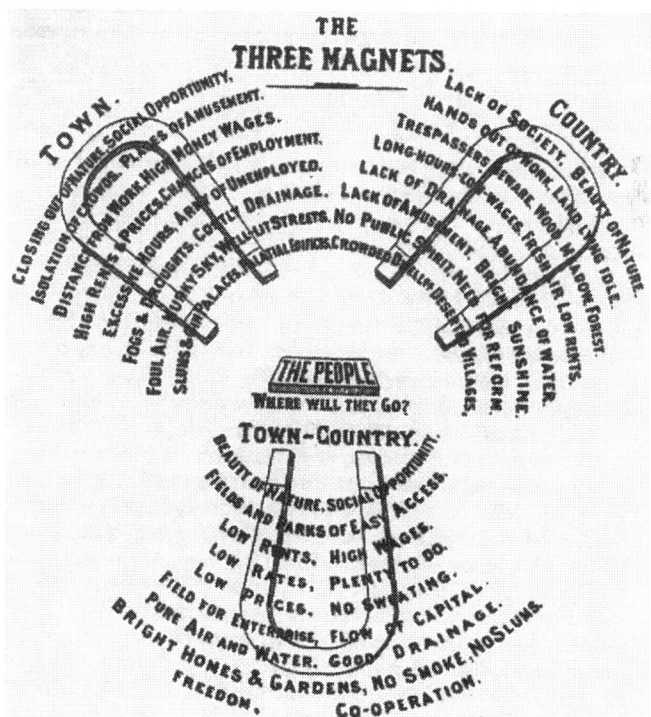

Figure 2.5 The Three Magnets of Ebenezer Howard.

in the suburb or country?"[20] This structurally convoluted but conceptually straightforward question, argues for the insertion of the delights of the countryside back into the heart of the city. McHarg's subsequent response suggests that decentralization via Garden Cities may not be the only remedy for the ills besetting the modern metropolis.

The article, published in *Architects' Year Book 8*, and in a condensed version in *Architectural Record*, began with the following praise for modern architecture:

> The free standing, single family house in suburb and countryside from the hands of Frank Lloyd Wright, Mies van der Rohe, Le Corbusier, Philip Johnson and Richard Neutra undoubtedly represents the greatest contribution of modern architecture to art and environment.[21]

Then, in a statement that established his argument, McHarg wondered why architects had yet to create housing for the city commensurate with the elegant villas they designed and built in the countryside. And since this was a rhetorical remark, he already had an answer and numerous supporting examples.

Rather than a house in a garden, as proposed by Howard in *Garden Cities of To-morrow*, McHarg called for a garden in a house. He was certain a townhouse with an enclosed courtyard – a courtyard house – was the urban counterpart to the freestanding country villa. The courtyard house would, he wrote in a reprisal of the themes explored in "Open Space and Housing," support an appropriate urban density, provide useable outdoor space, and maximize light, air, and privacy. Moreover, this typology exploited the cultural and economic advantages of the city and retained a close relationship with the natural landscape. The result was an environment "as humane as it is urbane."[22]

Throughout the essay, McHarg demonstrated an interest in data collection and cataloguing, as seen, for example, in his classification of the courtyard house into four regional groupings: a North American branch indebted to the work of Mies van der Rohe, Philip Johnson and Ludwig Hilberseimer; an Italian branch emanating from the rationalist theories of Giuseppe Pagano; a South American branch centered around the work of Josep Lluís Sert; and a North African branch exemplified by French architects emulating Le Corbusier. His taxonomy established the formal characteristics of the courtyard house, its place in history, its apparent timelessness, its adaptation to local context and culture, its contemporary relevance, and, equally important, its ability to accommodate the idiosyncratic tastes and prejudices of the designer. Supporting images reinforced these themes. A Mies van der Rohe design adjacent to a traditional Chinese house illustrated the typology's timeless, cross-cultural appeal. A canvas awning shading the courtyard of a Mexican house depicted adaptation to local climate. The Contraspatial House by Donald Olsen and the Case Study House #4 (Greenbelt House) by Ralph Rapson demonstrated the creative application of contemporary pre-fabrication techniques and depicted the courtyard as a space of suburban leisure and entertainment. Morse Payne's studies for the In-Line Court House connected the discussion to the grid, serial composition, and the imaginative possibilities of infinite expansion.

In light of his academic training at Harvard, McHarg's proposal for urban housing with interior courtyards is not as unexpected or intellectually novel as it at first seems. After all, Le Corbusier in *The City of To-Morrow and its Planning* referenced Howard and the Garden City as precedent for his Radiant City and its cellular system of housing units interspersed with garden terraces.[23] Less well known today, but equally popular among GSD students during McHarg's tenure there, was a similar argument presented by Paul and Percival Goodman in *Communitas: Ways of Livelihood and Means of Life*.[24] The Goodman's maintained, like Le Corbusier, that urban planners should refrain from promoting dispersion. But in a less aggressive countermeasure that called into question the *tabula rasa* posture of Le Corbusier, they argued the city could become more livable, inviting, self-expressive, and liberating through small-scale, decentralized, action that privileged the home and family. Robert Geddes, one of the architects who worked with McHarg on the collaborative master's thesis, would later note that the decision to revitalize downtown Providence, instead of designing a New Town in the suburbs, reflected the argument presented in *Communitas*.[25]

Once again drawing inspiration from the work of iconic designers and theorists, McHarg deployed the courtyard house to strategically reinsert key aspects of the Garden City back into the fabric of the existing city. He also borrowed ideas from *Communitas*, including the notion of the home as an "asylum for the personality," and based his scheme on small-scale, individualized action, separation, and seclusion.[26] Removed from the less than ideal reality of the outside world, the interior garden provided a safe refuge where it was possible to step back, reconnoiter, and construct a personally empowering environment, even if this meant retreating into an intensely private psychic realm. For McHarg, the courtyard house was the ideal urban dwelling precisely because it turned away from the "squalor, poverty, and ugliness" of the street. Conversely, the high walls restrained unruly manifestations of individual expression, no matter how "anarchic" or "vulgar" from intruding on the general public.[27] In this democratically ambiguous scenario, design operated within two equally problematic environments – the external one that people cooperatively inhabit and the internal one that people autonomously construct.

During his promotion of the courtyard house as a controlled zone of urban tranquility, McHarg presented the Rockefeller Guest House, a townhouse in Manhattan designed by the architect Philip Johnson, as an exemplary model. This praise no doubt reflected the fact that McHarg was at that time actively attempting to interest Johnson – a graduate of the GSD and director of the Department of Architecture at MoMA – in a potential housing project.[28] But self-promotion aside, his interest in The Rockefeller Guest House also reflected his appreciation of the design's organizational clarity and elegant minimalism.[29] In *Design with Nature*, he would refer to the space as an urban "oasis" of intelligence and art that accomplished so much with so little.[30]

The "Mrs. John D. Rockefeller III Guest House" (1949–1950) was Johnson's second commission for Blanchette Rockefeller, a prominent socialite and trustee of MoMA. Designed for entertainment and the display of art, the Guest House served as an adjunct salon for her New York apartment. The exterior façade ensured privacy, and the interior opened into a visually continuous space centered by a glass-walled garden open to the sky.[31] Although small in scale, this composition reflected the essence of architecture and landscape architecture – a hearth, sunlight, trees, and water – and it presented these design elements as essential aspects of the art of living. This was an intelligent and humane house, and it clearly emanated from wealth and power, which brings this discussion to a problematic aspect of McHarg's argument. Discrimination, which was nominally the singular domain of the educated and cultured, became after the social upheavals of the 1960s an elitist and exclusionary term, and rightly so. Nevertheless, it is a mistake to limit McHarg's interest in the Guest House to a dynamic of exclusion. Equally important, and as he indicated at the beginning of "The Court House Concept," he wanted middle class urban dwellers to live in houses commensurate with the custom-built villas of the wealthy. The housing examples in the essay accomplish this objective inexpensively and imaginatively. Discrimination, in this sense, represented innovative and pragmatic problem solving, and, somewhat ironically, the democratization of design. Similar motives were at play, as discussed earlier, when McHarg highlighted the Eames's molded plywood chair in the photographs of his student apartment.

In the opening chapter of *Design with Nature*, the Guest House appeared in a description of a 1949 student excursion to New York City that occurred on a sweltering summer day. As recounted, McHarg and his companions visited the Lever House, the United Nations, and the MoMA Sculpture Garden.[32] "Footsore, tired, sweaty, grubby, crumpled and thirsty," they arrived at the last project on their list – a brownstone conversion by Philip Johnson:

> We passed through the bland façade into a small vestibule and immediately left both heat and glare behind. We moved into a large and handsome living room, the end wall of glass subtending a small court defined by a guest wing. This was dominated by a pool with three stepping stones, a small fountain, a single aralia tree and on the white

painted bricks a tendril of ivy. We stood on the narrow terrace beside the pool, savoring the silence, then discovering below it the small noises of a trickling fountain, drips and splashes, the rustle of the delicate aralia leaves, seeing the reticulated patterns in the pool, the dappled light . . . sun and shade, trees and water, the small sounds under silence. What enormous power was exerted by these few elements in this tiny space.[33]

Serving as a touchstone for his ecological aesthetic, this account of the Guest House quickly moved past the utilitarian and into the poetics of unconscious and involuntary emotions that accompany unexpected pleasure. The façade concealed the interior from prying eyes and denied entry to the heat, noise, and grime of the street. Inside, delicate patterns and sounds created a soothing, evocative ambiance. The resulting internalization and intensification of perception was emotionally palpable, physically calming, and worthy of emulation. These same attributes made the encounter transitory, precious, and worthy of protection.

The relaxing, contemplative pleasure McHarg experienced when he entered this carefully crafted space mirrored a design discourse on the strictures of functionalism that had emerged in response to the rapid, overwhelming, and seemingly uncontrollable pace of mid-20th-century urbanization. Architecture, as a critical countermeasure, was here conceived as a restorative art that addressed the feelings of discomfort and alienation endemic to a world fractured by change.

The philosopher and social theorist John Dewey established an important framework for this discourse when he removed art from the rarified confines of the museum and situated it within the everyday as an object or activity, no matter how ordinary or commonplace that caused people to have heightened or intensified interactions with their surroundings.[34] Intrigued by the relationship of perception to shared cultural values and social action, Dewey transformed art into a participatory aesthetic where the events and objects of daily life defined how people experienced the world, how they inhabited the world they experienced, and how they treated the world they inhabited. To explain his ideas, he offered a tautology: "It [everyday experience], becomes a home and the home is part of our every experience."[35] In his mind, nothing transcended the commonplace in teaching people how to construct inclusive communities. And, by transforming everyday acts of social reproduction into expressive acts of hope, he was certain the environment would be aesthetically pleasurable and eminently practical.

Lewis Mumford provided an equally outspoken and more pessimistic defense of contemplative space and experiential practice. Similar to Dewey, he argued the environment, when sensibly organized, was a socially empowering and supremely democratic work of art that united people with their surroundings and each other. He likewise believed this interchange would only be meaningful when the viewer learned to appreciate and value what they saw and felt. In *Art and Technics*, he observed that art could:

> . . . widen the province of personality, so that feelings, emotions, attitudes, and values, in the special individualized form in which they happen in one particular person, in one particular culture, can be transmitted with all their force and meaning to other persons or to other cultures.[36]

Yet for Mumford, who fatalistically believed in fixed stages of cultural development – a biologically deterministic sequence of birth, growth, senescence, and death – the ultimate value of contemplative space lay in its ability to deter cultural decline.[37] It was his contention that physically impoverished environments hastened decline because they bred emotionally impoverished communities where it was common for relationships to unravel. Vibrant environments, in contrast, promoted vital interchanges that allowed people to establish living bonds with their surroundings and each other. To ensure the second outcome, he called for a contemplative aesthetic that slowed the frenetic pace of life. This action allowed the fleeting qualities of experience that "slip too quickly out grasp" or "sink too deeply into the human unconscious to be retrieved."[38] In a move that deliberately dismissed the cosmopolitan cars, yachts, and airplanes that populated the architectural renderings of European modernists, he promoted pastoral landscapes where stability, comfort, and simple pursuits overruled technological prowess, achievement, speed, and power. Mumford, in other words, accepted architectural modernity's heroic stance and universal quest for order, but not the accompanying technocratic insistence that progress was best achieved through the cogs and gears of mechanized production. He instead claimed, and this was critical for McHarg, that certain landscapes, such as the Greek forum and New England village, embodied time-honored values that fostered social prosperity and cultural longevity. Significant here, as it was for Dewey, was the indispensable role of the everyday in the development of common values and democratically progressive ideals.[39]

One of the most insightful discussions of perception, experience, and meaning as a living practice in artistic production, however, can be found in the argument made by Suzanne Langer in *Feeling and Form: A Theory of Art Developed from Philosophy in a New Key*.[40] Langer used poetry and music to position form as the materialization of a creative dynamic that merged intellect and emotion. In this union, interior and exterior space harmoniously coalesce, and the experience of life is symbolized by its setting.

McHarg's contribution to this aesthetic debate illustrated how natural sights and sounds were comparable in worth and beauty to architecture, art, poetry, and music. Indeed, and returning for a moment to the design principles disseminated by Gropius at Harvard, without the landscape there would be no gesamtkunstwerk, or total work of art, and, therefore, no such thing as complete building.[41]

"The Court House Concept" concludes with student work done under McHarg's tutelage. The housing in these projects employs a common trope in modernist architecture – an asymmetrical arrangement of walls and windows that dissolve the barriers separating interior and exterior space. This pattern of organization extends outward and forms a communal area sheltered by buildings and trees. The resulting spatial paradigm returns the essay to its rhetorical beginning and reinstates the garden in the city, honors architectural modernity, advances the aesthetic virtues of the landscape, supports the personal needs of the urban dweller, and touts the discriminating touches of skillful design. McHarg captured the generative potential of these ideas in a sketch that he drew on the back of the last page of the 1955 housing studio syllabus (Figure 2.6).

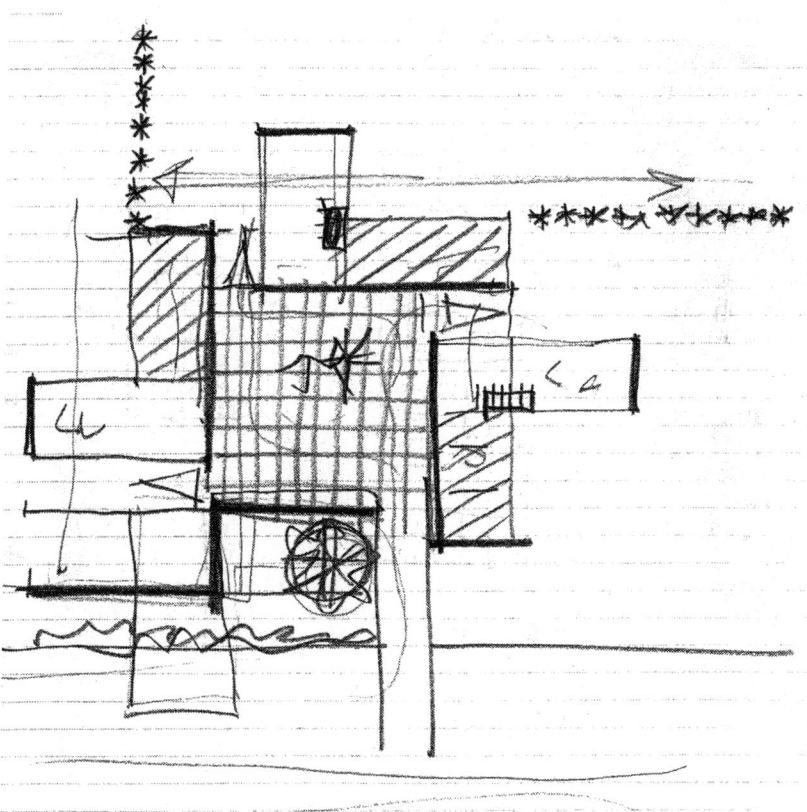

Figure 2.6 Sketch of courtyard housing by Ian McHarg. Ian L. McHarg Papers, The Architectural Archives of the University of Pennsylvania.

1958: The humane city

In 1957, McHarg delivered a talk at the GSD that explored the physical and social attributes of the ideal city and their origin: Did they come from the urban plan or the people who lived in the city? In a compilation of loosely structured observations that embraced civic idealism, transcendent form, and revelatory processes, he suggested it was a little of both. Throughout his presentation, personal insight challenged standard practice, aesthetics vied with function, and elegance and discrimination collided with a less than ideal urban reality. The resulting argument affirmed his desire to create an egalitarian environment and his acceptance of the civilizing power of discriminating elitism. The lecture appeared in *Landscape Architecture* in 1958 under the title "The Humane City: Must the Man of Distinction Always Move to the Suburbs?"[42]

McHarg began his remarks with a thinly veiled allusion to the prominence of the GSD, its design philosophy, and its illustrious teachers and students. The city, exemplified by the classic attributes civis (city, civilized, civilization), urbs (urbane), and polis (polite and elegant), was the historic home of "man's highest accomplishments," and its "most accomplished men." This "aristocracy of concern, ability, and art," he continued, served as "the vital vanguard and custodian of the development and conservation of the city, its changing form and expression, its social and cultural institutions, its history and artifacts." Energized by wealth, social stature, and philanthropic largesse these individuals opposed "greed and philistinism" and the abhorrent "rows of diners, gas stations, hot-dog stands, second-hand car lots, billboards and dump heaps." He also claimed that the enlightened guidance of these individuals would halt urban flight and transform the city into "a proud creation best representing the art, technology, wealth, and social and political aspirations of America."[43] His subsequent call for a pedestrian-friendly city with central services, parks, and plazas referenced ideas from his GSD thesis.

The convention-bound classicism that McHarg outlined at the beginning of his talk may seem passé, even for the 1950s, but this does not deny the fact that it aligned his argument with ideas sanctioned by the leading practitioners of post-war modernism, including Josep Lluís Sert, who was now the dean of the GSD and the chair of the architecture department, and a likely member of his audience. Sert, a former colleague of Le Corbusier and close associate of Gropius, was the president of CIAM when he assumed his position at Harvard. An avowed modernist and critic of decentralization, he believed it was possible to revitalize the heart of the city, and thus the heart of architectural modernity, through designs that emphasized community, spontaneity, culture, and art.[44]

Sert had earlier explained how to achieve this vision of redemptive civic design in the introductory essay of *The Heart of the City: Toward the Humanization of Urban Life*, a text, which, as previously noted included McHarg's thesis as a case study.[45] The essay that Sert wrote for this compendium of CIAM ideology referenced historical precedent to craft a timeless, and yet symbolically resonant and culturally superior acropolis, in the sense of Ortega y Gasset, made possible through physical and social circumstances that allowed people to congregate and events to happen.[46] A central square, modeled on the Greek forum, grounded his ideas and shaped the emotional life of the urban masses. A backdrop of "trees, plants, water, sun and shade," arranged to harmonize with the shape and color of the buildings, enhanced the humanism of the concept. The Piazza del Duomo in Milan, Rockefeller Center in New York City, and Sert's work in South America served as representative examples.[47]

When he spoke at the GSD, McHarg was similarly exploring the attributes of historically resonant and socially engaged urbanism through his involvement in the preparation of several research grants for The Rockefeller Foundation. One of the grants, endorsed by Jane Jacobs and funded, underwrote a visual survey of cities, small towns, highways, and farmland in the United States.[48] The architectural critic Ian Nairn conducted the survey and published his findings in *The American Landscape: A Critical View*.[49] Nairn's survey promoted a pedestrian-friendly and socially-oriented urbanism, and like Sert, he made classical traditions the foundation of innovative modernism.

With these thoughts in mind, it is important to note that McHarg looked to history for inspiration, but he did not seek to recreate the past. On the contrary, he argued that advances in science and technology had irrevocably altered demographics, employment, transportation, and communication systems. After he itemized these changes, he presented his main argument. It was easy to construct highways, assemble factories, and build homes, but it was difficult to instill these actions with imaginative inspiration:

> . . . it [the city] should be human – beyond that, it should be transcendental – that everyday activities of man should gain from and in turn be reflected in the physical environment. Man should be induced to rise above himself by the art which composes the city; and in turn the best philosophical, social and political thought should be represented in the art.[50]

McHarg found additional intellectual support for his vision of transcendent design in a discussion of the artist Morris Graves written by the left-wing essayist and social critic Kenneth Rexroth. While it is unclear how Rexroth came to McHarg's attention, his creative sensibility, lauded for its fusion of Western pragmatism and Eastern mysticism, contained themes of interest to McHarg.[51] These themes included the physical and psychological rejuvenation attained from intimate contact with the natural landscape and the belief that design should emulate life and its processes of becoming.[52] Rexroth also appeared in the lecture notes that McHarg prepared for his course at The Glasgow School of Architecture, next to the quote from Santayana, which suggests that the social criticism of this leftist guru, in some way, captured the quality of things and the inherent beauty of a particular way of life that mattered to him.

Graves had an evocative and recognizable painting style. He deployed a web of white calligraphic lines to bind imaginary animals, often wounded birds, to their surroundings.[53] A darling of pop culture, in 1953 *Life* magazine named him one of the United State's brightest and most original artists.[54]

"The Visionary Paintings of Morris Graves," written by Rexroth in 1955 for *Perspectives USA*, commended the reclusive, self-taught Graves for his primitive, intuitive brilliance and innate ability to express the "local color" of the Pacific Northwest where he lived. His art, in sympathy with this cultural heritage, entwined "indigenous" sensibility with the "non-plastic" simplicity and sophistication of oriental philosophy.[55] Rexroth also stated the spiritual communion of birds, trees, and rocks in these paintings symbolized a "pre-modern" knowledge in which "form emerges from formlessness" and miraculously "seems to bleed quietly into being."[56] Moreover, his use of the words "local color," "indigenous," "non-plastic," and "pre-modern" to describe Graves was a conscious ploy designed to make his readers (who for the most part either defined themselves as, or aspired to be members of the design intelligentsia) perceive the immersive, organic brushwork of this enigmatic west coast artist as a worthy, and inevitable, alternative to the abstract, hard-edged empiricism of cubist-inspired modernity. This stance, in keeping with the policy of the editorial board of *Perspectives USA* – an advocacy journal underwritten by the Ford Foundation to tout the superiority of American culture at the height of the Cold War – challenged the intellectual supremacy of the Euro-centric design community. As one might also suspect given his sponsor, Rexroth's strong endorsement of Graves's paintings of mystical birds was a sly, but recognizable rebuke of Le Corbusier.

In what is perhaps the most unexpected and interesting observation in "The Humane City," McHarg momentarily shed his east coast elitism (but not his officious tone), assumed the Zen-inflected posture of west coast anarchism, and quoted this passage from the Rexroth essay:

> The function of the artist is the revelation of reality in process, permanence in change, the place of values in a world of facts . . . The activities of men endure as long as they emanate from a core of transcendent calm. The contemplative, the mystic, assuming moral responsibility for the distracted, tries to keep his gazed fixed on that core. The artist uses the materials of the word to direct men's attention to it. When it is lost to sight, society perishes.[57]

This embrace of the cryptic patterns of Zen-thought allowed McHarg to use Rexroth's rhetoric of simplicity and beauty to promote contemplative designs that revealed the fundamental nature of life, and the enduring value and permanence of its processes. Timeless design, he informed his audience, was not timeless: change was inevitable and natural. But this reality did not preclude, and even enhanced the possibility of shaping civic, civil, and revelatory behavior. And then, after he positioned these ideals as the origin point of design, he noted that transcendent cities would materialize only when the fluidity of natural processes (feeling) counterbalanced hard-edged facts (thinking).

To express the importance of social relationships in this vision of design, McHarg returned to his intellectual roots and referenced The Pioneer Health Centre – a preventive medicine facility located in a poor section of London that was well known in CIAM architectural circles and profiled in *The Heart of the City*.[58] According to McHarg, the physical form of the building – which echoed the typology of the courtyard house – in conjunction with the spatial program – which included recreation rooms, nurseries, medical examination rooms, pre-marital counseling offices, a restaurant and theater, and a central atrium and swimming pool – represented an innovative, activist architecture that improved the lives of the people who frequented the facility. He appropriated a statement made by the Centre's founder, G. Scott Williamson, and issued a stern admonition to "prima donna architects" who focused their attention on the creation of monuments to their reputations and ignored the city and the people who live in the city: "The power of the architect [landscape architect and city planner] to fix the conditions in which life and living take place is tremendous, almost frightening."[59]

Ultimately though, McHarg's health-promoting, Zen-inspired, and socially-egalitarian design rebellion should not overshadow the fact that the moral certainty of his argument was inextricably bound to his Harvard education, and to ideas expressed by Gropius in the lecture "Apollo in the Democracy." This talk, which Gropius delivered in 1956 toward the end of his career following a trip to Japan, called for enduring values in a world where science removed magic from life and relativity negated eternal values:

> When we look at our task in its great diversity we see that it actually embraces the whole life of civilized man in all its essential aspects: the fate of the soil, the forests, the waters, the cities, and landscape; the sciences of man, biology, sociology, and psychology; law government, economics, art, architecture, and technology.[60]

In addition to considering the total environment as his professional domain, Gropius commanded architects to embed themselves within their "natural surroundings as the center point of all planning and building." The task required will, determination, and endurance. If designers persevered, they would generate "the dream, wonder, joy, and the illusion of human life in a new and magical beauty."[61]

The manifesto presented in "The Humane City" differed from that presented by Gropius in "Apollo in the Democracy" only in its promotion of landscape architecture. Displaying the same overarching ambition, McHarg sought a fundamental reorganization of priorities that allowed "trees and sky, rock and water, hills and valley, shrubs, flowers, and grass as they live and die and change through the seasons" to come first. To achieve this objective, he commanded landscape architects to include "light, granite and marble, steel and glass" in their design vocabulary.[62] Only then, he claimed, would they transcend the horticulture limits imposed by architects and capture the essence of nature in their art.

When McHarg seized the materials and operative principles of Gropius and Sert he opened his argument to the same criticisms. At this stage in his career, however, architectural modernity still reigned supreme and the advantages of appropriation outweighed the costs, particularly if, as he repeatedly claimed in this argument, the physical and psychological benefits of the landscape grounded his project. Moreover, by deploying the tools and tactics of architecture to his advantage, he could willfully, somewhat subversively, and with characteristic military bravado, embed his profession (and himself) at the center point of design, and through will and determination make it the organizing discipline of the total environment.

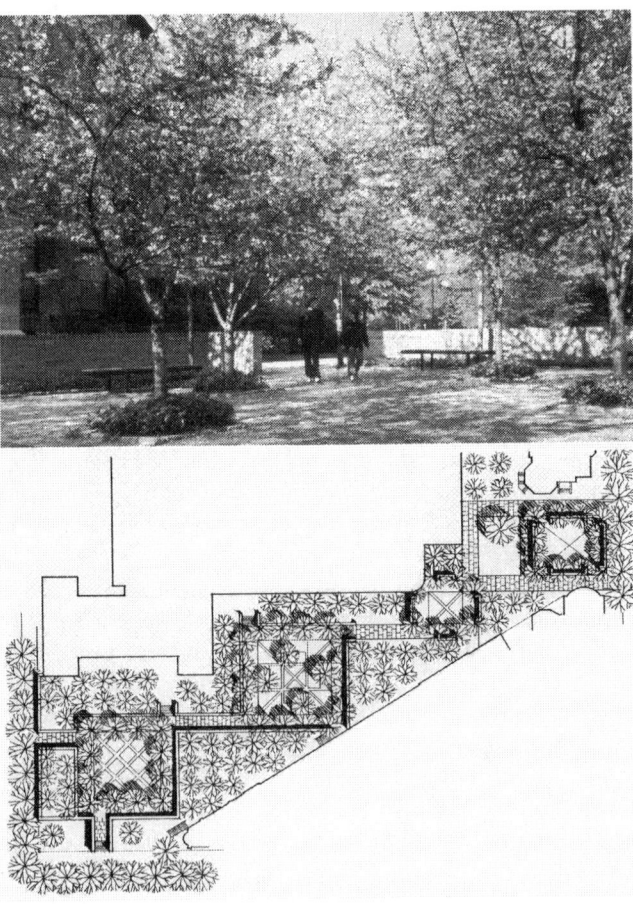

Figure 2.7 Woodland Avenue. Ian L. McHarg Papers, The Architectural Archives of the University of Pennsylvania.

McHarg's early acceptance and promotion of architectural modernity is notably represented in his design for Woodland Avenue, a pedestrian walkway at the University of Pennsylvania (Figure 2.7). Completed in the summer of 1957, and still in existence, the project occupies an abandoned trolley line that cut diagonally through the campus. The design consists of four rectilinear plazas paved in granite cobbles that increase in size as they move from the interior of the campus to a busy street intersection. Paths connect the plazas and demarcate their boundaries. Walls and trees further define these outdoor rooms. An intelligent response to an awkward site, the rhythm of elements, minimalist aesthetic, and clearly defined order provides a respite from the heat, noise, and grime of the city.[63]

Notes

1 See George Santayana, *The Philosophy of Santayana*, ed. Irwin Edman (New York: Random House, The Modern Library, 1936), 35. McHarg's personal library contained the book.
2 McHarg, *A Quest for Life*, 94.
3 Ibid., 95–19.
4 Ibid., 100–105.
5 The Architectural Archives of the University of Pennsylvania, Ian L. McHarg Collection, "Landscape Architecture Syllabus" (The Glasgow School of Architecture, The Royal Technical College, Session 1953/54). "Teaching Landscape Architecture at Glasgow School of Architecture," 109.I.B.1.18.
6 For further information on New Towns see William Ashworth, *The Genesis of Modern British Planning* (London: Routledge & Kegan Paul Ltd., 1954); Paul Kriesis, "A New Town Pattern is Born" in *Architects' Yearbook 6*, ed. Trevor Dannatt (London: Elek Books Limited, 1955): 53–74; and Ian McHarg, "Can We Afford Open Space? A Survey of Landscape Costs" *Architects' Journal* 123 (March 1956): 260–275.
7 The Neubühl housing development was built in 1930–1932, and designed by Max Ernst Haefeli, Carl Hubacher, Rudolf Steiger, Werner Moser, Emil Roth, Paul Artaria and Hans Schmidt. McHarg kept a plan of Neubühl in his personal files. See The Architectural Archives of the University of Pennsylvania, Ian L. McHarg Collection, "CIAM: CIAM étude présentée au CIAM 9 par le groupe CIAM Zurich" 109.I.B.2.4.
8 McHarg, *Quest for Life*, 117. Later, in *Design with Nature*, McHarg would reference an essay by C. H. Waddington during a discussion of form and rhythm in design. See McHarg, *Design with Nature*, 170; and C. H. Waddington, "The Modular Principle and Biologic Form" in *Module, Proportion, Symmetry, Rhythm*, ed. Gyorgy Kepes (New York: George Braziller, 1966), 20–37.
9 Ibid., 118.
10 Ibid.
11 Ian McHarg, "Open Space and Housing" in *Architects' Yearbook 6* ed. Trevor Dannatt (London: Elek Books Limited, 1955), 75–82.
12 Ibid., 79.
13 McHarg, "Open Space and Housing," 77, 80.
14 William Wurster, "The Outdoors in Residential Design" *Architectural Forum* 91 no. 3 (September 1949): 68–69.
15 McHarg, "Open Space and Housing," 77. See also Leon Festinger, Stanley Schachter, and Kurt Black, *Social Pressures in Informal Groups: A Study of Human Factors in Housing* (Stanford, CA: Stanford University Press, 1950).

16 McHarg, *A Quest for Life*, 95–99.

17 Ibid.

18 Ibid., 82.

19 Ebenezer Howard, *Garden Cities of To-Morrow* (Cambridge, MA: The M.I.T. Press, 1898, 1965).

20 Ian McHarg, "The Courthouse Concept" in *Architects' Year Book 8* ed. Trevor Dannatt (London: Elek Books Limited, 1957), 74–102.

21 Ibid., 74. See also Ian. L. McHarg, "The Court House Concept" *Architectural Record* 122 (September 1957): 193–200.

22 Ibid.

23 Le Corbusier, *The City of To-morrow and its Planning* (New York: Dover Publications, Inc., 1929, 1987).

24 Paul and Percival Goodman, *Communitas: Ways of Livelihood and Means of Life* (New York: Vintage Books, 1947).

25 Personal communication, Robert Geddes, August 2015.

26 Paul and Percival Goodman, *Communitas*, 160.

27 McHarg, "The Court House Concept," 75.

28 McHarg, *A Quest for Life*, 135. See also, Ian L. McHarg and Frederick R. Steiner, *To Heal the Earth: Selected Writings of Ian L. McHarg*, 157–159.

29 McHarg attended a meeting at the Philadelphia Redevelopment Commission to explore potential properties in Philadelphia where "my courthouse concept" could be developed. In addition to McHarg, attendees included Robert C. Wylie and David Wallace. McHarg's notes list Wylie as the partner of Mr. Philip Johnson, and Wallace, who would later become his professional partner, as an employee of the Philadelphia Redevelopment Commission. Johnson designed a private residence for Wylie (1953) and a prototype speculative home for Wiley Development (1955). See The Architectural Archives of the University of Pennsylvania, Ian L. McHarg Collection, "Court House Project, Philadelphia," 109.II.A.1.2.3. McHarg also invited Johnson to the University of Pennsylvania to serve as a guest critic for a studio on the design of Seagram Building's plaza, a project that Johnson was then working on with Mies van der Rohe. See McHarg, *Quest for Life*, 135.

30 McHarg, *Design with Nature*, 5.

31 See Stover Jenkins and David Mohney, *The Houses of Philip Johnson* (New York: Abbeville Press Publishers, 2001), 100–103.

32 The United Nations Building and the Lever House were built between 1947–1952 and 1951–1952, respectively. In an early draft of *Design with Nature*, McHarg noted that he traveled to New York in 1949 with two other students and the Lever House was completed. He would have seen the United Nations Building under construction. At the MoMA, he would have seen the 1944 design by Goodwin and Stone. See The Architectural Archives of the University of Pennsylvania, Ian L. McHarg Collection, "A Place for Nature in Man's World: Book Outline, January 19, 1967," 109.II.B.1.1; and The Architectural Archives of the University of Pennsylvania, Ian L. McHarg Collection, "Design with Nature Chapter Drafts and Related Research Material," 109.II.B.1.2.1. For further information on the MoMA sculpture garden by Goodwin and Stone, see Mirka Beneš, "Inventing a Modern Sculpture Garden in 1939 at the Museum of Modern Art, New York" *Landscape Journal* 13 no. 1 (Spring 1994): 1–20.

33 McHarg, *Design with Nature*, 5.

34 John Dewey, *Art as Experience* (New York: The Berkeley Publishing Group, 1934, 1980).

35 Ibid., 107–108.

36 Lewis Mumford, *Art and Technics* (New York: Columbia University Press, 1952, 2000), 16.

37 Patrick Geddes and Oswald Spengler influenced Mumford's concept of cultural development and decline. See Geddes, *Cities in Evolution*; and Oswald Spengler, *The Decline of the West: Perspectives of World-History Vol. 2* (Classic Reprint) (New York: Alfred A. Knopf, 1945).

38 Mumford, *Art and Technics*, 16.

39 See Lewis Mumford, *Sticks and Stones: A Study of American Architecture & Civilization* (New York: W. W. Norton & Company, 1924), and Lewis Mumford, *The Golden Day: A Study in American Experience and Culture* (New York: Boni and Liveright Publishers, 1926).

40 Susanne K. Langer, *Feeling and Form: A Theory of Art* (New York: Charles Scribner's and Sons, 1953). See also Susanne K. Langer, *Philosophy in a New Key* (Ambridge, PA: Harvard University Press, 1957).

41 Bayer, Gropius and Gropius, *Bauhaus 1919–1928*, 25.

42 Ian McHarg, "The Humane City: Must the Man of Distinction Move to the Suburbs?" *Landscape Architecture* XLVIII no. 2 (January 1958): 103–107.

43 Ibid.

44 See Alofsin, *The Struggle for Modernism*, 249–260.

45 Josep Lluís Sert, "Centres of Community Life" in *The Heart of the City: Towards the Humanization of Urban Life*, eds. J. Tyrwhitt, J. L. Sert, and E. N. Rogers (London: Lund Humphries & Co. Ltd., 1952), 3–16.

46 José Ortega y Gasset, *The Revolt of the Masses* (New York: W.W. Norton & Company, 1932), 7–28. Sert's essay in *The Heart of the City* contains an extended quote from *The Revolt of the Masses*.

47 Josep Lluís Sert, "Centres of Community Life," 27–28.

48 See Rockefeller Archive Center, Sleepy Hollow, New York, "Miscellaneous Correspondence." Rockefeller Foundation Record Group 1.2, Series 200, Box 44.

49 Ian Nairn, *The American Landscape: A Critical View* (New York: Random House, 1965).

50 McHarg, "The Humane City," 105.

51 The Architectural Archives the University of Pennsylvania, Ian L. McHarg Collection, "Landscape Architecture Syllabus" 109.I.B.1.18. The notes for the second lecture on "Open Space and Housing" situates Rexroth's work as an anodyne to the noise and squalor of the city. McHarg would present the same argument in subsequent essays on housing.

52 Linda Hamalian, *A Life of Kenneth Rexroth* (New York: W.W. Norton & Company, 1991), 374–375.

53 Ray Kass, *Morris Graves: Vision of the Inner Eye* (New York: George Braziller, Inc., 1983), 40.

54 Winthrop Sargent, "Mystic Painters of the Northwest" *Life* (September 28, 1953): 84–89.

55 Kenneth Rexroth, "The Visionary Paintings of Morris Graves" *Perspectives USA* 10 (Winter 1955): 58–66.

56 Ibid., 63.

57 McHarg, "The Humane City," 104. This passage from Rexroth also appears in the lecture notes that McHarg prepared for The Glasgow School of Architecture course.

58 See G. Scott Williamson, "The Individual and the Community" in *The Heart of the City: Towards the Humanization of Urban Life*, eds. J. Tyrwhitt, J. L. Sert, and E. N. Rogers (London: Lund Humphries & Co. Ltd., 1952), 30–35; Walter Gropius, *Rebuilding Our Communities* (Chicago, IL: Paul Theobald, 1945), 53. The Pioneer Health Center was conceived as a preventive-health initiative for a working class neighborhood in south-central London.

59 McHarg, "The Humane City," 105. McHarg inserted the words "landscape architect" and "city planner" in brackets into the Williamson statement.

A reference to "prima donna architects" also appears in the lecture notes that McHarg prepared for The Glasgow School of Architecture course.

60 Walter Gropius, *Apollo in the Democracy: The Cultural Obligation of the Architect* (New York: McGraw Hill, 1957), 6.
61 Ibid., 9.
62 McHarg, "The Humane City," 106.
63 The Architectural Archives of the University of Pennsylvania, Ian L. McHarg Collection, "Woodland Ave. (1958–1959)," 109.III.A.1.8, and "Woodland Ave. (1957–1958)," 109.III.A.I.9.

Part 2
The place of nature

3   Space, time, and being

When personal feeling transcends into Religion (not a religion but the essence of religion) and Thought leads to Philosophy, the mind opens to realizations.

Louis I. Kahn, Form and Design[1]

Early in his career, just when he began to formulate the environmental manifesto that he famously presented in *Design with Nature*, McHarg quoted the Puritan mystic Jonathan Edwards to describe his vision of design: "Space is necessary, eternal, infinite and omnipresent, but I have as good speak plain, I have already said as much as space is God."[2] These words embed the spiritual within the natural and they make divinity synonymous with insights obtained from direct observation. For McHarg, they also expressed a skeptical assessment of architectural modernity, which he considered closed to the world and trapped by an ossified, self-referential vision of itself and its truths. To rectify this situation, he turned to Edwards for inspiration and glorified the natural processes that lay between and beyond the walls of architecture's normative discourse. This desire to reveal and honor that which is commonly overlooked or culture conditions us not to see, such as the inextricable bonds that tie us to our surroundings, is key to understanding the environmental argument McHarg sought to make and its relationship to the modernist narrative. As later reconstituted in the hauntingly beautiful images of the sun and the earth that grace the front and back cover of the first edition of *Design with Nature*, he sought to make this celestial vision of space, time, and being so breathtaking and visceral that denial was not only blasphemous; it was impossible.

The phenomenon of space, and its importance as both an operative principle and mode of critique in 20th-century design discourse was, of course, not new or uncontested. Sigfried Giedion, for example, explored the issue and its relationship to technology, fabrication, and artistry in *Space Time and Architecture: The Growth of a New Tradition*. According to Giedion, the creative energies of modernity expressed a new visual dynamism that made it impossible to comprehend form and space from one vantage point. Consequently, designs that exemplified the singular view, such as the garden of Versailles, no longer reflected the spirit of the age. It was imperative, he argued, to shed the spatial confines of old absolutisms and ethical restrictions, and instead see the world "from all sides, from above and below, and from inside and outside," through the simultaneity and angled multiplicity of cubist space.[3]

McHarg readily agreed with Giedion's visual assessment and actively sought to understand the world from multiple viewpoints as a many-sided phenomenon, but as already noted he faulted his architecture colleagues for their general inability to see beyond built form. This is what brought him to Edwards, the omnipotent power of evidence, the spiritual resonance of space, and the immanence of nature. And, like Edwards, he used empiricism to embed the spiritual within the natural to sanctify his vision of the world.

When Edwards presented his observations on space in the essay "Of Being" he posed fundamental questions about knowledge and its relationship to what is observed and what is known by inference.[4] Empirical observation illustrated the hand of God in every point of space and time: inference presumed his universal presence in Nature and the order of things. McHarg likewise believed there were fundamental laws that animated the world, and these laws were tantamount to divinity. But in light of recent advances in science and technology, he also believed the divine will of creation was now in the hands of the designer and it was his duty to finish the work the gods of architecture deemed inconsequential.

McHarg's reference to Edwards appeared in a 1958 article titled "The Humane City." This quixotic essay, as previously discussed, explored the power of design to mold people and their actions, and it championed the landscape as the means to improve the human condition and urban life. If cities were not physically humane, McHarg maintained their citizens would not be benevolent and caring. To give further meaning to this claim, he quoted the Zen-inspired social critic Kenneth Rexroth to explain how truly revelatory designs employed the materials of the world to direct our attention toward "reality in process, permanence in change" as expressed by "trees and sky, rocks and water, hills and valleys, shrubs, flowers, grass-throughout the seasons."[5] He provided an additional glimpse of what he was after in another purloined quote, this time from Lâo-Tse: "It is on the space where there is nothing that the utility of the dwelling depends."[6] While it is hard to make sense of this bricolage of appropriated quotes and pilfered philosophies, they do begin to identify the evidences, inferences, and aspirations that McHarg would subsequently assemble and present in *Design with Nature*. As the first tentative steps of a larger spiritual journey, they illustrate how he sought to temper his intellectual dissatisfactions with Eastern wisdom. During this search for alternatives he accumulated more than one set of principles, and numerous assumptions, to fabricate his design ideology.

RIAS

McHarg acquired a critical component of his environmental manifesto in 1956 while assisting the architect Louis Kahn in the selection of a site for a new laboratory commissioned by the Research Institute for Advanced Science (RIAS).[7] During a visit to the existing facility, a chance encounter with one of the RIAS scientists opened his eyes to a vital but hidden dimension of the landscape that changed the way he read the land and perceived his relationship to it.

Established in 1955, RIAS was the research arm of the Glen L. Martin Company (now Lockheed Martin). The institution's mission statement declared allegiance to the phenomena of nature and the observation and investigation of the underlying basis of these phenomena, and to the application of this knowledge toward the improvement of human welfare (Figure 3.1). Areas of specialization included high-tech aeronautic materials, spacecraft guidance, and chlorophyll research. These scientific interests coincided with a larger strategic initiative launched by the federal government to advance the technical supremacy of the United States in its struggle against Cold War adversaries.[8]

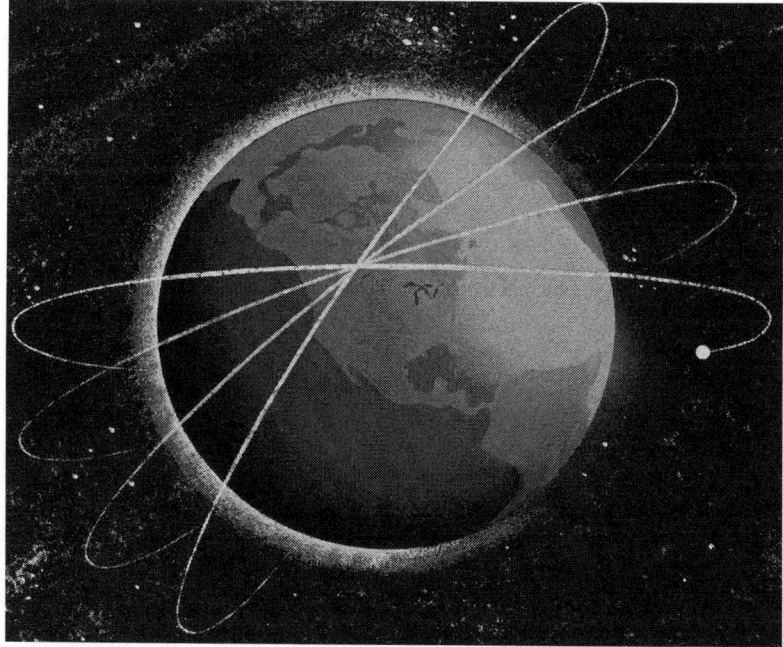

Figure 3.1 Research Institute of Advanced Study promotional material. Louis I. Kahn Collection, University of Pennsylvania and Pennsylvania Historical and Museum Commission.

The company's annual reports positioned the research conducted at the facility within a benign, but nonetheless expansionist vision of science and knowledge. The 1957 report, for example, noted the acronym RIAS was the plural of "ria," or a long narrow inlet formed by the partial submergence of a river valley. The intentional word play indicated the basic and applied research conducted by the institute's scientists, similar to the exploratory journeys of 19th-century natural historians, adventurously followed the channels of natural phenomenon "from their deltas on the ocean of technology to their fountain-heads in the rich hinterland of the rock-ribbed continent of pure science." A bust of the Greek scholar Democritus, who hypothesized that "all matter consists of innumerable tiny particles," positioned these explorations within the humanist tradition and reinforced the critical importance of RIAS research in the advancement of freedom, plurality, and democracy. An image of Sir Isaac Newton studying the prismatic mechanics of light referenced the fundamental nature of the work. A micrograph of a cosmic ray produced by one of the company's Nobel Laureates and a photograph of a hand cradling an array of semi-conductor transistors demonstrated the research produced at RIAS made important contributions to both the theoretical understanding of atoms and the practical application of this knowledge (Figure 3.2). The report also touted the new research facility and noted the strategic role it would play in future scientific and technical discoveries.[9]

Dr. Bessel Kok, an expert on photosynthesis, led the RIAS division that studied chlorophyll. His research on the molecular structure of this green pigment helped elucidate the biochemical reactions that enabled plants to absorb sunlight and transform it into sugars that other living organisms utilize as energy.[10] When McHarg observed Kok at work he was conducting a recycling experiment for the space program. The practical application of this research involved the use of algae as a food source in manned space

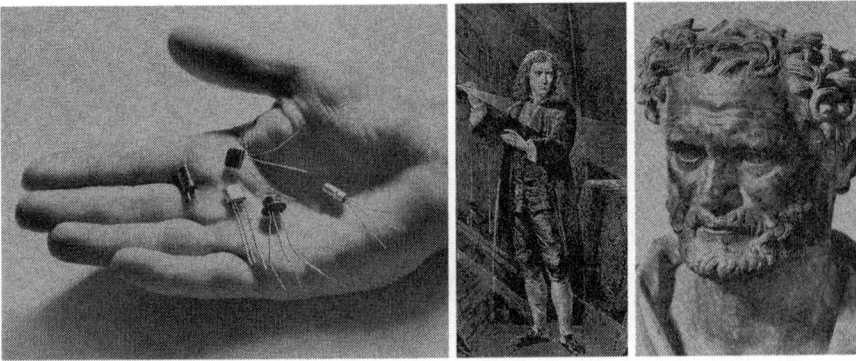

Figure 3.2 Research Institute of Advanced Study promotional material. Louis I. Kahn Collection, University of Pennsylvania and Pennsylvania Historical and Museum Commission.

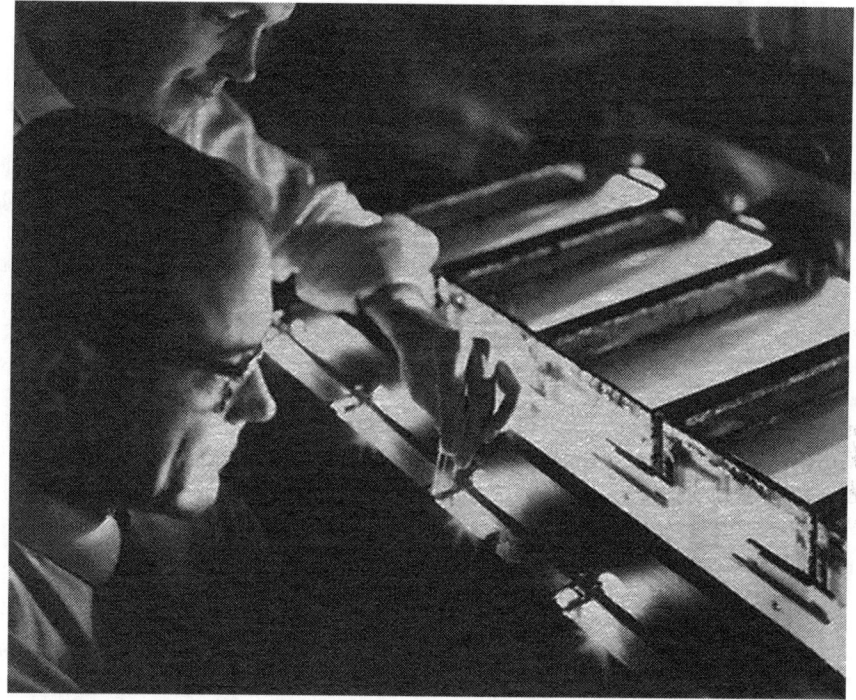

Figure 3.3 Dr. Bessel Kok, Research of Advanced Study promotional material. Louis I. Kahn, University of Pennsylvania and Pennsylvania Historical and Museum Commission.

exploration (Figure 3.3). This is how the encounter appeared in *Design with Nature*:

> Some years ago I spent a most instructive winter with Louis Kahn, searching for an appropriate site for a prospective temple of science. I learned much from my travels with this most perceptive architect, but my knowledge was even more enlarged by an encounter with a member of the research organization . . . his task was to send an astronaut to the moon with the least possible baggage . . . The experimental design required a plywood capsule, a fluorescent tube representing the sun, some air, water, and algae growing in water, some bacteria and man.[11]

Completely entranced by the notion that energy and material, as embodied by the plant, were one and the same, McHarg stated: "Suddenly I had an image of a green world, half turned toward the sun, leaves cupped to its light, encapsulating in their templates, into their beings, this modified

and ordered sunlight."[12] A spontaneous eruption of natural order –
something unspeakably old, pagan, and magical and yet utterly new, scientific, and rational – presented itself and he reveled in the power and the
glory: lightning struck; volcanoes erupted; hydrogen, oxygen and carbon
dioxide combined; DNA coiled upward into plants and animals; and the
gray-brown centers of cities, as they appeared from space miraculously
transformed into verdant gardens (Figure 3.4). By the time he completed
his description of the chlorophyll experiment, it was obvious this exercise
in applied science profoundly changed his thinking. Plants were not just
benign decorative elements: their ability to capture the light of the sun
reflected a natural, evolving order that he could utilize to realign design
practice. A diatom and a cross-section of a leaf represented the beneficial mutualism he had in mind: their shape, structure, and processes
made it easy to imagine the earth as simply a larger, albeit more complex
and sustainable version of the RIAS algae experiment (Figure 3.5). His
commentary, offered in soothing bio-chemical detail, conveniently overlooked Glen L. Martin's involvement in missile guidance systems and
nuclear warfare, and focused instead on the cooperative integration of
science and society.[13]

Though never realized, Kahn's proposed temple of science, designed with
Anne Tyng, combined Tyng's interest in precise, repetitive geometries with
ideas about the organization of space that Kahn recently gleamed from *The
Book of Tea* by Okakura Kakuzo.[14] This short text, which narrated the
history of tea consumption in Japan from its use as a medicinal herb to its
transformation into a refined ritual, argued for a way of life in harmony
with the proportions and rhythms of the universe. The philosophy of tea,
Kakuzo observed:

Figure 3.4 Images of the evolution of life from *Design with Nature*. Lightning
Louis I. Kahn Collection, The Architectural Archives of the University of Pennsylvania; Volcano Image # 122279, American Museum of
Natural History. Reproduced by permission; DNA helix reprinted by
permission of the estate of Bunji Tagawa.

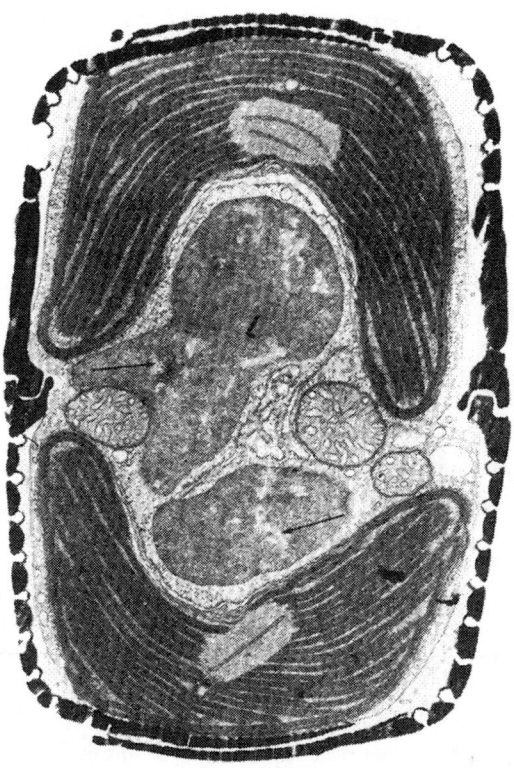

Figure 3.5 Electron micrograph of a diatom from *Design with Nature*. Diatom plate 563 from Drum et al., *Electron Microscopy of Diatom Cells* (Berlin: J. Cramer Publishing, 1969). Courtesy E. Schweizerbart Science Publishers, www.schweizerbart.de.

. . . expresses conjointly with ethics and religion our whole point of view about man and nature. It is hygiene, for it enforces cleanliness; it is economics, for it shows comfort in simplicity rather than in the complex and costly; it is moral geometry, in as much as it defines our sense of proportion to the universe.[15]

The underlying allure of *The Book of Tea* for Kahn, and what McHarg would duplicitously dismiss as inconsequential to his thinking, was the book's endorsement of the imaginative beauty of space, as expressed in the following adage that Kakuzo acquired from the *Tao* of Lâo-Tse: "The usefulness of a water pitcher dwelt in the emptiness where water might be put, not in the form of the pitcher or the material of which it was made." From this simple statement a moral philosophy and total design emerged. As Kakuzo further noted: "Perfection was everywhere if we only choose to recognize it."[16]

Intrigued by the generative simplicity of these universalizing pronouncements and their inversion of normative thought, Kahn would return to basics
and question his fundamental approach to architecture. During these explorations he examined the nature of space and the mystery of fabrication. His
discoveries, codified in a 1960 essay titled "Form and Design," were idealistic and practical and they involved the interplay of imagination and action.
Form – what a thing wants to be – reflected the immeasurable characteristics
and unifying essence of an idea and it served as the origin point of action:

> Form encompasses a harmony of systems, a sense of Order and that
> which characterizes one existence from another. Form has no shape or
> dimension. For example, in the differentiation of a spoon from a spoon,
> spoon characterizes a form having two inseparable parts, the handle
> and the bowl.[17]

Design – how a thing becomes what it wants to be – was all about measure-
ment, materials, technique, and conceptual realization, and it served as the
structural framework for action.

To resolve the dialectic tension between the idealism of aspiration (form)
and the pragmatics of realization (design) it was necessary, in Kahn's
words, "to go back to the beginning" and rediscover that singular moment
of inspiration when space was empty but nevertheless filled with "spirit
and resourcefulness."[18] Here, before action commenced and frustrations
emerged, it was possible to transcend the limitations of reality and intui-
tively know what a thing wanted to be. A year later, in a 1961 essay titled
"The Nature of Nature," he entwined his thoughts on form and design
with the elemental materials of the earth:

> Built into us is a reverence for the elements, for water, for light, for air –
> a deep reverence for the animal world and the green world. But, like
> everything which is deeply rooted in feeling and a part of our psychic
> existence, it does not come forward easily. There are times when we feel
> strongly, but the simple matter of doing daily chores and solving daily
> problems keeps us away from the feelings about such simple, wonder-
> ful, motivating things.[19]

In the case of RIAS, when Kahn returned to fundamentals and questioned
what a thing wants to be, he based his initial suppositions on the adage of
the water pitcher and embarked on an imaginative exploration of the mate-
riality and immateriality of space. This journey led to a checkerboard pattern of architecture and landscape that could theoretically accommodate
infinite expansion and still maintain the initial spirit and resourcefulness of
the design (Figure 3.6).

McHarg, deeply influenced by his discussions with Kahn regarding the
material order of things and the power of emptiness, would surrender to the

Figure 3.6 Proposed research facility, Research Institute of Advanced Study promotional material. Louis I. Kahn Collection, University of Pennsylvania and Pennsylvania Historical and Museum Commission.

seductive ambitions of this notable architect and follow a similar path in his promotion of urban housing with interior gardens. Also emboldened by the boundless vision of natural creativity that energized his thoughts when he witnessed the RIAS experiment, he moved beyond the circumstantial confines and crystalline certainties of architecture and began to explore the imaginative possibilities of spatial alternatives. The essential nature of this search led him to ecology, and, eventually, to the Quaker religion and the inner certainty of a spiritual energy – The Light Within – that arises from unmediated interactions with the environment.[20]

Notes

1 The Kahn essay was taped as a lecture titled "Structure and Form," for Voice of America prior to becoming "Form and Design." See Louis Kahn, "Form and Design" in *Louis Kahn: Essential Texts*, ed. Robert Twombly (New York: W.W. Norton & Company, 2003), 62–80.
2 McHarg, "The Humane City" 105.
3 Sigfried Giedion, *Space, Time and Architecture: The Growth of a New Tradition* (Cambridge, MA: The Harvard University Press, 1944), 356–357.
4 Jonathan Edwards, "On Being" in *Jonathan Edwards's Sinners in the Hands of an Angry God: A Case Book*, eds. Wilson H. Kimnach, Caleb J. D. Maskell and Kenneth P. Minkema (New Haven, CT: Yale University Press, 2010), 52–53. Edwards wrote the essay in 1721.
5 McHarg, "The Humane City," 105–106.
6 Ibid.
7 The Architectural Archives of the University of Pennsylvania, Louis I. Kahn Collection, "Research Institute for Advanced Science, Baltimore, MD; Unbuilt, 1956–1958."
8 William B. Harwood, *Raise Heaven and Earth: The Story of Martin Marietta People and their Pioneering Achievements* (New York: Random House, 1993), 283–284.

9 *Research Institute for Advanced Science, 1957 Annual Report*, 2, 7–9. The report provided courtesy the Glen L. Martin Maryland Aviation Museum.

10 Harwood, *Raise Heaven and Earth*, 184. See also J. Myers, "A Biographical Memoir of Bessel Kok 1818–1979" (Washington, DC: National Academy of Sciences, 1987). www.nasonline.org/publications/biographical-memoirs/memoir-pdfs/kok-bessel.pdf. Martin Marietta was under contract with NASA, and its Closed Environment Life Support Systems (CELSS) program, to conduct studies on the use of microalgae as a food source, supply of oxygen, and a catalyst for waste disposal in future human-crew planetary missions. Although the company discontinued this research in 1985, it eventually led to an algae additive in baby formula. See NASA Spinoff Technology Transfer Program, Consumer/Home/Recreation, "Nutritional Products from Space Research," https://spinoff.nasa.gov/spinoff1996/42.html.

11 McHarg, *Design with Nature*, 44.

12 Ibid., 46.

13 Martin Marietta eventually tested the experiment McHarg witnessed at RIAS in space. They sent the algae capsule into orbit as the piggyback payload of an intercontinental ballistic missile. See News Bureau Martin Company, A Division of the Martin Marietta Company. Press Release, Gravity Independent Photosynthetic Gas Exchanger (GIPSE) Release No. 3342–3163 (March 1, 1963), Courtesy of the Glen L. Martin Maryland Aviation Museum.

14 Okakura Kakuzo, *The Book of Tea* (Rutland, VT: Charles E. Tuttle Company, 1956). See also Peter Schneider, "Louis Kahn and the Little Book of Tea: Echoes of the Tao Te Ching in Louis Kahn's Thought" *International Journal of Humanities and Social Science* 14 no. 2 (Special Issue – July 2012): 22–27. The proposed research facility's checkerboard pattern reflected Tyng's interest in precise, repetitive geometries. In her Ph.D. thesis, Tyng described architecture as an additive process where built form encloses and shapes "in between" space. See Anne Griswold Tyng, "Simultaneous Randomness and Order: The Fibonacci-Divine Proportion as a Universal Forming Principle," University of Pennsylvania, Ph.D., 1975, Architecture, 106–108.

15 Ibid., 4.

16 Ibid., 45, 110.

17 Louis Kahn, "Form and Design," 64.

18 Ibid., 65.

19 Louis Kahn, "The Nature of Nature," in *Louis Kahn: Essential Texts*, ed. Robert Twombly (New York: W.W. Norton & Company, 2003), 119.

20 See Ian L. McHarg, *The Essential Ian McHarg: Writings on Design with Nature*, ed. Frederick R. Steiner (Washington, DC: Island Press, 2006), xxi. In the introduction of this text, Steiner noted that religion was central to McHarg's being and his decision to become a Quaker mirrored other momentous transitions in his life that included his change from soldier to scholar, his move to the United States, and his move from site-specific design to global planner. McHarg attended the Friends Meeting House in Chester County, Pennsylvania.

4   First principles

Self is the schoolmaster whose lessons are best worth his wages; and since the subject I am considering has not yet become a branch of formal instruction, those whom it may interest can, fortunately, have no pedagogue but themselves. To the natural philosopher, the descriptive poet, the painter, and the sculptor, as well as to the common observer, the power most important to cultivate, and, at the same time, hardest to acquire, is that of seeing what is before him.

George Perkins Marsh, 1864[1]

Shortly after he wrote the 1958 essay "The Humane City," McHarg refocused his design explorations to more directly engage the land and its processes of formation. This decision reflected his concern over the subordinate status of landscape architecture in the hierarchy of design professions, and his belief, following his observation of the chlorophyll experiment at RIAS, that his previous studies of housing and open space did not address important aspects of the environment that were crucial to health and prosperity. As he pursued this new line of inquiry, he became less interested in relating his work to the priorities of architecture and its operative codes of spatial organization and material detailing, and more intrigued by the complex environmental interactions that enable plants and animals to live together in relatively stable communities. He was soon drawn to the science of ecology and its study of the relationship of an organism to its surroundings, including all other organisms (Figure 4.1).

When McHarg first professed an interest in ecology, it was a relatively low-status science that lacked the public recognition it would achieve following the 1962 publication of *Silent Spring* by Rachel Carson. In light of his training at Harvard, however, his turn to ecology is not surprising, or unexpected. For example, in an *Approach to Design*, his GSD professor Norman Newton urged students to follow ecology's empirical path "as far as it goes" to gain deeper insight into the spatial arrangements that support the dynamics of life.[2] Even Walter Gropius, the chair of the school's architecture department during McHarg's tenure as a student,

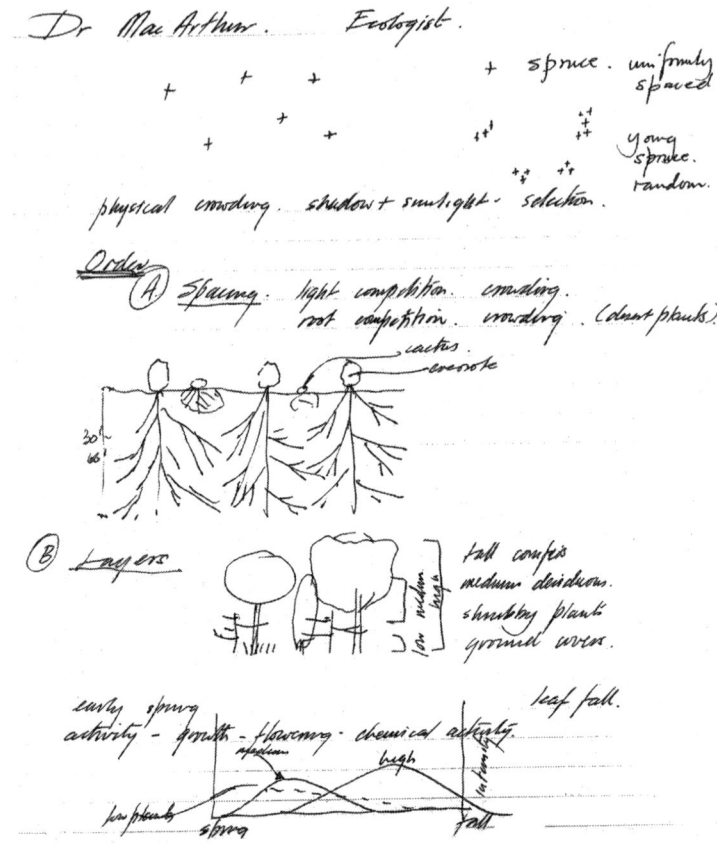

Figure 4.1 Sketch of the morphology of a deciduous forest by Ian McHarg completed during a lecture on ecology by Robert MacArthur. Ian L. McHarg Papers, The Architectural Archives of the University of Pennsylvania.

and a cause célèbre in modernist design circles, summarized his avant-garde design vision as an all-encompassing inquiry into "the fate of the soil, the forests, the waters, the cities, and landscape; the sciences of man, biology, sociology, and psychology; law, government, economics, art, architecture, and technology."[3]

A more immediate impetus for McHarg's turn to ecology came from the 1955 conference *Man's Role in Changing the Face of the Earth*. Sponsored by the Wenner-Gren Foundation and the National Science Foundation, and co-chaired by the anthropologist Carl Sauer, the biologist Marston Bates, and the social critic Lewis Mumford, this weeklong symposium celebrated the 19th-century statesman and natural historian George Perkins Marsh and the centennial publication of *Man and Nature; or, Physical Geography as Modified by Human Action.*[4] In this text, Marsh, for the most part, presented human action toward the land in less than stellar terms, and implicated the advance of Western civilization with denuded forests, soil erosion, and degraded water supplies. Distressed by "the dangers of impudence," he urged caution "in all operations which, on a large scale, interfere with the spontaneous arrangements of the organic and inorganic world."[5] Sharon Kingsland, in her biography of Marsh, observed that he wrote *Man and Nature* to persuade his countrymen to quit clear-cutting forests, depleting soils, destroying water supplies, and to instead settle down and build "a well-ordered and stable commonwealth" appropriate for a country committed to progress.[6] The 1955 conference updated the scientific context of Marsh's inquiry, globalized its scope, and reinforced the importance of natural harmonies in the maintenance of a healthy environment and a well-ordered society. A critical topic of speculation was the agency of "man" in geologic change, and why this anthropogenic action was often detrimental.[7]

In 1959, four years after the conclusion of the conference, McHarg developed a seminar course, titled Man and Environment, to advance the principles proffered by Marsh in *Man and Nature.*[8] He divided the course into sections devoted to science, religion, and social attitudes. The use of the term "environment" in the course title indicated the inquiry would be interdisciplinary, and it would present the world as a dynamic entity with its own laws and processes. Course readings included the conference proceedings from *Man's Role in Changing the Face of the Earth.*[9]

Ecology

Each of the guest speakers in Man and Environment provided useful information that could be applied to design, but the three lectures delivered by the ecologist Robert MacArthur, an assistant professor in the Department of Zoology at the University of Pennsylvania, proved fortuitous. The principles of ecology that this young, and at that time little known member of the scientific profession presented to the class would become central to his older colleague's theory of environmental design. The following discussion of these lectures is highly descriptive due to the complexity of the material, but this detail provides important insight into McHarg's understanding of ecology and his subsequent application of this knowledge in design and planning.

MacArthur studied ecology at Yale University under the guidance of G. Evelyn Hutchinson and received a Ph.D. from that institution in 1957. One year later he assumed a teaching position at The University of Pennsylvania.[10] His research examined population dynamics and species spatial distribution, and like many ecologists in the mid-20th century his analyses of these phenomena stressed competition between and within species, and the optimization of material and energy flows.[11]

When MacArthur spoke to McHarg and his students in the fall of 1959, he presented what was at that time a standard overview of species distribution in plant communities.[12] His lecture on "Ecology" began with a comparative description of a forest biome in the north of Canada and a desert biome in the southwest of the United States, as they appeared from an airplane.[13] The forest biome displayed little species diversity, and consisted of tightly packed and uniformly spaced spruce trees. The desert biome also displayed little species diversity and consisted of creosote plants separated by large patches of bare ground. MacArthur explained that both landscapes looked the way they did because they reflected selective pressures generated by their physical location. The distribution of spruce trees in the far north, where there was little sunlight for much of the year, reflected an adaptive response that optimized available light. The distribution of creosote plants in the desert where there was little rainfall reflected an adaptive response that optimized the availability of water. In this scenario, widely spaced plants with extensive root systems that spread deep into the soil in search of water had the best chance of survival.

MacArthur moved from the discussion of selective pressures in these extreme conditions to the temperate climate of United States in order to explain how the complex, three-dimensional pattern of species in a deciduous forest biome – a top layer of large canopy trees, a middle layer of understory trees and shrubs, and bottom layer of low shrubs, herbs, and flowering plants – reflected an adaptive response to sunlight that enabled multiple species of plants to co-exist. In addition to resource competition, time was an important factor in this discussion. The selective pressures at play in the deciduous forest resulted in plants that commenced metabolic activity (germination, and the appearance of leaves and flowers) and food production (photosynthesis) at different times in the spring in a sequential pattern from low to high as the season progressed. As a result, the layered organization of the forest reflected a pattern of competitive dominance in which the plants on the forest floor were subordinate to the plants in the understory, and the plants in the understory were subordinate to the canopy trees.

To complete his discussion of resource competition and spatial dominance in the deciduous forest biome, MacArthur introduced the concept of succession using an abandoned farm field as a model. His explanation began with the cessation of human disturbance and the subsequent sprouting of trees in the field – a few in the middle in a random pattern, but most along the edge of the field, near existing trees (Figure 4.2). This pattern of growth continued until the trees completely filled the field. In parallel with this process, the composition of plant species changed. Trees that tolerated shade replaced the early colonizers that required light. When the observed changes in distribution and species composition became so slow that they never appeared to alter, he further explained the forest had reached a climax state, and it would remain in that state as long as there was no major disturbance.[14]

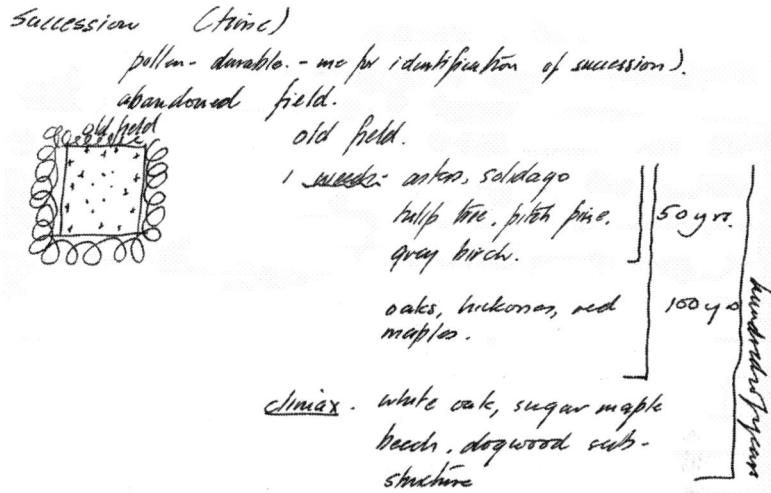

Figure 4.2 Sketch of plant succession in an old field by Ian McHarg completed during a lecture on ecology by Robert MacArthur. Ian L. McHarg Papers, The Architectural Archives of the University of Pennsylvania.

MacArthur's last example (selected in honor of McHarg's homeland) involved a type of succession that occurred high on the slopes of the Cairngorm Mountains of Scotland that cycled through stages and never reached a mature climax state. In this landscape, succession began with bearberry, which was forced by the prevailing winds to grow in one direction. Heather eventually grew in the soil that accumulated around the bearberry roots. But it, too, was blown in one direction by the wind and overtopped the bearberry. Reindeer moss covered the roots of the heather, which, in turn, was blown away by the wind. Once this happened, succession recommenced with the rooting of the bearberry in the bare soil.

In December of 1961, MacArthur delivered two additional lectures to McHarg's class that discussed spatial distribution and population dynamics in animal communities. There is no official transcript, but McHarg made extensive notes under the heading "Robert MacArthur Ecologist: Patterns of Natural Communities."[15] The notes, which contain the diagrams that MacArthur drew along with explanatory annotations, make it possible to recreate the presentations. The course syllabus lists the lectures under the titles "Symbiosis" and "Entropy."

Both lectures adhered to the mathematical model of population dynamics that MacArthur had theorized in "Fluctuations of Animal Populations, And a Measure of Community Stability."[16] This paper, which reads as a statement of logic and mathematical proof, borrowed from the system theorems outlined by the cyberneticist Claude Shannon in "A Mathematical Theory of Communication." In this paper, Shannon devised a sequence of differential equations to illustrate how multiple communication pathways increased the possibility of successful message transmission. The equations also explained how energy maintained the orderly structure and operations of these communication pathways, and how noise, or the faulty transmission of messages within this network, was a function of entropy and the dissipation of energy into a non-useable form.[17] Inspired by Shannon's elegant mathematical formulations, MacArthur created a logical sequence of differential equations to explain population dynamics in terms of energy, order, information, and communication.[18]

In regard to the topic of communication pathways, it is important to note that McHarg and his students revisited the concept of message transmission and discussed the theory of "Organized Complexity" proposed by the social scientist Lawrence K. Frank. It was Frank's contention that an organism's ecological relationship with its surroundings reflected a process of selective signaling and message filtering, and these "To Whom It May Concern" messages allowed different species to co-exist within a given territory in complex and diverse communities:

> Through evolution each organism has developed a concern for those messages which are essential to its living function and survival as a species, while ignoring what is not biologically relevant or useful. Accordingly, in any geographical area, many different species, bacteria in the soil, worms, insects, fish, reptiles, birds, amphibians, and the array of mammals, carry on their life careers, selectively receiving and responding to signals that are of concern to each species, while unaware of the many other messages that are being concurrently transmitted.[19]

Later, in *Design with Nature*, McHarg would state, in an unacknowledged reference to Frank, that the world is "a great voice of 'to whom it may concern messages,' clothed in form."[20]

MacArthur began the "Symbiosis" lecture with a description of cyclical fluctuations in animal populations. A simple two-species predator-prey model and the Lotka-Volterra equation explained how animal populations tended to oscillate around optimum conditions – balanced numbers of individuals and sufficient food sources.[21] If conditions were such that the numbers of predators and prey remained stable, the formula produced a double sine wave pattern of uniform size and duration. If conditions were such that there were more prey than predators to consume them, or more predators than prey to feed on them, the oscillations became unstable and fluctuated outwards and inwards respectively, in increasingly erratic, but nonetheless predictable, self-destruction (Figure 4.3). McHarg's notes contained sketches of the territory diagrams and the predator-prey graphs. The words "Competition," "Cooperation," "Commensalism," and "Co-existence" appear

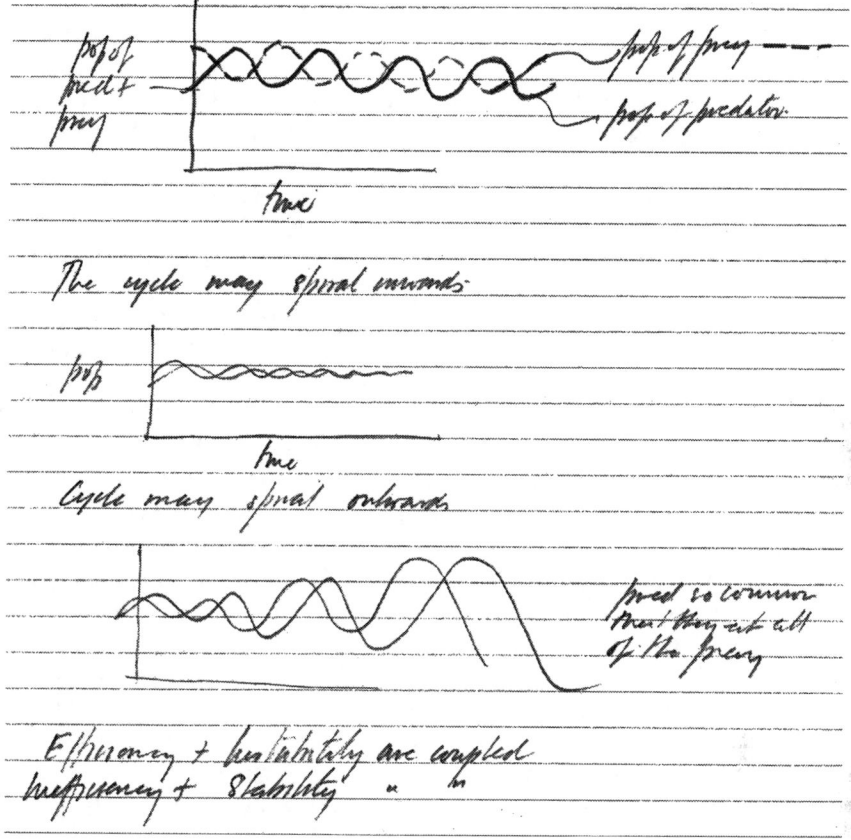

Figure 4.3 Sketch of predator and prey interactions by Ian McHarg completed
during a lecture on ecology by Robert MacArthur. Ian L. McHarg
Papers, The Architectural Archives of the University of Pennsylvania.

next to the territory diagrams in capital letters. Mimicking MacArthur's
mathematical treatment of the material, he also wrote: "Stability in diversity +
numbers of species."[22]

The "Entropy" lecture described the energy dynamics of species distribution and evolution, and it reflected MacArthur's recent focus on the interaction of song birds in forest trees, and the relationship of this behavior to food
sources, competition, territory invasion, and habitat complexity.[23] He began
the lecture with an explanation of the concept of natural selection and the
ability of an organism to adapt to its surroundings and successfully produce
offspring. This brief, but thorough, introduction to the basic terminology and
operations of Darwinian evolution included definitions of fitness, genetics,
and environment-induced mutations using references to R.A. Fisher and

J. B. S. Haldane.[24] As the preamble to a discussion of habitat and adaptive response, these remarks explained how natural selection worked at the cellular level, and who and what did the selecting. His subsequent comments described how the physical characteristics of the environment produced selective pressures that led to evolutionary change, and how the movement of animals to new habitats, in search of food and mates in their struggle to survive and reproduce, altered the selective forces acting upon them. Collectively, this information provided a concise explanation of how the environment and the organism both serve as agents of evolutionary change, while also demonstrating how resource abundance, population pressures, and territorial struggles enliven and increase the complexity of this iterative relationship. In his notes, McHarg wrote (and underlined) the "fitness of a population – its success at leaving progeny – is at the core of evolution."[25]

During the lecture, MacArthur also discussed the critical role of energy in the structure of ecosystems. To explain this topic he used the food web (an organization of organisms based upon predation) and habitat (the place where an organism lives) model proposed by Charles Elton in 1927, and later modified in 1942 by Raymond Lindeman to incorporate population dynamics and energy flows.[26] He illustrated his remarks with a diagram of the ecosystem energy transfer process that began with plants, photosynthesis, and the transformation of sunlight into a food source that could be utilized by animals. Following this essential first step, the flow of energy continued upward in the food chain through the herbivores (prey) that fed on the plants, to the primary carnivores (predators) that fed on the prey, and secondary carnivores that fed on the primary carnivores, and so on until it was totally consumed. Notations in the diagram documented the dissipation of energy into heat at each level of the food chain as mandated by the Second Law of Thermodynamics and the principle of entropy.[27] The notations indicated that living organisms absorbed and utilized approximately 10 percent of the energy during each step of the exchange, and this negentropic capture enabled them to maintain structure and function and avoid decay.[28] At the conclusion of this discussion MacArthur observed, in a statement that supported his thesis of diversity and stability, that the environment contained many interconnected food chains, and this web of energy and nutrition was so structured that an animal could potentially survive even when food was scarce. An important corollary to this remark was the understanding that energy capture enabled organisms to diversify and evolve.

MacArthur summarized the lecture as a series of verifiable propositions: available energy controls the number of individuals found in a geographic area; species abundance reflects adaptation to climate, latitude and environmental complexity; energy from the sun absorbed by the earth's surface moves up the ecosystem through a network of nutrition pathways; complex nutrition networks provide more opportunities for animals to feed and greater partitioning of habitat; habitat partitioning reduces the total number of individuals in a given area but increases the number of species; species diversity provides more opportunities for adaptive response when the environment changes. In addition to the energy diagram and summary principles, McHarg's notes contained the statement that competing species could co-exist if they utilized slightly different resources – land, water, light, nutrients, and shelter. And, in an editorializing comment that indicated how he interpreted MacArthur's thesis of diversity and stability, he wrote: "Wherever man goes instability follows."[29]

Material from these lectures proved critical to McHarg's nascent concept of environmental design, and he subsequently incorporated commensurate interactions, communication pathways, habitat partitioning, adaptation, entropy, stability, dominance, and diversity into the language of his arguments. Correspondingly, he considered land planning in terms of pattern distribution, signal transmission, and territorial dynamics as he sought to understand how the forces, events, elements, and organisms that constitute the environment adjust to each other, as they in turn respond to new circumstances.[30] In fact, the next spring, McHarg and his students used ideas from these lectures in a study of housing on a barrier island that they commenced following a major storm that caused significant damage to the shore communities of New Jersey. The development plan detailed in this study removed the houses from the primary dunes to minimize disturbance to the pioneering plant communities. As McHarg and his students learned from MacArthur, diverse ecosystems in which organisms optimize resources through the selective partitioning of habitat are the most successful and enduring.

McHarg would ultimately entwine sunlight, energy, entropy, order, equilibrium, and evolution into the structure of the argument presented in *Design with Nature*. His statement of principles included fact, faith, aspiration, and certitude:

> Entropy is the rule, it demands its price; all energy is destined to be degraded but physical systems are becoming more ordered on earth, while life systems continue to evolve toward greater order, greater complexity, less randomness – toward negentropy . . . Is this not a description of life and the direction of evolution – it is negentropic – creative.
>
> Can one then imagine negentropy as a tide of ordering, moving differentially against the force of entropy, paying its tithe, evolving from the order of the non-living into life, from simple to complex life. From uniformity to diversity, from a small number to an infinite number of species, aspiring to dynamic equilibrium, always imperiled as evolution moves forward? Within this path the cycles of life, death and decay recycle the increasing storehouse of ordered matter . . . evolving – as Teilhard de Chardin has suggested – towards increasing consciousness.[31]

McHarg's overriding objective was not that difficult to comprehend – if life creatively evolved and became more complex and ordered in response to the energy of sunlight, then societies, cities, landscapes, and people had to embrace this energetic principle if they, too, were to creatively evolve and become more complex and ordered. To stress the essential nature of these ideas, he culminated his remarks with an idealistic reference to the Jesuit priest Pierre Teilhard de Chardin and his prophetic vision of evolution.[32] This action, however, endowed ecosystem energy flow and thus his vision of environmental design with vital purpose, and by so doing transformed the provisional, experimental nature of the science that grounded his ideas into a messianic pursuit with a problematic teleology.

Religion

When McHarg introduced the subject of religion to the students in his seminar course Man and Environment, he explained that his interest in the topic reflected a conflict in his personal beliefs. He had been taught as a child that "man was made in the image of God" and that morality existed only in "man's acts to man."[33] But his study of natural history led him to believe the cosmos was united by a common order, and the earth, as an expression of this order, was in some way an aspect of God. A morality limited to relations among people, he continued, may advance social justice, but it did nothing to address the pollution of the air and water, the disorder of the urban environment, or the desecration of the countryside caused by poorly planned urban expansion. In their race to gain the functional and the necessary, people had somehow lost respect for nature. He then told the students, in a sentence that revealed his debt to Albert Schweitzer and his philosophy of reverence for all life, that it was possible to rectify this situation through the institution of a morality that extended beyond the "church and synagogue" and included all of nature and all of nature's creations.[34]

McHarg had clearly determined his religious stance and moral aspirations, but he needed to convince his students to also adopt a design covenant that was modern in its application of science, just in its actions toward people, ethical in its behavior toward the land, and visionary in its thinking. Using the class as a platform for change he stated his purpose:

> We heard the natural scientists talk about an order that can be observed, a natural order. Philosophers and theologians are also concerned with a natural order which can be seen, which can be explained, and a supernatural order which engenders belief, which cannot be seen . . . both these quests are united in an attempt to find meaning for life, for acts; knowledge can satisfy part of this quest, but man must always inquire beyond the limits of knowledge to find purposefulness and meaning.[35]

Underlying this ambitious agenda were two interrelated but disparate visions of Nature. On one hand, it was a natural process that could be isolated, divided into manageable pieces, quantified, and put to good use. On the other hand, it was a life principle that existed in a deified realm beyond scientific substantiation. For McHarg, both visions prompted one overarching question: How shall I use my knowledge and talent to promote good?

To situate these thoughts within the historical practice of landscape architecture, McHarg invited Morse Peckham, a professor in the English department at the University of Pennsylvania and scholar of Victorian literature and Romanticism, to lecture on the concept of Nature in Western garden tradition. Peckham received his Ph.D. in comparative literature from Princeton, and he was strongly influenced by the philosophers Charles Sanders Peirce, William James, and John Dewey. His scholarly research explored the semiotics of symbols and signs, the pragmatics of their use, the meanings they communicated, and the role of personal experience in their interpretation. A master of extemporaneous speaking, Peckham was known for the adroit inventiveness of his intellectual style. When he lectured, he introduced the topic, defined important terms, and then sought to educate and persuade through comparative language analysis. He often concluded his remarks with a statement that returned the discussion to the introductory proposition.[36] In 1958, Peckham, at McHarg's behest, discussed the influence of religion in garden design at the National Conference of Instructors in Landscape Architecture (now CELA). The conference, which was held at the University of Pennsylvania and organized by McHarg, examined religious, social, and scientific attitudes to the environment, and Peckham presented a paper in the first session titled "Christian Attitudes to Environment."[37]

In the fall of 1959, Peckham delivered two lectures to McHarg's class.[38] The first lecture, derived from his conference talk and titled "Man-and-Environment in the Christian Tradition" discussed biblical symbolism in garden design. The second lecture, titled "18th and 19th Century Philosophies," connected the material presented in the first lecture to the theory of natural selection detailed by Charles Darwin in *The Origin of Species*. Peckham derived the material for the second lecture from his recent study of the revised editions of the text that Darwin published in response to social and scientific criticism.[39] Both lectures illustrated how ideas were never discarded, but instead amended in response to social and scientific change.

Peckham began his discussion of Christian attitudes toward the environment with the argumentative claim that the visual squalor endemic to the modern world was due in large part to the absence of cultural values that either failed to prevent this physical debilitation, or did not agitate for a revolt against it. After a short digression into the concepts of worldliness (the temptations of the here and now) and otherworldliness (the heavenly rewards of the afterlife), he continued to court controversy and stated the physical and imaginative squalor that plagued contemporary life was a manifestation of the fall of humankind in the book of Genesis as recounted in the biblical tale of the Garden of Eden.

Nature in Judaism and Christianity, he explained, had been forged by the will of God to create order out of chaos. The biblical paradise garden, as a symbol of this order, was a land of milk and honey given to humans for their benefit and innocent enjoyment. This gift, however, came with rigorous conditions tied to obedience, temptation, and the determination of right (what is good) from wrong (what is non-good). In this parable, God endowed humans with free will to test their righteousness. When Adam and Eve tasted the apple of knowledge despite God's warning, they failed the test and were banished to a wilderness of alluring enticements and worldly pleasures that only the virtuous could resist. In terms of garden design, he stated the issue was not a question of right or wrong, but rather the cultural significance of the symbols that emerged in response to the tale of expulsion, and how they expressed nature, redemption, and the earthly restoration of paradise.[40] His intent, in keeping with his interest in textual analysis, was not to trace morality back to a singular source, but to demonstrate how ideas were represented and transmitted from generation to generation.

A reference to the epic poem *Paradise Lost* illustrated how the humanist tradition resolved the dialectic of good (order) and evil (chaos) through the merger of intellect (man) with emotion (woman). Quickly extrapolating his semantic parsing of the terms "good" and "evil" to architecture, Peckham explained how the structure of a gothic cathedral expressed the humanist synthesis of thinking (intellect, order, and man) and feeling (emotion, chaos, and woman) through mathematic proportions that recreated the harmonic balance of the world prior to the biblical fall from grace.

As he continued to make his case, Peckham noted the 17th-century gardens of Versailles achieved harmonic balance through an intricate geometry of paths, vistas, and precisely clipped plants that were carefully separated from worldly disorder by walls and rows of trees. To enter the "dance of lines and angels" of this paradigmatic expression of enlightenment was to get "as close to paradise as a living human could get." He then explained how this design, built for a king by the royal gardener André Le Nôtre, achieved further divinity and cultural authority through the symbolic appropriation of ancient pagan references to Apollo the Sun God[41] (Figure 4.4).

But there was more to this story. Congruent with the construction of Versailles, science asserted itself, enduring Laws of Nature were formulated, and a new material universe emerged. God, in this vision of life, created the atoms, set them in motion, and then stepped back to view his handiwork. The mathematician and natural philosopher Isaac Newton, entranced by the operations of these phenomena, created complex formulas to explain the way the colors of the world, planetary motion, and the cyclical regularity of the tides physically (and magically) expressed the manifold power and wisdom of divine law. All of a sudden, Peckham exclaimed, members of the upper echelons of culture considered their surroundings through this

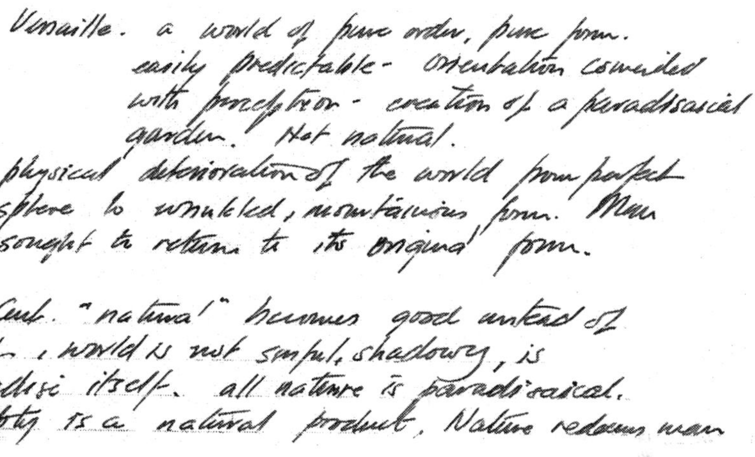
Versaille. a world of pure order, pure form.
easily predictable - Orientation coincided
with perception - creation of a paradisaical
garden. Not natural.
physical deterioration of the world from perfect
sphere to wrinkled, mountainous form. Man
sought to return to its original form.

18th Cent. "natural" becomes good instead of
bad. world is not sinful, shadowy, is
paradise itself. all nature is paradisaical.
identity is a natural product. Nature redeems man

Figure 4.4 Notes by Ian McHarg produced during a lecture on garden design by Morse Peckham. Ian L. McHarg Papers, The Architectural Archives of the University of Pennsylvania.

spectrum of knowledge and the earth became a garden and the natural landscape a paradise. These enlightened individuals not only felt at home in the world, they also glorified the beauty of an inherent order that extended from the distant stars to the sparkling atoms of the soul. He quoted Anthony Ashley Cooper, Third Earl of Shaftsbury to explain:

> O mighty Nature! wise substitute of Providence: impowered creature! Oh thou impowering Deity, supreme creator! thee I invoke and thee alone adore! To thee this solitude, this place, these rural meditations are sacred; whilst thus inspired with harmony of thought, though unconfined by words, and loose in numbers, I sing of Nature's order in created beings, and celebrate the beauties that resolve in thee, the source and principle of all beauty and perfection![42]

It was, therefore, only natural that the landscape gardener Capability Brown, in sympathy with this divinely ordained unity of the human microcosm with the planetary macrocosm, removed the intricate parterre garden that surrounded Blenheim Palace and replaced it with a meadow that swept up to the building's walls. The great beauty of the English landscape as it still existed, Peckham further asserted in one of many editorial asides, reflected this unified vision of "man in nature." But reason alone did not fully reveal the creative genius of the 18th-century English garden tradition. Intuition, feeling, emotion, and introspection were equally important sources of illumination. "Through an act of symbolic contemplation the individual discovered both his self and his God, since neither can be apprehended directly, empirically, or deductively."[43] Peckham's next statements reveal that he was less pleased with subsequent developments. As this discourse matured, he noted, it also ossified and the intimate soul-searching of the Romantic imagination coalesced into an exclusionary connoisseurship that limited contemplative transcendence to those conversant in the fine-grained distinctions of discriminating taste. In his opinion, it was imperative for everyone, not just the privileged, to feel at home in the world. Only then, he argued, would people emotionally respond to the land, perceive it to be beautiful, and maintain it in the state that it deserved; and with more than a hint of elitist consternation and religious skepticism, he concluded:

> When you see our hideous, dirty, and torturing cities, when you see the highways and the countryside ruined by gas-stations and bill-boards, when you watch our beaches destroyed by thousands of squalid cottages and hot-dog stands, you can say to yourself, Ah the cross has passed this way.[44]

The second lecture on "18th and 19th Century Philosophies" began where the first ended, and commenced with the proposition that people, in general, created visions of the world consistent with the dominant cultural discourse. Peckham again referenced Newton to explain how science had displaced God from everyday existence and supplanted his oversight with mathematical proofs and scientific deductions that illuminated the divine majesty of the Laws of Nature. In this world view, the enlightened belief that all was knowable and would eventually be discovered, fostered the assumption that an ideal society perfectly adapted to the environment could be obtained if human action conformed to reason.[45]

But science progressed, Peckham continued, and the Newtonian view was superseded when Charles Darwin had the temerity to suggest that organisms were not perfectly adapted to their surroundings. Indeed, Darwin's carefully documented observations in *The Origin of Species* indicated that this was impossible because the world itself was constantly changing. In sympathy with this new, uncertain, and immersive view of man in nature, landscape design was no longer bound by the need to provide recognizable symbols of divine intent – though that remained important. The new imperative was the creation of imaginative structures so "cunning" in their mimesis of nature that the hand of the designer was all but "invisible," as notably accomplished by Frederick Law Olmsted in the design of Central Park.[46] At the close of the lecture, he suggested it was once again time to channel the Romantic spirit, emulate the sentiments of Olmsted, and craft living works of art that challenged the dominance of mass-produced visual squalor.

Peckham's overview of garden history provided the confirmation that McHarg needed to remain convinced that his interpretation of the limitations of Judeo-Christian scripture was indeed correct. People were easily tempted to behave badly, at least where the environment was concerned, and they required instruction in the nuances of the Laws of Nature as reconstituted in the guise of ecology. In his notes, McHarg wrote that landscape architects had lost their metaphysical role and had forfeited their capability of creating order.

Stewardship

The lectures in Man and Environment provided McHarg with an ancillary education. When he listened to his invited guests speak, his own ideas evolved and he began to discern how he could critically shape culture. An intriguing passage in *Design with Nature* that referenced ideas presented by Robert MacArthur, however, suggests, at least at face value, that he harbored a personal uneasiness with the extent of his formal training. Considering the fact that he obtained three degrees from Harvard, in five years, his remarks are somewhat disingenuous. Nevertheless, by relinquishing academic superiority and letting his readers know that there were gaps in his knowledge, he reassured them that it was all right if they, too, were unaware of the critical role of plants in their lives:

> As I never obtained a college degree and entered graduate school without this dispensation, I never acquired the illusion of being educated which these diplomas often confer. Teaching is that device whereby I assemble a fragmentary, ragged and belated education. As it occurs at no cost to the student and cannot be terminated with a degree, it clearly has certain advantages. But it does also have embarrassments, not the least of which is confronting the commonplace of knowledge as total novelty late in life. I well remember that occasion, when I first heard that all life, with minor exceptions, is now, and forever has been, entirely dependent upon photosynthesis and the plant. I recall looking around me, searching for other eyes equally overwhelmed by this revelatory statement. I found only the dead faces of those who had long since absorbed this information and for whom it had no moving power.[47]

But if we look at this statement another way, dig a little deeper, and relate his pedagogy to both his environmental agenda and passion for ecology (and his impatience with those who did not share his enthusiasms), then his reference to teaching becomes an acknowledgment that logic and facts would not promote the behavior he sought. To implement his environmental agenda, his ideas had to be easily communicated, catch the attention of his audience, and spark imagination.

The botanist Paul Sears, who helped McHarg understand the importance of ecology as a way of life, provided guidance on how to instill a sense of environmental purpose, and a few common sense tricks on how to get people to listen. Sears received a Ph.D. in botany from The University of Chicago in 1922. In his long and nomadic career as an educator, he held positions at Ohio State University, The University of Nebraska, The University of Oklahoma, Oberlin College, and Yale University where he served as the first Director of the Conservation Program.[48] Sears retired from Yale in 1960, but he continued to lecture and promote land stewardship. It was in this capacity as an informal ambassador of conservation that he spoke to McHarg and his students.

Early in his career, Sears authored the best-selling book *Deserts on the March*. This factual and yet poetic account of the drought and dust storms that plagued the southern plains of the United States in the 1930s detailed how drifting dunes of wind-blown soil buried fences, houses, and machinery and caused many farming families to migrate in search of better living conditions. In his history of this environmental disaster, Sears linked soil erosion to climate, the destruction of prairie grasslands, and ill-conceived practices of dry-land farming.[49] Later, in *The Living Landscape*, he outlined the knowledge and skills that were needed to avoid similar disasters. His recommendations included a basic knowledge of the way organisms draw upon the energy and materials of their surroundings to produce their own substance, and how these actions correspondingly change the environment. It was only natural, he stated, that humans altered their surroundings, as all living things did the same thing. This was not the problem. The real issue was the rapid increase in the rate and scale of human-induced change that followed the advent of modern technology, which, he claimed, was an even more pressing issue because most people had no idea how the environment actually worked. In his judgment, practical measures to conserve soil and water would never occur and environmental disasters would become increasingly common if people failed to understand the "tightly woven fabric of activity" that united the operations of climate, air, geology, water, soil, and life into a living biosphere. He touted the science of ecology, and its knowledge of living communities, as the solution.[50]

Sears had come to McHarg's attention through his participation in the conference *Man's Role in Changing the Face of the Earth* and the essay "The Processes of Environmental Change by Man" that he wrote for the conference proceedings. In this essay, Sears argued, in a slight change of tactics, that science did an exemplary job defining the physical characteristics that enabled *Homo sapiens* to assume the position of dominant species on earth, however, this research tended to ignore the fact that humans required other organisms for survival, and for maintaining habitat conditions that made survival possible. To amend the situation, he again called for instruction in ecology, which he now defined, and this was the critical change, as both a social and physical science. Most people, he explained, agreed the lessons of ecology applied to the world apart from people – to nature – but they tended to resist applying these concepts to their own actions. Sears claimed this refusal reflected a dogmatic adherence to the norms of behavior and religious beliefs that separated people from their surroundings. He also faulted laboratory scientists unfamiliar with natural history, and considered them culpable as well.[51] People, he lamented, never realized that when they harmed the land they harmed themselves. The following analogy explained his thinking and what happened when people denied this reality and heedlessly used and abused the environment to satisfy their wants and desires:

> Any species survives by the virtue of its niche – the opportunity afforded it by the environment. But in occupying its niche it also assumes a role in relationship to its surroundings. For further survival it is necessary that the role at least not be a disruptive one. Thus one generally finds in nature that each component of a highly organized community serves a constructive, or at any rate, a stabilizing role. The habitat furnishes the niche, and, if any species breaks up the habitat, the niche goes with it.[52]

Sears delivered two lectures to McHarg's students that recapped the message of commensurate living that he had presented in the "Processes of Environmental Change by Man." The talks were delivered in a rambling down-to-earth style that belied their simple sophistication.[53]

The first lecture, titled "From the Physical Environment to the Ecology of Man," outlined the two basic roles of science in human life: as an applied practice, it provided the technological conveniences that made life easier; as a source of knowledge, it provided perspective on applied practices and their environmental impact. According to Sears, people were, in general, familiar with and supportive of technology, especially when it improved their lives. Knowledge, however, was harder to substantiate but more important because it explained why it was necessary to quit thinking of the land as "an inert stockroom" of goods and services, and why it was crucial to pay more attention to the land's patterns and processes and the role they played in their lives and livelihood.[54]

The second talk, titled "An Ethic for Man and Environment," chronicled the benefits that accrue when environmental conservation was practiced as a social and physical science.[55] He quoted several sentences written by the ecologist Aldo Leopold to indicate what he meant and how he sought to reconfigure society:

> An ethic ecologically is a limitation in the freedom of action in the struggle for existence. And I think philosophically is a differentiation of social from anti-social conduct. These are two definitions of one thing. The thing has its origin in the tendency of independent individuals or groups to evolve modes of co-operation. The ecologist calls these syndioces; politics and economics are advanced syndioces in which the original free-for-all competition has been replaced in part by cooperative mechanisms with an ethic of conduct.[56]

Sears was by no means a radical, subversive socialist in the sense that he wanted an all-out revolution in behavior. He was a subversive social conservative in the sense that he endorsed time-honored stewardship practices that preserved the ecological integrity of the land. Beginning in 1935 with *Deserts on the March*, he consistently criticized "scientific" management practices that maximized production but depleted the soil, and he consistently promoted agriculture practices that left the land in the same, or better condition as it passed from one generation to the next. The Amish farms of Pennsylvania and traditional Japanese agriculture were, in his estimation, superior models of stewardship. Indeed, for Sears, it was perfectly acceptable to emulate old ways, particularly when they enriched the land and life. This was just plain common sense. As Leopold noted in his land ethic: "A thing is right when it tends to preserve the integrity, stability, and beauty of the biotic community. It is wrong when it tends otherwise."[57]

The concept of land stewardship presented by Sears upended McHarg's conception of modernity and design. From Sears he learned that innovation did not require an avant-garde refutation of history. To be a forward thinking steward of the land, it was more important to look backwards and critically assess the field of practice. In terms of artistic validity, this meant it was necessary to discard any style that ignored the processes of nature and failed to protect the land.

Equally important, Sears introduced McHarg to the writings of Aldo Leopold. McHarg subsequently read *A Sand County Almanac* and discovered ideas that supported his environmental ambitions. Numerous excerpts from this text appear in a hand-written outline that he prepared for a book with the provisional title *Towards a New Landscape*. This outline formed the beginning of *Design with Nature*.[58] The first excerpt consisted of a passage from "The Land Pyramid" that explained, in layman's terms, the concept of trophic levels and the energy flow discussed by MacArthur:

> Plants absorb energy from the sun. The energy flows through a circuit called the biota, which may be represented by a pyramid consisting of layers. The bottom layer is the soil. A plant layer rests on the soil, an insect layer on the plants, a bird and rodent layer on the insects, and so on up through various animal groups to the apex layer, which consists of the larger carnivores.
>
> The species of a layer are alike not in where they came from, or in what they look like, but rather in what they eat. Each successive layer depends on those below it for food and often for other services, and each in turn furnishes food and services to those above. Proceeding upward each layer decreases in numerical abundance. Thus, for every carnivore there are hundreds of his prey, thousands of their prey, millions of insects, uncountable plants.[59]

The second excerpt, also from "The Land Pyramid," linked ecosystem energy dynamics to the soil:

Land, then, is not merely soil; it is a fountain of energy flowing through a circuit of soils, plants, and animals. Food chains are the living channels which conduct energy upward; death and decay return it to the soil. The circuit is not closed; some energy is dissipated in decay, some is added by absorption from the air, some is stored in soils, peats, and long-lived forests; but it is a sustained circuit, like a slowly revolving fund of life. There is always a net loss by downhill wash, but this is normally small and offset by the decay of rocks. It is deposited in the ocean and, in the course of geologic time, raised to form new land and new pyramids.[60]

He copied the following passage to relate ecosystem energy flow to human occupation of the land:

The process of altering the pyramid for human occupation releases stored energy, and this often gives rise, during the pioneering period, to a deceptive exuberance of plant and animal life, both wild and tame. These releases of biotic capital tend to becloud or postpone the penalties of violence.[61]

And he copied the following summary and questions posed by Leopold to define what he wanted design and planning to address:

1   That land is not merely soil.
2   That the native plants and animals kept the energy circuit open; others may or may not.
3   That man-made changes are of a different order than evolutionary changes, and have effects more comprehensive than is intended or foreseen.

These ideas, collectively, raise two basic issues: Can the land adjust itself to the new order? Can the desired alterations be accomplished with less violence?[62]

At a fundamental level, Leopold saw the protection of the land as a scientific necessity and moral obligation, and, correspondingly, his language embedded people in nature and his ideas required sacrifice.[63] A lifetime of observations led him to believe that cooperation was not a natural human trait, particularly when it required effort. To prompt behavioral change, he promoted ecological literacy and the institution of environmental regulations. It was his contention, or at least his hope, that reasonable restrictions would hinder competitive instincts and replace "us *versus* them" propositions with acts of "love, respect, and admiration." This sublimely simple premise would, if properly implemented, enlarge the human community and its economic imperatives to include the soil, water, plants, and animals. "An ethic ecologically, is a limitation in the freedom of action in the struggle for existence," he stated.[64]

Leopold's argument supplied McHarg with a scientific and ethical framework for environmental action. Sears had done something similar in his lectures, but Leopold went further and indicated how to analyze the problem through ecological energy dynamics and management surveys that indicated where and why the land remained productive following human interference. Again, Sears had done the same when he praised Amish and Japanese farming practices, but he made no reference to quantifiable methods. McHarg readily adopted Leopold's ethical stance and made it a fundamental principle of landscape design; and, true to his nature, he had no trouble imposing limits on actions that he considered morally suspect. He would simultaneously begin to formulate a design method that mapped the critical ecological functions of the land.

Notes

1 George Perkins Marsh, *Man and Nature: Or, Physical Geography Modified by Human Action* (New York: Charles Scribner & Co., 1864), 10.
2 Newton, *An Approach to Design*, 144.
3 Gropius, *Apollo in the Democracy*, 6.
4 See Marsh, *Man and Nature*. The conference was held in Princeton, New Jersey on June 16–22, 1955. See Wenner-Gren Foundation, History, "Man's Role in Changing the Face of the Earth," www.wennergren.org/history/mans-role-changing-face-earth.
5 Ibid., v.
6 Sharon E. Kingsland, *The Evolution of American Ecology 1890–2000* (Baltimore, MD: The Johns Hopkins University Press, 2005), 8.
7 See Clarence J. Glacken, "Changing Ideas of the Habitable World" in *Man's Role in Changing the Face of the Earth*, ed. William L. Thomas, Jr. (Chicago, IL: University of Chicago Press, 1956), 89.
8 The Architectural Archives of the University of Pennsylvania, G. Holmes Perkins Papers, "1959 Syllabus Man and Environment," call #:054.268. McHarg would cite *Man and Nature* in the 1963 essay "Man and Environment." See also McHarg and Steiner, *To Heal the Earth: Selected Writings of Ian McHarg*, 10–23.
9 William L. Thomas Jr., ed. *Man's Role in Changing the Face of the Earth* (Chicago, IL: The University of Chicago Press, 1956).
10 Edward O. Wilson and Hutchinson, G. Evelyn, *Robert Helmer MacArthur 1930–1972* (Washington, DC: National Academy of Sciences, 1989), www.nasonline.org/publications/biographical-memoirs/memoir-pdfs/mac-arthur-robert-h.pdf.
11 Personal communication, Peter Morin, June 27, 2018. Morin, an expert in community ecology and species interactions, provided insight into ecological theory in the 1950s and 1960s. As noted by Morin, MacArthur's hypothesis of stability has been modified by subsequent research, but his desire to understand why particular geographic areas contain a specific number of species remains a fundamental question in the science of ecology.
12 G. Holmes Perkins Papers, "1959 Syllabus Man and Environment."
13 A biome is a geographically specific assemblage of plants and animals determined by climate and physiography. In the mid-20th century, with the rise of ecosystem studies, biome replaced the word "formation" in the ecological literature. For a description of this nomenclature change as presented in a textbook from this era see Colinvaux, *Introduction to Ecology*, 57–59.

14 The Architectural Archives of the University of Pennsylvania, Ian L. McHarg Collection, "Dr. Robert MacArthur, Ecology," 109.II.E.2.43.

15 Ibid.

16 Robert MacArthur, "Fluctuations of Animal Populations and a Measure of Community Stability" *Ecology* 3 no. 36 (July 1955): 533–536.

17 C. E. Shannon, "A Mathematical Theory of Communication," *The Bell System Technical Journal* 27 (July, October 1948): 379–423, 623–656. In this essay, Shannon mathematically theorized how communication signals allow messages to move from a point of origin to a destination. MacArthur used this same approach to mathematically theorize how species interactions determine spatial distribution. See also James Gleick, *The Information: A History, A Theory, A Flood* (New York: Pantheon Books, 2011), 221–234.

18 At the end of the lecture, MacArthur summarized his stability and diversity argument as follows: species abundance indicated the amount of energy that entered and existed a system; in some communities species abundance remained constant, in others it fluctuated; stability in the pattern of fluctuation arose from, and reflected interactions between species and the physiology of the organism in question; interactions between species could be graphed as networked food webs that convey solar energy upward in trophic (nutrition) levels from plants to herbivores and predators; the number of choices an organism had in following the paths of the food web indicated overall community stability.

19 There is no transcript of this lecture, however, an essay by Frank in the Vision + Value series edited by Gyorgy Kepes explains the concept of "Organized Complexity." Frank's essay begins with the following sentence: "The world, as Norbert Weiner once remarked, may be viewed as a myriad of To Whom It May Concern Messages." See Lawrence K. Frank, "The World as Communication Network" in *Sign, Symbol, Image* ed. Gyorgy Kepes (New York: George Braziller, 1966), 1–14.

20 See McHarg, *Design with Nature*, 168. McHarg included several images from the Vision + Value Series in *Design with Nature*, including an electron micrograph of a platinum atom, an x-ray of a nautilus shell, a beehive, and diatom. See McHarg, *Design with Nature*; 164, 169–170: and Gyorgy Kepes, *Module, Proportion, Symmetry, Rhythm* (New York: George Braziller, 1966), 17, 40, and 53.

21 See Alfred J. Lotka, *Elements of Physical Biology* (Baltimore, MD: Williams & Wilkins Company, 1925).

22 See "Dr. Robert MacArthur, Ecology," accompanying notes.

23 Peter Morin provided the brief one-sentence synopsis of MacArthur's research. Personal communication, Peter Morin, June 27, 2018. The material presented in this lecture would reappear in *The Theory of Biogeography*, a text that MacArthur co-wrote with E. O. Wilson. The introduction to this text states that territory dynamics and landscape partitioning would become a central concern of ecology due to the fragmentation of "formerly continuous natural habitats now being broken up by the encroachment of civilization." See Robert M. MacArthur and E. O. Wilson, *The Theory of Island Biogeography* (Princeton, NJ: Princeton University Press, 1967), 3–4.

24 The references to Fisher and Haldane indicate that MacArthur discussed the role of genes and genetic change in evolution as formulated in the Synthetic Theory of Evolution. See: R. A. Fisher, *The Genetical Theory of Natural Selection* (Oxford: Oxford University Press, 1930); J.B.S. Haldane, "A Mathematical Theory of Natural and Artificial Selection, Part V: Selection and Mutation" *Mathematical Proceedings of the Cambridge Philosophical Society* 23 no. 7 (1927): 838–844.

25 "Dr. Robert MacArthur, Ecology," accompanying notes.

26 C. Elton, *Animal Ecology* (New York: Macmillan Co., 1927); and Raymond Lindeman, "The Trophic-Dynamic Aspect of Ecology" *Ecology* 23 (October 1942): 399–418.

27 The laws of thermodynamics describe two fundamental principles of thermal energy, or heat. The First Law of Thermodynamics, or Conservation Law, states that the total energy in the universe stays constant: it cannot be created or destroyed; it is only transformed. The Second Law of Thermodynamics, or Entropy Law, states when energy is consumed to either maintain or create order it dissipates over time. Entropy is the measure of the amount of energy that is no longer available to work, and thus it is a measure of randomness, disorder, and the absence of pattern. The physicist Erwin Schrödinger used the term "negentropy" to explain the ability of living organisms to avoid the negative effects of entropy and decay. See Jeremy Rifkin, *Entropy: A New World Order* (New York: Bantam Books, 1981); Vaclav Smil, *Energy a Beginners Guide* (Oxford, UK: One World Publications, 2006); Tom Butler, Daniel Lerch, and George Wuerthner, *The Energy Reader: Over development and the Delusion of Endless Growth* (Sausalito, CA: The Center for Deep Ecology, 2012); E. Schrödinger *What is Life?* (Cambridge: Cambridge University Press, 1944); and Science in Society Archive, Dr. Mae-Wan Ho, "What is (Schrödinger's) Negentropy" (Modern Trends in BioThermoKinetics 1994, www.i-sis.org.uk/negentr.php.

28 See Eugene P. Odum, *Fundamentals of Ecology* (Philadelphia, PA: W. B. Saunders Company, 1953).

29 "Dr. Robert MacArthur, Ecology," accompanying notes.

30 General systems theory is concerned with the parts, or elements of complex interactive systems, and it studies how these elements adjust to each other as they, in turn, adapt to new circumstances. See Ludwig von Bertalanffy, *General System Theory: Foundations, Development, Applications* (New York: George Braziller, 1968), 23.

31 McHarg, *Design with Nature*, 53.

32 See Pierre Teilhard de Chardin, *The Phenomenon of Man* (New York: Harper & Row Publishers, 1959); and Pierre Teilhard de Chardin *Man's Place in Nature* (New York: Harper & Row Publishers,1956). McHarg's personal files contained a news clip on Teilhard de Chardin that described this Jesuit priest, who was forbidden by the Vatican to publish his writings on evolution, as the progenitor of a philosophy in which "things do not die," but are instead recycled as part of the continuing work of God. As noted in the article, this included "the industry of bees as they make juices scattered in so many flowers – these are but pale images of the continuous process of elaboration which all the forces of the universe undergo in order to become spirit." See The Architectural Archives of the University of Pennsylvania, Ian L. McHarg Collection, "Passionate Indifference," (n.a.), 109. II.C.87.2.

33 The Architectural Archives of the University of Pennsylvania, Ian L. McHarg Collection, "Religious Attitudes to Man and Environment," 109.II.C.87.2, 2.; and The Architectural Archives of the University of Pennsylvania, Ian L. McHarg Collection, "Man and Environment (1959), the 1961 copy," 109. II.E.2.2, 1.

34 Schweitzer argued that ethics had to extend beyond relations of "man-to-man" and must include plants and animals. McHarg cited Schweitzer in *Design with Nature* in an argument that states reverence for life must not stop "at those creatures having a utility to man, but encompasses all matter and all creatures." As noted in Chapter 1, McHarg heard Schweitzer speak at the Aspen Institute and Music Festival in the summer of 1949. See *Design with Nature*, 125; and Albert Schweitzer, *Out of My Life and Thought: An Autobiography*, trans. C.T. Campion (New York: Henry Holt, 1946).

35 "Religious Attitudes to Man and Environment," 109.II.C.87.2, 1.
36 See H. W. Matalene, *Romanticism and Culture: A Tribute to Morse Peckham ad a Bibliography of his Work* (Columbia, SC: Camden House, 1984), xv–xxi; and Harry Smoak, *Meaning as Response: Experience, Behavior, and Interactive Environment Design*, Ph.D. Diss., Concordia University (2015), 10.
37 The Architectural Archives of the University of Pennsylvania, Ian L. McHarg Collection, "Conference on Instruction on Landscape Architecture, Philadelphia 1958," 109.II.A.1.67. Conference attendees included Grady Clay, J.B. Jackson, Louis Kahn, Dan Kiley, G. Holmes Perkins, David Crane, Hideo Sasaki, Peter Walker, and Stanley White.
38 G. Holmes Perkins Papers, "1959 Syllabus Man and Environment."
39 See Charles Darwin, *The Origin of Species: A Variorum Text* ed. Morse Peckham (Philadelphia: University of Pennsylvania Press, 1959); and Morse Peckham, "Darwinism and Darwinisticism" *Victorian Studies* 3 no. 1 Darwin Anniversary Issue (September 1959): 19–40.
40 The Architectural Archives of the University of Pennsylvania, Ian L. McHarg Collection, "Morse Peckham, Man-and-Environment in the Christian Tradition," 109.II.E.2.53, 5.
41 Ibid., 6.
42 Ibid.
43 Ibid., 7.
44 Ibid., 8.
45 The Architectural Archives of the University of Pennsylvania, Ian L. McHarg Collection, "Morse Peckham 18th and 19th Century Philosophies," 109. II.E.2.52.
46 Ibid., 17–18.
47 McHarg, *Design with Nature*, 46.
48 An equally astute politician, Sears served as the President of the American Association for the Advancement of Science (AAAS) and the President of The Ecological Society of America. See Archives at Yale, Guide to the Paul Bigelow Sears Papers MS 663, "Paul Bigelow Sears, 1891–1990." https://archives.yale.edu/repositories/12/resources/4448.
49 Paul B. Sears, *Deserts on the March* (Norman: University of Oklahoma Press, 1935).
50 Paul B. Sears, *The Living Landscape* (New York: Basic Books, 1962), 17, 33, 104.
51 Paul B. Sears, "The Processes of Environmental Change by Man" in *Man's Role in Changing the Face of the Earth*, ed. William L. Thomas, Jr. (Chicago, IL: University of Chicago Press, 1956), 471–484.
52 Ibid., 472.
53 Paul Sears is listed in the Fall 1961 "Man and Environment" syllabus under the title "The Land Ethic." Syllabus courtesy The University of Pennsylvania, School of Design Department of Landscape Architecture.
54 The Architectural Archives of the University of Pennsylvania, Ian L. McHarg Collection, "Paul Sears, from the Physical Environment to the Ecology of Man," 109.II.E.2.63.
55 The Architectural Archives of the University of Pennsylvania, Ian L. McHarg Collection, "Paul Sears, an Ethic for Man and Environment," 109.II.E.2.62.
56 Ibid., 1. See also Aldo Leopold, *A Sand County Almanac and Sketches Here and There* (New York: Oxford University Press, 1949), 202.
57 Leopold, *A Sand County Almanac*, 224.
58 The Architectural Archives the University of Pennsylvania, Ian L. McHarg Collection, "Towards a Modern Landscape," 109.II.V.C.21.2.
59 Leopold, *A Sand County Almanac*, 214.

60 Ibid., 216.
61 Ibid., 218.
62 Ibid.
63 For further discussion of the religious strains in Leopold's writing and his attempts to blend these beliefs with pragmatic conservation see Roderick Frazier Nash, *Wilderness and the American Mind* (New Haven: Yale University Press, 1967), 183–199.
64 Leopold, *A Sand County Almanac*, 202–203, 223.

5   The House We Live In

The evolution of a land ethic is an intellectual as well as an emotional process. Conservation is paved with good intentions which prove to be futile, or even dangerous, because they are devoid of critical understanding either of the land, or of economic land –use. I think it a truism that as the ethical frontier advances from the individual to the community, its intellectual content increases.

Aldo Leopold, The Land Ethic[1]

Beginning in the fall of 1960 and extending through the spring of 1961, McHarg moderated a nationally syndicated television talk show titled *The House We Live In* that reprised his seminar course Man and the Environment for the general public.[2] The overall tone of the program, which was produced by WGBH in Boston and filmed in Philadelphia, reflected a mid-20th-century confidence when people in the United States still believed, however briefly, that they could do or overcome anything. Speaking as a cultural arbiter and moral authority of the moment, McHarg renounced dominance and control and called for a new environmental order governed by restraint and interdependence. As the impresario responsible for compiling the guest list and orchestrating the interviews, he was equally determined to position himself as an intellectual leader conversant in a number of disciplines, and a bracingly subversive idealist who challenged convention (Figure 5.1).

Over the course of twenty-two interviews with scientists, social critics, and theologians, McHarg promoted a broad reading of ecology that extended an organism's relationship to its environment beyond science and quantification and into a boundless domain of philosophical ideas, ethical propositions, and imaginative suppositions. The scientists explained how biophysical systems worked. The social critics explained why these systems were not operating the way they should. The theologians unearthed moral correctives. As later reconstituted in the starkly dramatic photographs of the sun and the earth that graced the front and back cover of the first edition of *Design with Nature*, this was a breathtaking vision that embraced the sublimity of the natural world, the insignificance of the human

Figure 5.1 Promotional photograph *The House We Live In*. Ian and Carol
McHarg Collection, The Architectural Archives of the University of
Pennsylvania.

presence, the modernist project for universal civilization, and humanity's
capacity for immense destruction.

McHarg's decision to host *The House We Live In* reflected his participation in the Rockefeller Conference on Urban Criticism, held in Rye, New
York in October of 1958. Organized by David Crane under the auspices
of the Institute for Urban Studies at the University of Pennsylvania, the
conference assembled a talented group of practitioners, academics, and social critics to discuss deficiencies in urban design in the United States and
to propose solutions. A memorandum by Crane outlined the agenda. This
working paper claimed the dismal state of the country's urban design, exemplified by "antiseptic, dull, meaningless at best, and, at worst, garish,
pretentious, and inhumane" products, was due to the dismal state of the
country's urban design criticism. The artistry of European civic design, in
contrast, reflected a history of artful design critique. But this laudatory
statement did not mean that European design and design criticism could,
or should, become the model for the United States. As Crane cautioned,

European civic design and design criticism was artful, however, its emphasis on the formal arrangement of buildings in city centers did not address the social, economic, and physical problems that plagued urbanization in the United States. Multiculturalism, unprecedented technological development, the 1949 Housing Act and the rapid growth of suburbia, and the 1956 Federal-Aid Highway Act and the rapid increase in automobile ownership, were of particular concern. To deal with these culturally specific issues, the memorandum called for a critical approach equivalent to Henry James's plea for an "American intelligence" that matched the country's dynamic energies. Commensurate with this author's voluminous productivity and his European inflected but nonetheless American perspective, Crane cited Lewis Mumford's "Skyline" essays in the *New Yorker* as "the single most brilliant contribution to the field of American criticism." What Crane and his fellow conference organizers sought (in addition to self-promotion) was a professional coming-of-age in which a pragmatic and provocative American social realism produced a comparable, and perhaps superior urban vision. Associated with this pursuit was the desire for an easily accessible and technologically savvy literary intelligence – a "Lewis Mumford of television, who would look at urban planning through the eyes of people in architecture, landscape architecture and city planning," but "hopefully with fewer commercials."[3] In his capacity as host of *The House We Live*, McHarg responded to both recommendations and positioned himself as an American-inspired European and the Lewis Mumford of television. Numerous ideas from the program appear in *Design with Nature*.

Season one

The *House We Live In's* format was simple. The program began with the disembodied voice of a studio announcer stating over angst-ridden, downbeat music: "Man, Man's God, and the World In Between in this Post Atomic Age." Following this theatrically ominous beginning the camera moved in for a close-up of McHarg comfortably ensconced in an American designed and manufactured Eero Saarinen tulip chair, surrounded by curling tendrils of cigarette smoke. Once his face was firmly fixed in the center of television screen, he proceeded to introduce the scope of the program as "the evolution of matter, life and man," attitudes toward "nature and God," and human "psychological and physiological" needs.[4] When the camera moved out to reveal his guest, McHarg leaned forward and with a penetrating stare asked this individual to relate their field of specialty to these themes. The resultant conversations were lively, informative, and serious.

The first three guests, the physicist Harlow Shapley, the biologist David Goddard, and the physical anthropologist Carlton Coon provided the program's scientific foundation. Shapley's description of the newly formulated Big Bang theory allowed viewers to catch a glimpse of the birth of the

universe and the beginning of time.[5] Goddard described the common attributes that link organisms to each other and to their surroundings, and
he explained the genetic mechanisms of inheritance and mutation.[6] Coon
chronicled human evolution from its animal origins.[7]

In contrast, the fourth guest, the anthropologist Margaret Mead, argued the way an individual comprehends the world was as much a cultural construct as it was a scientific fact. Everything a person sees, she
stated, they see through the institutions and ways of thinking that defined
their particular culture. Using case studies to illustrate her point, Mead
detailed how actions toward the land were entwined first with technology; second with customs that taught people to perceive the environment
as either friendly or hostile; and third with notions of punishment and
reward. She also warned against reductive generalizations and cautioned
McHarg when he claimed pre-modern cultures were more in tune with
their surroundings, and therefore less environmentally destructive. She
countered his statement with several examples in which the introduction
of weapons into these cultures enhanced perceptions of power and control, and fomented social changes that led to the subsequent deterioration
of the environment.[8]

Mead's commentary not only redirected the program's discussion away
from its initial consideration of natural balance and toward an inquiry into
cultural imbalance and a broader discourse on power, ethics, and control, it
also reflected a general skepticism among mid-20th-century environmental
commentators regarding the ability of governing institutions to supervise
the technology that emerged during and immediately after World War II.
As McHarg stated in his opening remarks to Shapley, his *modus vivendi*
for rethinking the ethical principles that guided contemporary environmental relationships – what he referred to in an unacknowledged reference to
George Perkins Marsh as the newest problem of "man and nature" – was
his anxiety over the very real threat of thermonuclear annihilation.[9] This
fear, along with his conviction that spiritual improvement was needed, set
the stage for religion.

The five speakers following Mead – Rabbi Abraham Heschel, Father
Gustave Weigel, Paul Tillich, Swami Nikhilananda, and Alan Watts –
represented the major religions of the world – Judaism, Catholicism, Prot-
estantism, Hinduism, and Zen Buddhism respectively. Once again, the
conversation was lively and informative and it touched upon topics as
diverse as Wordsworth and Romanticism, Teilhard de Chardin and evolution, and Beatniks and Saint Francis.[10]

But unlike earlier discussions, McHarg's attitude contained a note of
defiance. He pressed these guests to explain (and then defend) the moral
precepts of their religions. Softly, but repeatedly, he called out doctrines
that he believed promoted environmentally controlling and exploitive attitudes. Although the theologians responded with provocative and compelling answers, McHarg's numerous references to Genesis 1:28 and its charge

to be "fruitful, multiply, and fill the earth and subdue it" revealed he already had his truth. In his judgment, Judeo-Christian doctrine was the antagonist of enlightened environmental design.[11] His determination to have his guests validate this supposition influenced the questions he asked and the landscapes he preferred.

McHarg, who could not resist preaching, even to the converted, believed, as he stated to Heschel, that everyday actions must transcend the utilitarian. He observed to Weigel that people must humbly see something of God in all forms of life.[12] To Tillich, he spoke of an inclusive ethic where the "micro-organisms of the soil, sun, star, wind, reptiles, amphibians, mammals exist in one system."[13] With Nikhilananda, he talked of matter and spirit, sacred rivers and mountains, and discussed the grand melody of the great web of being that draws everything toward the creator.[14] Like Watts, he sanctioned the shock tactics and mental gymnastics of Zen because it forced people to look beyond themselves.[15] There had to be some type of mediating entity that united these thoughts into a singular world-view, and this is where he entwined ecology, which taught the interdependence of all things, with religion. Going back to the etymological root of the word – *oikos* or household – he domesticated his beliefs in the program's title *The House We Live In*.

During his conversations with these eminent theologians, McHarg's agenda, never wavered. He was in search of a reasoned, socially relevant, and yet spiritual belief system that would provide objective evidence of his place in nature. The sanctioned doctrines of Judeo-Christianity for the most part, failed to provide a complete answer. True, these dictums had allowed him to humbly see something of God in all forms of life: but, in his mind, they were constrained by authoritarian injunctions, prone to aggressive dominance, and thus the antithesis of verdant growth and change. His desire to equate spiritual truth to something less assertive and more enduring preserved his sermonizing from tiresome banality.

When Paul Tillich presented a metaphysical discourse on transcendence that conceived of life as an ontological choice between mindless servitude (the desire to have) and personal affirmation (the ability to be), McHarg used the opportunity to relate this ethical speculation to design. In full agreement with Tillich, but situating his remarks within physical rather than social space, McHarg suggested these alternative modes of existence were readily apparent in Western garden tradition. The Renaissance Garden (later exemplified in *Design with Nature* by André Le Nôtre and Versailles) was merely the symbolic décor of a centralized authority that imposed "human patterns upon a reluctant landscape," and it signified deference to stultifying rules of conduct, central authority, servitude, and the desire to have. Conversely, the English Landscape Garden (later exemplified in *Design with Nature* by Capability Brown and Blenheim Palace) took its cues from the natural ecology of plant communities and allowed the land to be what it wanted to be.[16] McHarg, in other words, promoted the organic plasticity of life over the crystalline stasis of the complete object, and he embraced the grand gesture and total design. This stance reflected the remarks made by Morse Peckham on science, symbolism, and humanist garden traditions in the course Man and Environment.[17] It also aligned his thinking with the argument of Sigfried Gideon in *Space, Time and Architecture*, but with a not-so-subtle shift in emphasis away from built form and toward the land and its generative processes.[18]

McHarg, it seems, knew what constituted a harmonious man and nature relationship. The problem involved how to convince others to follow. He found a kindred spirit and one possible path to salvation in the interconnected "way of things" provided by Alan Watts in his description of Zen Buddhism.[19] Most striking to McHarg was the claim by Watts that the shock tactics and mental gymnastics of Zen lifted the veil of illusion that clouded reality by forcing people to look beyond themselves and ask Nature what it wanted to be, and what shape it wanted to become. This argument, in keeping with the teachings of Lâo-Tse, balanced the polarities of existence – good and evil, creativity and destruction, and wisdom and foolishness – with the intent of turning people away from the desire to master the world and toward the greatness of little things.[20] Picking up on these ideas, while simultaneously promoting the vision of land stewardship advanced by the ecologist Paul Sears, McHarg praised the productivity of Japanese agriculture and claimed that it could "only come from a people who have an extraordinary acuity to the processes of nature."[21] The veil of illusion in this instance, at least as witnessed by McHarg, was Western society's general inability to perceive the benefits of natural processes and to honor humble landscapes that exemplified those processes.

Yet McHarg confessed to the next guest, the social critic and psychologist Erich Fromm, that he was unable to feel a sense of belonging, or know himself as part of a larger, living order. This revelation was made in response to Fromm's assertion that modern society excelled in the creation of alienating barriers that separated people from their surroundings and each other. Fromm argued reason alone would not topple these barriers, but would in and of itself create an ideological structure that increased their height. To "feel at home in this world" also required "love," which Fromm defined, like Watts, as a positive engagement with, and active immersion in the experiences of life. McHarg brought these existential musings down to earth by noting ecology's science of interdependence makes it impossible to think of humanity as separate from its surroundings.[22]

The last two guests of the first season returned the discussion to science. The physiologist Hans Selye warned of the deleterious impact of stressful environments on the body, and irrevocably demolished the physical boundaries that separated people from their surroundings.[23] Selye not only described the physiological mechanisms of stress syndromes, he also called for altruistic behaviors that advanced the wellbeing of others. The geneticist Julian Huxley, in contrast, assumed a distancing-objectivity and conceived of religion as a self-correcting hypothesis. If we intellectualize religion this way, Huxley argued in words that echoed the opening remarks of Shapley and foreshadowed the front cover of *Design with Nature*, humanity would learn to see itself as part of the natural world "made of the same matter and the same energy as the most distant star." [24] Then, in a comment faintly tinged with the taint of eugenics, he claimed the higher order of abstraction made possible by a secular "ecological theology" would allow people to contemplate the unlimited possibilities of nature – their own nature and the rest of nature – as they dreamt of, and planned for, higher and higher levels of organization.[25] Huxley also observed, in a statement McHarg would repeat in *Design with Nature*, that science was essential to an ecological world-view because it was, like life, reproducible and capable of development. Less than satisfactory, however, was Huxley's claim that the soaring spires of gothic cathedrals, a beloved symbol of architectural modernity, were an apt expression of transcendence. McHarg hoped Huxley would have instead affirmed the intrinsic value of sunlight in the nourishment of the microorganisms, insects, animals, and plants of the earth. In his judgment, this was a divinity that transcended intellectual speculation and issues of control, for it involved the heat of the sun, the warmth of the soil, the growth of plants, and the diversity of life.

Season two

The second season, which aired the following spring, further bemoaned the state of the environment and the dire consequences of the status quo. Conversations similarly explored natural balance and cultural imbalance, and their relationship to power, ethics, and control. The urban environment, rather than religion, served as the main focus of attention. Throughout these discussions there was an overriding sense, particularly in the questions McHarg posed, that the earth's ecological limits were knowingly transgressed for the sole purpose of economic gain. To illustrate the adverse consequences of this action, the discussions delved into the flow of material, energy, and thought when it became encoded within the body as a living, but not necessarily life-affirming residue of cultural aspiration. Related discussions of density, disease, and stress provided the opportunity to link economic and social politics to an ecological paradigm of dominance, sub-dominance, territorial aggression, and spatial injustice. The threat of extinction opened and closed each episode. Considering his personal history, it is not surprising that McHarg also equated overcrowding and the lack of sunlight and fresh air to tuberculosis, pneumonia, and bronchitis. The invited guests provided compelling and provocative answers, and McHarg continued to use the opportunity to promote ecology.

The first guest of the second season, the physical anthropologist Loren Eiseley, supplied a sweeping narrative of evolutionary history that established the framework for subsequent discussions. His remarks were notable for the way they co-joined natural processes and cultural development through a paradigm of action and response.[26] Eiseley began his argument with a description of the emergence of life and its maturation into a living web of plant and animal communities that were adaptively linked by chance, calamity, and unimaginable creative possibilities. The novel products of this interactive system included the human mind, technology, and cities. But here, he cautioned, danger lurked, as evidenced by the increasingly elaborate mechanical constructs, both real and imagined, that separated people from the world and its "green leaves and ancestral waters." To re-establish contact with "the earth from which we sprang and upon which we are still dependent," he stated it was necessary to balance scientific and technological progress against the knowledge of ourselves as organisms embedded within, and responsive to the world like all other living creatures. Eiseley further observed that it was precisely because the world was naturally unfinished, and the environment of the human mind was unconstrained by physical limits, that eveyone must become "a disciple of history, not its servant." If modern society failed to alter its policies and actions, he foretold a future in which the concrete highways and suburban housing endemic to urban growth would leave a trail of destruction comparable to the rhizomatous tracks of an invasive fungus as they spread over the surface of an orange.[27]

The second guest of the spring season, the British historian Arnold Toynbee, called upon years of carefully collected empirical evidence to discuss the evolution of civilization. Aristocratic and erudite, he was famous in both academic circles and popular culture, as evidenced by his exhaustive twelve-volume magnum opus *A History of Civilization* and his profiles in *Time* and *Life*.[28]

Toynbee began his historical overview with the observation that advanced cultures developed from primitive societies when "a surplus over and above the necessities of life" enabled the emergence of an elite minority who had the time and resources to retire from society, reflect, and plan.[29] To be successful, however, the leadership of this newly emergent leisure class required the concomitant development of political, economic, and social institutions that supported their proposed dictates, as well as the willingness of the rest of society to voluntarily follow these directives.

McHarg, who had announced at the beginning of the broadcast that Toynbee would "look at the degree to which man has adapted to his environment to survive" and where he has failed to adapt, "having been extinguished," was less than happy when this distinguished historian (after repeated prompting) failed to agree with his assessment that the maintenance of a healthy physical environment was the most important variable in cultural longevity.[30] Toynbee, it must be noted, did agree that the environment as the material setting in which a particular history unfolds could either enhance or hinder historical development. Nevertheless, he kept restating that cultural longevity was generally determined by the response to the challenges posed by exterior forces. Civilizations prospered, he argued, when the elite members of society led by example, and they declined when these individuals failed to do so. Decline, therefore, more often than not reflected internal social deficiencies and self-satisfaction, rather than external threat or the limitations imposed by the physical environment. Unhappy with this response, McHarg kept pressing his guest until Toynbee did admit there were times in history when environmental challenges were so great that it was impossible to mount an effective response.

Although Toynbee repeatedly discounted environmental determinism, in line with the general theme of the program he did suggest that advances in technology eroded the distinctions that traditionally separated "man and nature." But he qualified this statement, and, similar to Margaret Mead, used case studies to argue that the challenges posed by technology were relative. Uranium, he observed, with an academic detachment born from years of practice, was useless to people in a pre-atomic society, just as the plow was useless to people in pre-agrarian cultures, or oil to the Arabian herdsman. And, then, reiterating his central theme of challenge and response and its attendant dialectic of good and evil, he stated it was not necessarily the tools of technology that threatened civilization, but how well the cultural elite responded to the changes they engendered.[31] Of particular interest in this regard (though not explicitly mentioned during this conversation), was Toynbee's ancillary argument that many of the technological challenges besetting the world could be attributed to Judeo-Christian theology and its vision of human dominion.[32]

Despite, or not inconceivably because of, the difference in their positions, Toynbee's paradigm of action and response allowed McHarg to voice (in another unacknowledged reference to Paul Sears) an environmental ethic in which the land expressed "an order, a pattern and a process" governed by a natural system of checks-and-balances.[33] In the brave new world of his imagination, natural harmony reigned supreme, critical ecologies were restored, work was pleasurable, the land was beautiful, and everyone lived in a clean, orderly environment. This aspiration possessed something of the web-like naturalism of a Morris Graves painting and the providential supernaturalism of a Jonathan Edwards sermon, as well as an avant-garde posturing not unlike that exhibited by the elite CIAM architects in their defense of the material practices enumerated in *The Heart of the City*.

The five speakers following Toynbee – the conservationist Fairfield Osborne, the physiologist John Christian, the geneticist Theodosius Dobzhansky, and the epidemiologists A. M. M. Payne and Leonard Duhl – inserted a Malthusian specter of over population and resource scarcity into the discussion. Osborne detailed the factors that limited agricultural production and led to mass starvation.[34] Christian, who based his commentary on the study of laboratory rats, observed that overcrowding magnified stress, promoted territorial aggression, increased infant mortality, and enhanced susceptibility to disease.[35] Dobzhansky described the deleterious impact of environmentally induced genetic mutations.[36] Payne argued that heart, lung, and mental health diseases were primarily associated with dense urban populations.[37] Duhl outlined practical actions that reduced the severity of these maladies, such as the removal of smog, noise, and infectious disease. But in an intriguing counterclaim, Duhl observed that human nature was reactive and opportunistic, therefore, progress required stress and crisis. Besides, he argued, in a remark that reinforced Toynbee's thesis of challenge and response, the ills that beset the city would never be fully resolved as this terrain would inevitably be confronted by new problems.[38]

The seventh guest of the second season, William L. C. Wheaton, who had been one of McHarg's planning professors at Harvard and was now his colleague and director of the Institute for Urban Studies at the University of Pennsylvania, expanded the discussions of urban density and dysfunction to include suburbia. He began by stating the biggest challenge the United States faced in the near term was the enormous growth of the urban population, coupled with the outward flow of people from city centers. According to Wheaton, these interrelated circumstances reflected two contradictory impulses engrained in American culture – the democratic pastoralism of the country's agrarian ideals and the laissez-faire commercialism of its mercantile history. To address these divergent forms of growth, and their attendant living preferences, he called for two environments – a high-density, urban landscape for those who enjoyed the noise and commotion of the city, and a low-density suburban landscape for those who preferred the quiet rusticity of country life. Aware that this proposition did not completely address the issue of urban sprawl, Wheaton reasoned that it was still preferable to mass-produced uniformity, which, he stated, did not satisfy the physical and psychological necessities of anyone.[39]

Convinced of the need for planned dispersion, Wheaton called for urban centers and surrounding greenbelts that emulated British New Towns. But he also stated, in a nod to the reality of contemporary life in the United States, that it was necessary to account for the personal autonomy, imaginative freedom, and dynamic energy of the county's nomadic car culture. To demonstrate how to achieve both objectives, he called attention to the survey of regional land use patterns that the Institute for Urban Studies, with the assistance of McHarg, was then conducting on the Delaware River watershed.[40]

The last two guests of the season, the writers and social critics Kenneth Rexroth and Lewis Mumford, returned the discussion to science. Rexroth claimed the reductive logic of the scientific method separated people from the world, while Mumford observed the inductive observations of science provided valuable insight into humankind's interaction with the environment. Rexroth correspondingly insisted that facts did not build livable cities, while Mumford built his case for livable cities around carefully selected facts. Both men qualified their nominally factual remarks with decidedly subjective opinions.

Rexroth's relaxed, Eastern-inflected meditation on oneness with the world reprised the counter-cultural stance of Alan Watts in the first season. He began his remarks with the observation that modern science made important strides when it began to think of people as part of the environment. But he tempered this observation by noting that the tendency of science to break the world into easily managed parts inhibited the perception of hard to quantify truths and holistic modes of action. "We are not rats in mazes," he stated. He also cautioned that it was important for humans to remain cognizant of their inherent self-interest, as seen in the following assertion: "We don't live for others, for the universe, for some competing species." This did not mean that he sanctioned destructive behavior or thought of "man as over or against the environment." On the contrary, he considered this behavior to be a habit of mind in need of serious revision. A quick reference to Joseph Needham, a developmental biologist and historian of Eastern religions, added scientific stature and a touch of contrarian spiritualism to his next remark. We should think of ourselves, he stated, as part of "a homogeneous, dense universe that stretches out to intergalactic space."[41]

Rexroth clearly held a pluralistic, anti-authoritarian worldview, which prompted him to see the environment, and by extension human existence, as a web of dynamic interactions that defied hierarchy, order, and easy categorization. In keeping with these beliefs, he called for intellectual flexibility and the abandonment of fixed principles and certainty. He also opposed the self-aggrandizement of actions based on one person's absolutism, and instead encouraged permissiveness. For design purposes, this meant designers had to refrain from trying to fit the world into a pre-determined mold, and they had to stop imposing their personal vision on others, as Le Corbusier had done in the Radiant City. According to Rexroth, this intellectual generosity would give rise to a "fantastic and quite insane" number of creative designers – "hundreds of thousands of Frank Lloyd Wrights all creating buildings that are quite good."[42]

McHarg, unconvinced that intellectual leniency would propagate either design genius or a life-enhancing environment, responded by noting the issue was not the lack of pluralism or permissiveness, but instead substandard social and environmental values. Tolerance of this moral deficiency, he claimed, promoted the construction of squalid, disreputable cities that lacked vitality. Even worse, this enervating milieu had spread outward and debilitated the countryside. After stating people had "lost the capacity to consider what is gracious and contemplative," he returned to the issue of creative genius. "You have said," he observed to his guest, "the ecological view is not enough, that nature is not an inevitable process, what then of this man, the artist and the cities he will build?" Using a quote from Rexroth's essay on Morris Graves to bolster this position, he then asked how the promotion of pluralism and permissiveness advanced "the place of value in the world of facts?"[43]

Rexroth readily conceded that cities were technologically broken, but he also stated (in a remark that tacitly assumed they were both members of the creative elite) that the ordinary man did not have to "live like us." Indeed, and after referencing the grace and elegance of Haussmann's Paris, Aix-en-Provence, and Vincenzo, he stated it was acceptable for cities to be dense and crowded, but only if areas of "creative, contemplative life" were built into their physical form. Then graciously modifying his stance on pluralism and permissiveness to acknowledge his host's argument, Rexroth noted that cities should, "within reason," allow for eccentricity, diversity, and the fulfillment of personal desires, whether or not these actions were considered to be of good or bad taste. After he removed a cigarette from the package on the table, he observed this stance might cause discomfort, but it was still necessary to emphasize "the infinite variety of human beings," rather than tasteful uniformity or stultifying perfection.[44] Here again his words suggested the way designers (particularly European modernists) were trained to act, and the things they were taught to believe had nothing at all to do with what made cities great.

When McHarg introduced Lewis Mumford, he called him an "oracle for man and environment," and "the man best advised in the Western world on urbanization and the development of its resources."[45] This glowing accolade, unlike any other introduction in the program, indicated that in this particular interview McHarg would assume the stance of pupil and quietly listen to his eminent guest. Mumford feigned modesty but was more than happy to assume the paternalistic role of teacher. In response, McHarg leaned forward and avidly listened as his illustrious guest provided enlightened solace and cause for hope.

During his thirty-minute presentation, Mumford crafted an argument that transfigured modern science into a progressive metaphysic, which, if properly pursued, would ensure the American dream of life, liberty, happiness, and prosperity. The lesson he sought to impart embraced the positivist belief that science could explain the nature of things through precisely measured properties, but only if people retained a healthy skepticism toward technological prowess and its promise of power and control. By outlining the limits of human understanding, and the fallacy of policies that promote power over the natural world, he positioned himself (and by association McHarg) as a knowledgeable philosopher of science and society.

As he began to speak, Mumford asked McHarg if he had read *The Fitness of the Environment* by the biochemist Lawrence Henderson.[46] After McHarg replied that he had not, Mumford assured his young host there was no shame in this admission, as few people had done so. But, he continued, it was a very important book that taught essential lessons about the disposition of the major elements of the earth – carbon, hydrogen, oxygen, and nitrogen – and how these chemicals interacted over time to create an environment "favorably disposed toward life." He further noted that Henderson's work not only made it impossible to think of the earth as a hostile environment, it also demanded that we stop acting like competitive Social Darwinists and instead advance "what is essential in any definition of life," which Mumford defined as cooperative social norms that balanced continuity with change, stability with growth, and security with adventure.[47]

After a quick glance at his notes, Mumford presented a brief synopsis of the natural history of urban civilization that he was currently documenting in *The City in History*.[48] His narrative began with small bands of hunter-gatherers and the capture of fire, moved to agricultural communities and the domestication of plants and animals, and ended with the internal combustion engine and industrial cities. When he concluded, he asked McHarg to both agree with his assessment that modern society's unwarranted faith in mechanization fostered social degeneration and the death of diversity, and to concur with his judgment that this malady prevented people from seeing "the great extraordinary choice life gives you every moment, the unexpected things that come from any organic situation." Everyone knew, he sadly continued, that the physical impoverishment of the environment encouraged a litany of destructive behaviors that included the deadly explosions of nuclear energy, the incendiary menace of overpopulation, and the socially annihilating impact of highways, automobiles, and suburbia. This suicidal behavior would stop, he reassuringly stated, if people acknowledged the fundamental importance of the "sun, air, water and salt" and "the green leaf on which all organic life depends."[49] Echoing Rexroth's fecund vision of creativity, he claimed that organic co-existence would lead to:

> . . . hundreds and thousands of species, all cooperating with each other, all necessary to each other for their own development and for the further development of life itself. All of man's culture depends on this immense organic variety – his own development and creativity would be unthinkable if Nature were not infinitely more creative than man has ever proved himself to be.[50]

As he continued to speak, Mumford continued to rail against the unnatural power of technology and its destructive influence. People no longer believed in limits, and worse, they were not aware of the inherent dangers of their actions, he stated. It was still possible, however, to address this situation through the dissemination of proper information, as was done, for instance, in the conference *Man's Role in Changing the Face of the Earth* that he had co-chaired. Once this was accomplished, people would discern the benefits of limits and they would work to restore the "symbiosis between city and countryside" disrupted by the myth of technological power.[51]

Nature was clearly Mumford's greatest ally, and mechanization his greatest adversary. But he was not against modernization. Instead, he wanted progress to be civil, cooperative, and organic. This was why he urged McHarg to read Lawrence Henderson and learn an alternative approach to living in which human action, and by extension design, was an adaptive transaction that fitted the environment to life and life to the environment. Mumford considered this biologically deterministic vision of global order to be supremely democratic because it enabled people to freely pursue knowledge, oppose authority, and use reason to discern legitimate from illegitimate action. The challenge, as he saw it, was how to advance social progress without falling prey to a model of struggle in which only the dominant and selfish survive. He based his solution on organic cooperation, as biochemically demonstrated by Henderson. When human action aligned with this naturally equilibrated will to be, society would materially and spiritually flourish, functional architecture would go hand-in-hand with transcendent urbanism, and the city and countryside would unite in regional harmony.[52] Similar to Tillich's philosophical speculations, this paradigm of cooperation was tantamount to an existential revolt against a standardized, mechanized, and artificially conditioned life:

> Organism, communities, human persons are means of regulating energy for the purposes of life. Too much and too little are equally fatal. Power without love, in other words, power without response becomes murderous and suicidal. And so the final lesson is the ultimate lesson of life – love not power is the means by which life prospers.[53]

The worldview that Mumford promoted embraced the biochemical energetics of life, and it coincided with the boundary-toppling vision of togetherness proposed by Eric Fromm. It also deftly united many of the ideas expressed by other guests on the program and shaped them into a communal ethic where the natural world provided the organizing principles, science the inductive logic, and people the operative actions.

Perhaps, then, the best way to summarize *The House We Live In* is to note that McHarg's discussions with his eminent guests raised probing questions about moral authority, alienation, and love, and gave forceful presence to an emancipatory ethic that saw the whole world as divine. But this examination was not without ironies and mixed-messages as evidenced by his desire to present ecology as both the elite purview of specialists and a rebellious critique of existing norms. Moreover, the select guest list that he assembled represented a community of prominent, and by that time, firmly entrenched and relatively old-fashioned gurus of bourgeois culture. As much as he sought to extricate himself from the bonds of tradition, he was unable or unwilling to do so. Then, there is the fact that he repeatedly ignored, or failed to grasp, the subtlety of his guest's remarks, and instead used their words to reinforce his point of view. It all came down to whose philosophy, what religion, and which scientific data best articulated his vision.

The House We Live In, like the later argument presented in *Design with Nature*, was not an open-minded, unsparing look at scientific knowledge and cultural norms for what they reveal about our environmental interactions and values. Rather, it was a quixotic search by McHarg for a world he would rather have than the one in which he found himself. As mentioned at the beginning of this chapter, his vision embraced the sublimity of the natural world and the insignificance of the human presence, as well as the modernist project for universal civilization and humanity's capacity for immense destruction. Seen from this perspective the earth was both a tiny speck in a vast unknowable universe and a safe place of refuge. The program's discussions reinforced this sentiment by promoting a scientific secularism in which actions toward the world, not the kingdom of God or a paradisiacal afterlife, were the essential truth. Bolstered by this ecological morality, he sought to convert as many as possible to his point of view. His comments and questions, however, indicated that he never understood, or would not admit, that his righteous indignation came with its own inherent biases, insensitivities, and dishonesties. In emulation of Zen philosophy, McHarg sought to lift the veil of illusion that clouded reality, but he remained blind to his own prejudice. Steeped as he was in the Protestant teachings of his youth, and by nature unable to admit ambiguity, he could never quite see, or accept, all of the alternatives he discovered when he decided to jettison dogma and embrace the wondrous diversity of life. What he failed to address, and this would become a major criticism of his work, were the tumultuous changes that would soon propel a more diverse social vision to the forefront of environmental dialogues. McHarg's reaction, as noted by his critics, was a redoubling of his efforts to convince others of the moral sanctity and authoritative power of ecological design.[54] And yet, despite the criticism that can be leveled against his use (and abuse) of ecology, his quest to redefine social values and environmental actions was eminently human, and true to life, in the sense that it captured in miniature all the aspirations, dreams, and self-deceptions that are essential to the nature of design.

Notes

1 Leopold, *A Sand County Almanac*, 225.
2 The Architectural Archives of the University of Pennsylvania, Ian and Carol McHarg Collection, "The House We Live In, Related Materials," 365.II.71.
3 The Architectural Archives of the University of Pennsylvania, Ian L. McHarg Collection, "Rockefeller Conference on Urban Criticism Oct. 3–4, 1958," 109.II.A.1.98. In addition to McHarg, conference attendees included William Wheaton, David Crane, Louis Kahn, G. Holmes Perkins, Arthur C. Holden, Leslie Cheek, Jr., Chadbourne Gilpatrick, Eleanor Larrabee, Grady Clay, J. B. Jackson, Catharine Bauer, Kevin Lynch, Lewis Mumford, and Jane Jacobs. For further information see, Peter L. Laurence, "The Death and Life of Urban Design: Jane Jacobs, The Rockefeller Foundation and the New Research in Urbanism, 1955–1965" *Journal of Urban Design* 11 no. 2 (June 2006): 145–172; and Peter L. Laurence, *Jane Jacobs, American Architectural Criticism and Urban Design Theory, 1935–1965*, PhD diss., (The University of Pennsylvania, 2009).

4 The Architectural Archives of the University of Pennsylvania, Ian L. McHarg Collection, "Program 1: Dr. Harlow Shapley," 109.II.B.2.10; and Moving Image Research Center at the Library of Congress, *The House We Live In* Harlow Shapley MAVIS: 1831490.

5 Ibid.

6 The Architectural Archives of the University of Pennsylvania, Ian L. McHarg Collection, "Program 2: David Goddard (edited)," 109.II.B.2.12; and Moving Image Research Center at the Library of Congress, *The House We Live In* David Goddard MAVIS: 1831542.

7 The Architectural Archives of the University of Pennsylvania, Ian L. McHarg Collection, "Program 3: Dr. Carlton Coon (edited)," 109.II.B.2.12; and Moving Image Research Center at the Library of Congress, *The House We Live In* Carleton Coon MAVIS: 1831500.

8 The Architectural Archives of the University of Pennsylvania, Ian L. McHarg Collection, "Program 4: Dr. Margaret Mead (edited)," 109.II.B.2.13; and Moving Image Research Center at the Library of Congress, *The House We Live In* Margaret Mead MAVIS: 1831501.

9 "Program 1: Dr. Harlow Shapley," 1; and *The House We Live In* Harlow Shapley.

10 The Architectural Archives of the University of Pennsylvania, Ian L. McHarg Collection, "Program 5: Rabbi Joshua Heschel (edited)," 109.II.B.2.14; Moving Image Research Center at the Library of Congress House We Line In Abraham Heschel MAVIS 1831502; The Architectural Archives of the University of Pennsylvania, Ian L. McHarg Collection, "Program 6: Dr. Paul Tillich (edited)," 109.II.B.2.25; Moving Image Research Center at the Library of Congress, *The House We Live In* Paul Tillich MAVIS: 1831516; The Architectural Archives of the University of Pennsylvania, "Program 7: Father Weigel (edited)," 109.II.B.2.16; Moving Image Research Center at the Library of Congress, *The House We Live In* Gustave Weigel. S.J. MAVIS: 1831517; The Architectural Archives of the University of Pennsylvania, Ian L. McHarg Collection, "Program 8: The Reverend Swami Nikhilananda (edited)," 109.II.B.2.17; Moving Image Research Center at the Library of Congress, Swami Nikhilananda MAVIS: 1831518; and The Architectural Archives of the University of Pennsylvania, Ian L. McHarg Collection, "Program 9: Professor Alan Watts (edited)," 109.II.B.2.18; and Moving Image Research Center at the Library of Congress, *The House We Live In* Alan Watts MAVIS: 1831519.

11 McHarg recites this passage from Genesis as follows: "Man, exclusively, is made in the image of God, man is given dominion over birds and beasts, fish, fowl, every creeping thing that creepth, and man is licensed to subdue the earth." See "Program 7: Father Weigel, 4. In the King James Bible the passage reads: "And God blessed them [man and woman], and God said to them, Be fruitful and multiply, and fill the earth and subdue it; and have dominion over the fish of the sea and over the birds of the air and every living thing that moves upon the earth." See Division of Christian Education of the National Council of Churches in the United States of America. *The Holy Bible*, translated from the original tongues being the version set forth in A. D. 1161, and revised A. D. 1952 (New York: Thomas Nelson & Sons, 1952), 1.

12 "Program 7: Father Weigel," 7, and *The House We Live In* Gustave Weigel.

13 "Program 6: Dr. Paul Tillich," 9, and *The House We Live In* Paul Tillich.

14 "Program 8: The Reverend Swami Nikhilananda," and *The House We Live In* Swami Nikhilananda.

15 "Program 9: Professor Alan Watts," and *The House We Live In* Alan Watts.

16 "Program 6: Dr. Paul Tillich," 6, and *The House We Live In* Paul Tillich.

17 "Morse Peckham, "Man-and-Environment in the Christian Tradition," 5.
18 Giedion, *Space, Time and Architecture*, 356–357.
19 "Program 9: Professor Alan Watts," 6–7, and *The House We Live In* Alan Watts.
20 See Lâo-Tse, *Tao Te Ching*, translated by D. C. Lau (New York: Penguin Books, 1963). See also Alan W. Watts, "Beat Zen, Square Zen, and Zen" *Chicago Review* 12 no. 2 (Summer, 1958): 3–11.
21 "Program 9: Professor Alan Watts," 6–7, 12, and *The House We Live In* Alan Watts.
22 The Architectural Archives of the University of Pennsylvania, Ian L. McHarg Collection, "Dr. Eric Fromm (edited)," 109.II.B.2.19, 9; and Moving Image Research Center at the Library of Congress, *The House We Live In* Erich Fromm MAVIS: 1831520.
23 The Architectural Archives of the University of Pennsylvania, Ian L. McHarg Collection, "Program 11: Dr. Hans Selye," (number not assigned); and Moving Image Research Center at the Library of Congress, *The House We Live In* Hans Selye MAVIS: 1831521.
24 The Architectural Archives of the University of Pennsylvania, Ian L. McHarg Collection, "Program 12: Sir Julian Huxley," 109.II.B.2.21, 6–7; and Moving Image Research Center at the Library of Congress, *The House We Live In* Julian Huxley MAVIS: 1831523.
25 Ibid., 3. See also Paul Weindling, "Julian Huxley and the Continuity of Eugenics in Twentieth-century Britain" *Journal of Modern European History* 10 no. 4 (November 1, 2012): 480–489.
26 The Architectural Archives of the University of Pennsylvania, Ian L. McHarg Collection, "Series 2, Program 1: Loren Eiseley (edited)," 109.II.B.2.23; and Moving Image Research Center at the Library of Congress, *The House We Live In* Loren Eiseley MAVIS: 1831524.
27 Ibid., 3, 4, 9.
28 See Arnold J. Toynbee, *A Study of History Abridgement of Volumes I–VI* (New York: Oxford University Press, 1974). See also John Mackintosh Shaw, "Arnold Toynbee's Evaluation of the Place of Christianity in Religious History," *Time Inc.* 69 no. 9 (9 June 1958): 274–277; Edward T. Gargan, "Toynbee Revisited," *Time Inc.* 77 no. 20 (12 May 1961): 99–100; Henry Anatole Grunwald, "Toynbee's Best Seller" *Life* 24 (23 February 1948): 118–124, 126, 128, 130, 133; and Henry Grunwald, "Arnold Toynbee: Mapping of a Great Mind" *Life* 37 (November 29, 1954): 87.
29 Moving Image Research Center at the Library of Congress, *The House We Live In* Arnold J. Toynbee MAVIS: 1831530.
30 Ibid.
31 Ibid.
32 During a series of lectures at Edinburgh University in 1952 and 1953, Toynbee argued that the Judeo-Christian concept of human dominion abetted the reckless consumption of resources and the development of destructive technology. McHarg was living in Edinburgh and teaching at Edinburgh University when Toynbee delivered his remarks, but there is no evidence that he attended the lectures. See Arnold Toynbee, *An Historian's Approach to Religion* (New York: Oxford University Press, 1956).
33 *The House We Live In* Arnold J. Toynbee.
34 The Architectural Archives of the University of Pennsylvania, Ian L. McHarg Collection, "Series 2, Program 3: Fairfield Osborn (edited)," 109.II.B.2.24; and Moving Image Research Center at the Library of Congress, *The House We Live In* Fairfield Osborn MAVIS: 1831539.

35 The Architectural Archives of the University of Pennsylvania, Ian L. McHarg Collection, "Series 2, Program 4: Dr. John Christian (edited)," 109.II.B.2.25; and Moving Image Research Center at the Library of Congress, *The House We Live In* John Christian MAVIS: 1831541.

36 Moving Image Research Center at the Library of Congress, *The House We Live In* Theodosius Dobzhansky MAVIS: 1831531.

37 Moving Image Research Center at the Library of Congress, *The House We Live In* A.M.M. Payne MAVIS: 1831532.

38 The Architectural Archives of the University of Pennsylvania, Ian L. McHarg Collection, "Series 2, Program 9: Dr. Leonard Duhl (edited)," 109.II.B.2.27; and Moving Image Research Center at the Library of Congress, *The House We Live In* Leonard Duhl MAVIS: 1831534.

39 The Architectural Archives of the University of Pennsylvania, Ian L. McHarg Collection, "Series 2, Program 7: Dr. William Wheaton," 109.II.2.26; and Moving Image Research Center at the Library of Congress, *The House We Live In* William L. C. Wheaton MAVIS: 1831540.

40 See William L. C. Wheaton, "The Institute for Urban Studies" in *The Book of School: 100 Years*, eds. Ann L. Strong and George Thomas (Philadelphia: University of Pennsylvania Graduate School of Education, 1990), 168–169. See also Richard C. Albert, *Damming the Delaware: The Rise and Fall of Tock's Island* (University Park: The Pennsylvania State University Press, 1987), 51–67.

41 Moving Image Research Center at the Library of Congress, *The House We Live In* Kenneth Rexroth MAVIS: 1831537.

42 Ibid.

43 Ibid.

44 Ibid.

45 The Architectural Archives of the University of Pennsylvania, Ian L. McHarg Collection, "Series 2, Program 8: Lewis Mumford (edited)," 109.II.2.20, 1; and Moving Image Research Center at the Library of Congress, *The House We Live In* Lewis Mumford MAVIS: 1831533.

46 Lawrence J. Henderson, *The Fitness of the Environment: An Inquiry into the Biological Significance of the Properties of Matter* (Boston, MA: Beacon Press, 1913, 1958).

47 Ibid.

48 Lewis Mumford, *The City in History: Its Origins, its Transformations, and its Prospects* (New York: Harcourt Brace Jovanovich, Publishers, 1961), 1.

49 "Lewis Mumford (edited)," 2; and *The House We Live In* Lewis Mumford.

50 Ibid.

51 Ibid., 6.

52 Ibid., 2.

53 Ibid., 10–11.

54 See Anne Whiston Spirn, "Ian McHarg, Landscape Architecture, and Environmentalism: Ideas and Methods in Context" in *Environmentalism in Landscape Architecture*, ed. Michel Conan (Washington, DC: Dumbarton Oaks Research Library and Collection, 2000), 97–114; and Marc Trieb, "Nature Recalled" in *Rediscovering Landscapes: Essays in Contemporary Landscape Architecture*, ed. James Corner (New York: Princeton Architectural Press, 1999), 29–44.

6   The Ecology of the City

But what distinguishes the worst architect from the best of bees is this, that
the architect raises his structure in imagination before he erects it in reality.
Karl Marx, *Capital: A Critique of Political Economy*[1]

In the summer of 1961, shortly after the conclusion of the television program *The House We Live In*, McHarg and the ecologist Robert MacArthur
jointly submitted a grant proposal to the United States Department of
Health, Education and Welfare for funds to support a two-week workshop that would study the impact of air and water pollution on the health
of people, plants, and animals in the urban ecosystem. The ecologist
G. Evelyn Hutchinson (MacArthur's Ph.D. thesis adviser) and the botanist
David R. Goddard (McHarg's guest on *The House We Live In* and the
provost of The University of Pennsylvania) compiled the list of workshop
participants. The grant application, which was titled "The Ecology of the
City," claimed the resulting data would allow architects, landscape architects, planners, and engineers to remodel the "distorted" morphology of
the city to conform to the dynamics of natural systems[2] (Figure 6.1).

   The proposal did not receive funding, but undeterred, in 1962 McHarg
delivered a talk at the Cranbrook Academy of Art titled "The Ecology of
the City" in which he promoted urban ecosystem studies as the means to address overcrowding, uncontrolled growth, and deficiencies in air and water
quality.[3] During these remarks, he displayed his newly acquired scientific
literacy, and interspersed his commentary with references to photosynthesis, entropy, nutrient cycling, and succession. A description of the temporal
morphology of a deciduous forest, taken verbatim from MacArthur, indicated what he sought to achieve:

   Recall that hepaticas and snowdrops and all the small plants tend to
   flower earliest in the spring. The reason for this is that they can have
   at that time all the sunlight they need, because the herbaceous plants,
   shrubs, and trees are not in leaf. The next group of plants to flower
   and seed are either herbaceous plants or shrubs. They, in turn, while

Figure 6.1 Photograph of a deciduous forest from the personal files of Ian McHarg. Ian L. McHarg Papers, The Architectural Archives of the University of Pennsylvania.

casting their shade upon the smallest plants, do not yet receive any shade from the taller trees. They are then followed by dogwoods and redbuds, which are still unshaded by the larger trees, and can absorb the necessary amount of sunshine. Finally, the tallest trees of the forest spread their canopy and cast their shade on the plants and trees beneath them, while having access to the sunshine they themselves need. The morphology of the forest can thus be understood in terms of cycles of plants related to periods of sunlight . . . The forest, which is at first seen as an undifferentiated mass of green, in fact, has a specificity and is an extraordinarily expressive statement of an ordered system.[4]

The city was not a deciduous forest, but McHarg wanted his colleagues to copy the stratigraphic complexity of this biome in their urban designs. This objective required an in-depth examination of material and energy flows, the selection of physical attributes that aligned with these temporal patterns, and the insertion of the most stable and enduring of these traits into the structure of the city. It all came down to studying, as MacArthur had done, the spatial dynamics of movement, territory, succession, and sunlight.

The concept of design as a mimetic expression of natural forms and processes not only adhered to the ecological principles of MacArthur, it also recalled McHarg's master's thesis and, in particular, his appropriation of ideas from James Marston Fitch regarding the use of plants to mediate air temperature, filter atmospheric pollutants, and reduce glare.[5] In this sense, he was not asking his colleagues to totally rethink their modernist principles, but he was asking them to refrain from imposing preconceived forms on the land, and to instead look around, observe nature, and incorporate this knowledge in their work. Excited by the imaginative possibilities, he stated that ecological design transformed the architectural manifesto "form follows function" into the dictum "form expresses process" or, "even better, process is expressive."[6]

That same year, McHarg received a request from the epidemiologist Leonard Duhl to prepare an essay for *The Urban Condition: People and Policy in the Metropolis*.[7] The essays in this text were to examine those aspects of the urban environment that cause stress – noise, overcrowding, pollution, poverty, physical fragmentation, and crime – and propose comprehensive solutions. To ensure the essays dealt with the complexity of urbanization in all of its political, economic, social, and spatial dimensions, Duhl asked each author to forgo a reductionist cause and effect approach, which, he explained, meant that they must consider the city ecologically, as "an open system, always in flux, never returning to the point at which it starts." Many of the essays originated as talks delivered during a 1962 meeting of The Orthopsychiatric Association devoted to the examination of alienation in the urban environment. The work of the physiologist John Calhoun on the behavior of rats in overcrowded cages provided the impetus for, and organizing theme of the conference, and by extension the impetus for, and organizing theme of the text.[8]

McHarg titled his essay "Man and Environment," and he began his argument with contrasting descriptions of the earth as it appeared from space.[9] In the first description, unchecked urban growth rapidly consumed resources and the land and lifeless gray tissue spread ominously from the city. In the second description, urban growth remained within bounds, the land stayed green, and plants, animals, and people symbiotically shared resources. He then posed a question: Was humankind a planetary disease or a beneficial organism? Appropriating the thoughts of Loren Eiseley, he concluded, in words that echoed the providential warnings of George Perkins Marsh, that humans were indeed an insatiably hungry and rapidly spreading pestilence:

> In this period we have seen the despoliation of continental resources accumulated over eons of geologic time, primeval forests destroyed, ancient resources of soil mined and sped to the sea, marching deserts, great deposits of fossil fuel dissipated into the atmosphere. In the country, man has ravaged nature; in the city, nature has been erased and man assaults man with insalubrity, ugliness and disorder.[10]

The essay's less than comforting aerial synopsis of human settlement patterns and angry assessment of human action toward the land announced McHarg's intent to present an expansive theory of design rather than the point-by-point analysis of stress in the urban environment. As he pressed forward with his argument, he proposed an overarching, and purposefully shocking, synthesis of operative dualities that reconciled man and nature, West and East, white and black, brains and testicles, Classicism and Romanticism, St. Thomas and St. Francis, Calvin and Luther, and anthropomorphism and naturalism. Holism was in order, and, in his opinion, this necessitated the infusion of Eastern philosophy into the bloodstream of Western cultural canons:

> Conceptions of man and nature range between two wide extremes. The first, central to the western tradition, is man-oriented. The cosmos is a pyramid erected to support man on its pinnacle, reality exists only because man can observe it, indeed God is made in the image of man. The opposing view, identified with the Orient, postulates a unitary all-encompassing nature within which man exists, man in nature.[11]

Later, in *Design with Nature*, McHarg would illustrate the all-embracing and geographically synoptic synthesis of man and nature outlined in this essay using images of the Acropolis in Athens, Greece and the Katsura Imperial Villa in Kyoto, Japan as paradigmatic examples of ecological unity. These images reflected his personal observations and life experiences. He knew both landscapes from time spent in Greece during the war and later travels in Japan with his wife Pauline. And it is perhaps no coincidence that Gropius praised the dynamic spatial qualities of the Katsura Villa in *Apollo in the Democracy* and Mumford praised the sublime power of the Acropolis in *The City in History*[12] (Figure 6.2).

From these broad generalities, he moved to specifics and targeted the Judeo-Christian doctrine of otherworldliness for censure. If earthly existence was simply probation before the glories of the afterlife, he bluntly stated, it was preordained that ugliness would disfigure the land.[13] He followed this statement with a discussion of the gardens created by André Le Nôtre for Louis XIV and his royal ministers in the 17th century.

Figure 6.2 Photograph of the Acropolis from *Design with Nature* that illustrates the ideal relationship of "man and nature." Courtesy Foto Marburg/Art Resource, NY. Reprinted by permission.

His argument took its cues from Morse Peckham, but provided a slightly different biblical interpretation that emphasized arrogance, power, and control (Figure 6.3). Here Cartesian rationality and the divine right of kings imposed patterns of behavior, which, in keeping with Genesis 1:28 and its charge to be fruitful, multiply, and have dominion countenanced a self-centered, insensitive, and domineering disregard of everything that existed outside the carefully prescribed boundaries of the *ancien régime* and its perfectly manicured conception of life.[14]

A true renaissance in thinking, he alternatively claimed, occurred in the country estates of 18th-century England. Conveniently ignoring the aristocratic display of wealth and power inherent to the enclosed hills and vales of these vast parklands, he proceeded to read this pastoral aesthetic as a sensitive and practical melding of physiography, science, philosophy,

Figure 6.3 Delagrive plan of Versailles (top) and photograph of Blenheim Palace (bottom) from *Design with Nature*. Delagrive plan courtesy Historic Urban Plans collection. Reprinted by permission; Photograph Blenheim Palace © 1992 by John Wiley & Sons, Inc. Reprinted by permission of John Wiley & Sons, Inc.

poetry, and art that spoke to sublime transcendence, creative imagination, pragmatic stewardship, and benign coexistence. The sylvan meadows and scenic lakes of these landscapes were not just pleasing to the eye and stimulating to the mind, they were impeccable examples of a natural economy that required minimal maintenance to be self-perpetuating (Figure 6.3). In contrast to the rigid patterns of Le Nôtre, which he stated usurped the power of God, the organic arrangement of plant and animal communities in the "proto-ecological" landscape of the English tradition "spoke to man of God."[15] Divinity, in this understated, albeit grand expression of harmonious habitation was of the world, not external to it, and therefore worthy of emulation.

The enticing charm of pastoral beauty was only a partial explanation of what McHarg sought to achieve. To complete his argument, he looked to the forms of nature for guidance. Building protective shelters was not an exclusive human skill, he explained. The chambered nautilus constructed spiral shells. Bees built beehives, and coral incorporated the minerals of the earth into the substance of their bodies, and in the process created reef habitats that sheltered a myriad of plants and animals. Each of these structures, he continued, was built with an economy of means and they had "survived periods of evolutionary time vastly longer than the human span."[16] The implications were staggering. Not only did McHarg downgrade *Homo sapiens* from its apex position in the grand scheme of life, he also challenged the supremacy of Judeo-Christian doctrine. If animals did what people did, and did it better, then the claim of dominion in Genesis was at best dubious. He borrowed a passage from the essay Paul Sears wrote for *Man's Role in Changing the Face of the Earth* to defend his position:

> Any species survives by the virtue of its niche, the opportunity afforded it by environment. But in occupying this niche, it also assumes a role in relation to its surroundings. For further survival it is necessary that this role at least be not a disruptive one. Thus, one generally finds in nature that each component of a highly organized community serves a constructive, or, at any rate a stabilizing role. The habitat furnishes the niche, and if any species breaks up the habitat, the niche goes with it.[17]

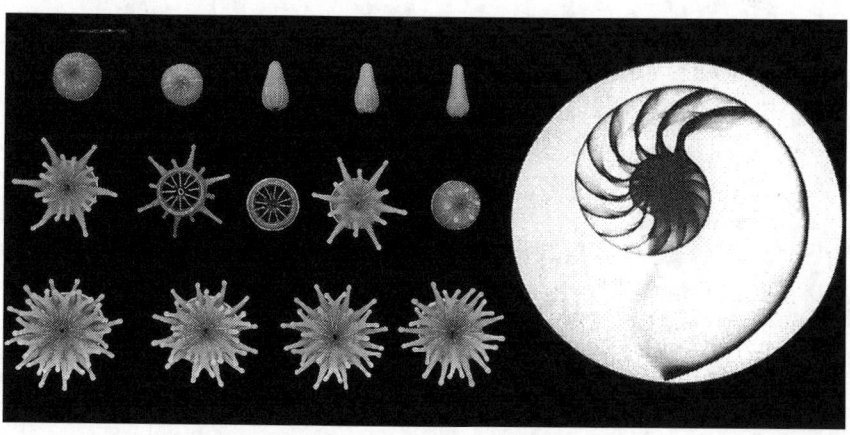

Figure 6.4 Coral polyps and a nautilus shell from *Design with Nature*. Coral image #12616 American Museum of Natural History. Reprinted by permission; Nautilus © 1992 by John Wiley & Sons, Inc. Reprinted by permission of John Wiley & Sons, Inc.

Again, the implications were clear. Humans resided within a community of species on a finite planet with limited resources, and they had to live commensurately, behave with a modicum of courtesy, quit fouling their nests, and build with nature. McHarg would include images of a nautilus shell, beehive, and coral in *Design with Nature* (Figure 6.4), and he would pair these images with photographs of the Taos Pueblo, a communal adobe house located in the desert southwest that was built by Native Americans prior to the European settlement of North America, and Fallingwater a private residence designed by the architect Frank Lloyd Wright and located in the heart of a deciduous forest (Figure 6.5). Situated in dramatic settings and constructed with an economy of means, both of these structures exist commensurately with their surroundings and appear to be timeless extensions of the land.[18]

Figure 6.5 Taos Pueblo (top) and Fallingwater (bottom). Photograph of Taos Pueblo from *Design with Nature*. © 1992 by John Wiley & Sons, Inc. Reprinted by permission of John Wiley & Sons, Inc; Fallingwater photograph courtesy Arianne Giudicelli. Reprinted by permission.

Notes

1 Karl Marx, *Capital: A Critique of Political Economy Volume I, Part I, The Process of Capitalist Production*, trans. Friedrich Engels (New York: Appleton & Co., 1889), 157.

2 The Architectural Archives of the University of Pennsylvania, Ian L. McHarg Collection, "Ecology of the City Participants," 109.II.A.1.28. In addition to McHarg, MacArthur, Hutchinson, and Goddard the proposed "Ecology of the City" workshop participants included J. B. Ovington, Dr. Fritz Went, G. N. Plass, H.T. Odum, and David Kendall. Loren Eiseley authorized the grant application for the University of Pennsylvania.

3 G. Holmes Perkins, J. B. Jackson, and Jacqueline Tyrwhitt also spoke at the conference. The papers were later compiled into an edited volume. See: Ian McHarg, "The Ecology of the City" in *The Architect and the City: Papers from the AIA-ACSA Teacher Seminar Cranbrook Academy of Art*, June 11–22, 1962, ed. Marcus Whiffen (Cambridge, MA: The M.I.T. Press, 1966), 53–65; and Ian McHarg "The Ecology of the City" *Journal of Architectural Education* 17 no. 2 The Architect and the City. The 1962 AIA0ACSA Seminar Papers Presented at the Cranbrook Academy of Art. Part I (November 1962): 101–103.

4 McHarg, "Ecology of the City," 56–57.

5 Fitch, *American Building*, 295.

6 Ibid.

7 Leonard Duhl, *The Urban Condition: People and Policy in the Metropolis* (New York: Basic Books, Inc., 1963).

8 Ibid., ix–xii. Duhl worked for The National Institute of Mental Health and he subsequently developed a federal initiative titled the Model Cities program to co-ordinate government services for the nation's poorest and least-served urban communities. In 1966, President Johnson enacted this proposal as part of the Great Society initiative. North Philadelphia, a poor, predominately African-American community adjacent to the University of Pennsylvania, was one of the first communities to receive Model City funding. See Joe Flower, *Healthcare Forum* "Building Healthy Cities: Excerpts from a conversation with Leonard J. Duhl, M.D," https://people.well.com/user/bbear/duhl.html; Jason T. Bartlett, "Model Cities" *The Encyclopedia of Greater Philadelphia*, https://philadelphiaencyclopedia.org/archive/model-cities/; and John H. Strange, "Citizen Participation in Community Action and Model Cities Programs" *Public Administration Review* 32 Special Issue Curriculum Essays on Citizens, Politics, and Administration in Urban Neighborhoods (October 1972): 655–669.

9 Ian L. McHarg, "Man and Environment" in *The Urban Condition: People and Policy in the Metropolis*, ed. Leonard Duhl (New York: Basic Books, Inc., 1963), 44–58.

10 Ibid., 45.

11 Ibid.

12 In *Design with Nature*, the photograph of the Katsura palace is attributed to an anonymous source. The photograph of the Acropolis is attributed to Walter Hege, a German photographer associated with National Socialism. McHarg likely found this photograph in *The Acropolis*, a 1930 text that was reissued in 1957. It is also more than likely that he was unaware of Hege's problematic history. Instead, he would have been attracted to the rhetoric of natural harmony, perfection, purity, and simplicity that permeated this text's discussion of the Acropolis, including the observation that this architectural ensemble "rises above the tumult of the town, bearing high up into the clear air the temples of the sacred precinct, an immense and firm foundation for the most beautiful edifices ever raised by the hand of man." See Gerhardt Rodenwaldt, *The Acropolis* (Oxford: Blackwell Publishing, 1957). See also Gropius, *Apollo in the Democracy*, 124; and Mumford, *The City in History*, 160.

13 McHarg, "Man and Environment," 52.
14 The passage reads as follows in the King James Bible: "And God blessed them [man and woman], and God said to them, Be fruitful and multiply, and fill the earth and subdue it; and have dominion over the fish of the sea and over the birds of the air and every living thing that moves upon the earth." See Division of Christian Education of the National Council of Churches in the United States of America. *The Holy Bible*, 1.
15 McHarg, "Man and Environment," 53.
16 Ibid., 54.
17 Ibid., 57. See also Sears, "The Processes of Environmental Change by Man," 472.
18 The comparison of modern and indigenous housing and settlement patterns was a common trope in mid-20th-century architectural discourse. Similar to the science of ecology, it was seen as a way to re-energize the modernist project. This idea appeared, for example, in an essay written by Aldo van Eyck for a student publication sponsored by the University of Pennsylvania Department of Landscape Architecture. This text also contains a discussion of succession and a description of McHarg's work. See Aldo Van Eyck, Paul Parin and Fritz Morgenthaler, "A Miracle of Modernization" in *Via 1, Ecology in Design*, eds. Rolf Sauer, et al. (New York: Grossman Publishers, Inc., 1968), 96–125. See also, Sigfried Giedion, *The Beginnings of Architecture: The A. W. Mellon Lectures in the Fine Arts, 1957, The National Gallery of Art, Washington, DC* (Princeton, NJ: Princeton University Press, 1964).

7   Towards a New Landscape

The architect must think of his responsibility – his responsibility to create something which is always true to the nature in man and to the Laws of Nature, and which is conscious of water, of air, of light, of the animal world and the green world.

Louis Kahn, The Nature of Nature[1]

Each person who lectured in *Man and Environment* and each guest who appeared on *The House We Live In* provided thought provoking appraisals of knowledge, perception, choice, and action. But they did not provide definitive answers. Life, unfortunately, was more complicated than McHarg wanted it to be. And this is what prompted, or perhaps it is better to say reinforced, his belief that ecology and its study of the relationship of an organism to its environment could serve as a synthesizing model. Besides, as several of his guests on *The House We Live In* indicated, this way of knowing and acting provided a bit of karmic hope that nature was resilient and it could bounce back from the harmful actions that people inadvertently and thoughtlessly inflicted.

In the creative burst of energy that followed the conclusion of *The House We Live In*, McHarg delivered talks, composed essays, prepared grants, expanded his teaching faculty, and began his professional collaboration with David Wallace. He also outlined the chapters of a potential book under the heading *Towards a New Landscape*. The outline consisted of a series of notes on tablet paper in which he puzzled over the meaning and role of landscape architecture in contemporary society. His musings reprised questions he posed in Man and Environment and *The House We Live In*: What are the canons of landscape architecture? How did they develop over time? How were they expressed in different cultures? How did they relate to scientific knowledge? How were they influenced by modernist architecture? Did they honor all forms of life? Could they make people perceive the world as it is, and as something better than it is? And, if so, what did this portend for the future? He annotated his notes with a timeline of culturally specific garden styles – Japanese, Middle Eastern, Renaissance (Italian and French), and

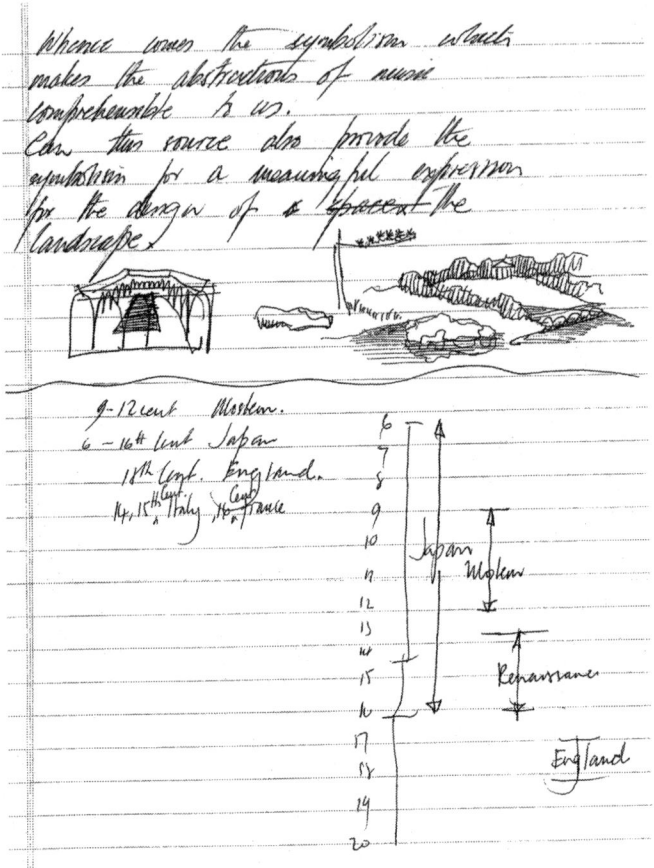

Figure 7.1 Garden history timeline and associated sketches by Ian McHarg. Ian
L. McHarg Papers, The Architectural Archives of the University of
Pennsylvania.

English – that in many ways paralleled the historical chronology of civilizations documented by Arnold Toynbee. He cross-referenced the timeline against quick but elegant sketches that captured the spatial attributes of the traditions he deemed worthy of emulation (Figure 7.1). He omitted the Renaissance garden from this drawing vocabulary based on a long-standing antipathy to clipped hedges and straight lines, and instead delineated the sanctity of water in the gardens of the Middle East, the dialogue of rock and tree in the Japanese garden, and the rolling hills, meadows and hedgerows of the English estate landscape. This selective annotation of garden history, however, reflected only one aspect of his aspirations. Eagerly reaching beyond the confines of tradition to capture the broad scope of his ambitions, he created a comprehensive definition of mid-20th-century landscape architecture that included urban parks, land conservation, science, architecture, and regional planning. The outline and sketches, as mentioned previously, evolved into the environmental manifesto *Design with Nature*.[2]

More critically, and in contradiction to his provocative all-or-nothing rhetoric and the immaculate vision of the sun and the earth on the cover of *Design with Nature*, during this intense period of exploration McHarg developed an approach to practice that reflected less ideal realities and more immediate fears. The attributes of this landscape are seen in a collage that was initially proposed as the cover of *Design with Nature*. The collage consists of two panels. Each panel contains an image of a hand: one hand reaches toward the city as if to heal it, the other reaches toward the countryside as if to save it. The hand hovering over the city consists of pictures of the countryside, and the hand hovering over the countryside consists of pictures of the city (Figure 7.2). The fragmented structure of the image acknowledges the impossibility of completely repairing past mistakes,

Figure 7.2 Proposed cover *Design with Nature*. Ian L. McHarg Papers, The Architectural Archives of the University of Pennsylvania.

but the mosaic of objects and events also implies, in keeping with McHarg's unwavering faith in ecology, that a judicious mix of science, technology, conservation, and stewardship could heal, or at least ameliorate, most environmental wounds.

If we extend this thinking, McHarg's struggle to construct a design approach from an assortment of ideas, images, and conversations corresponds to the concept of paradigm shift developed in the early 1960s by the philosopher Thomas Kuhn in The *Structure of Scientific Revolutions*.[3] Indeed, when this comparison is made, McHarg's skeptical assessment of architectural modernity, his search for alternatives, and his turn to ecology not only assumes added meaning, it also becomes easier to position his critical stance within the discourse of the era. According to Kuhn, revolutions in scientific thinking, particularly in the early stages of formation, were not *tabula rasa* transformations, but instead multifaceted and often contradictory accumulations of evidence that prompted a "crisis" of thought. This crisis subsequently provoked additional inquiries designed to rectify disparities, resolve incompatibilities, and fill conceptual voids. The resulting evidence irreversibly revolutionized thinking and produced a paradigm shift that precluded a return to what existed before. Equally important (and contentious), was Kuhn's claim that the question-oriented basis of science was not, as its practitioners would have you believe, an objective undertaking. Scientists may seek truth, their methods may be testable, their compilation of information may make persuasive arguments, their work may solve problems, but they were not impartial. McHarg's quest to resolve discrepancies, critical questioning, decidedly partial defense of ecology, and meticulous compilation of evidence from multiple sources exemplified, in spirit and practice, a paradigm shift that made it impossible to return to strictly formalist designs that failed to consider the agency of natural processes and change over time.

Support for this speculation can be found in "An Ecological Method for Landscape Design," an essay written concurrently with *Design with Nature* in which McHarg presented the evidence he had accumulated over the years to justify a new, and in his mind, revolutionary system of practice. His proposal appropriated modernist faith (progress was real) and ecological science (action was contingent, causal, and relative). It rejected religious strictures (Judeo-Christian doctrine was not the only truth) and economic determinism (value was more than a monetary good). His design proposition utilized the language of architectural modernity to subversively claim place is because, form and process are indivisible, nature always comes first, and design is more than the application of style.[4] To explain his intent, he returned to his Lâo-Tse inspired conversations with Louis Kahn and stated:

> Cup is form and begins with the cupped hand. Design is the creation of the cup, transmuted by the artist, but never denying its formal origins. As a profession, landscape architecture has exploited a pliant earth,

tractable and docile plants to make much that is arbitrary, capricious, and inconsequential. We could not see the cupped hand as giving form to the cup, the earth and its processes as giving form to our works. The ecological method is then also the perception of form. . .which is to say design, and this, for landscape architects, may be its greatest gift.[5]

Earlier in "The Nature of Nature," Kahn had made a similar statement about form and design using a spoon to create an analogy between the essence of an idea (the bowl of the spoon), and its materialization into something useful (the handle of the spoon). The final product – the consummate utensil or tool – required the designer to select, shape, and arrange the elements of the design – both physical and metaphysical – until they fit together in perfect harmony. According to Kahn, this way of thinking and acting was multifunctional and scalable: it could define the formal attributes of singular objects, and it could revitalize obsolete systems that no longer functioned as originally intended due to social and technological change. As a method of practice, this approach was a means to return to origins – brick, stone, arch, and wall in architecture and water, air, light, plants, and animals in landscape architecture – and a means to create connections among places – buildings, streets, people, cars, and cities – which, he claimed, made it authentic and real. Kahn imaginatively deployed this way of thinking to dismantle architectural compositions into their constituent parts prior to reconstructing the pieces in dramatic new ways, which, in this case he claimed, transformed the willful act of design from the singular, selfish "I" to the plural, inclusive "thou." Indeed, for Kahn this sensibility turned everyday actions and objects into things of great value that advanced the ideas and actions of others:

> Find the form and from it many designs can come – many notions many personal acts. Design is a personal act, it is how you see it. But the principles, the unique characteristics, are something which do not belong to you at all. They belong to the activity of man of which you happen to be a part and which you must discover.[6]

In the concluding section of *Design with Nature*, McHarg would refer to his quest for an authentic design paradigm – one that broke with the past and yet retained traces of continuity – as an "extemporaneous inquiry" that stitched together "a ragbag of memories and notes" cut from "the vestments of wiser men." His ecological revelation came later, after he assembled the pieces of this fabric into a design proposition and discovered the evidence made larger claims than he had anticipated. Proud of his achievement, but aware that it required refinement, he acknowledged with feigned modesty that his work followed no singular plot and it contained "incongruities, the seams were imperfect, but finally, although the product is only a patchwork quilt, is it not one piece of cloth?"[7]

Notes

1 Kahn, "The Nature of Nature," 121.
2 The Architectural Archives of the University of Pennsylvania, Ian L. McHarg Collection, "Towards a Modern Landscape," V.C.21.2. The outline for the proposed text was ambitious and eclectic and it consisted of five sections: Historical Legacies, Modern Precursors, Modern Influences, Modern Solutions, and The Future. Historical Legacies discussed the visual qualities of landscapes and their generating attitudes. Modern Precursors included the garden traditions of 18th-century England, China, and Japan, the work of Fredrick Law Olmsted, and the conservationists John Muir, Theodore Roosevelt, and Gifford Pinchot. Modern Influences discussed the "Naturalist Tradition" and the work of Charles Darwin, Thomas Huxley, Harlow Shapley, Patrick Geddes, and G. Scott Williamson. Modern Influences and Modern Solutions discussed "The Functionalist Tradition," and included Mies van der Rohe, Le Corbusier, Walter Gropius, Alvar Aalto, Louis Kahn, Richard Neutra, Frank Lloyd Wright, Lewis Mumford, Thomas Church, Christopher Tunnard, Garret Eckbo, Lawrence Halprin, James Rose, Dan Kiley, Hideo Sasaki, and Roberto Burle Marx. The Future discussed river valleys and metropolitan regions, and called these landscapes a "national treasure and scenic art."
3 Thomas S. Kuhn, *The Structure of Scientific Revolutions* (Chicago, IL: The University of Chicago Press, 1962, Third Edition 1996). My discussion of Kuhn is indebted to the following essay, Matthew C. Rees, "The Structure of Scientific Revolutions at Fifty" *The New Atlantis* 37 (Fall 2012): 71–86.
4 Ian L. McHarg, "An Ecological Method for Landscape Architecture" *Landscape Architecture Magazine* 57 no. 2 (1967), 105–107.
5 Ibid., 107.
6 Kahn, The Nature of Nature, 120. Kahn likely appropriated the analogy of the spoon from Ernesto Rogers who used the phrase "from the city to the spoon" to define the scale of architectural production and illustrate its importance in everyday life. As noted by the architecture historian Michelangelo Sabatino, Rogers sought to counter the dogma of architectural modernity through a return to tradition. See Michelangelo Sabatino, *Pride in Modesty: Modernist Architecture and the Vernacular Tradition in Italy* (Toronto: University of Toronto Press, 2011): 165–167.
7 McHarg, *Design with Nature*, 196.

Part 3
Implementing order

8   City and countryside

The hope of the city lies outside itself. Focus your attention on the cities – in which more than half of us live – and the future is dismal. But lay aside the magnifying glass which reveals, for example, the hopelessness of Broadway and Forty-second Street, take a reducing glass and look at the entire region in which New York lies. The city falls into focus. Forests in the hill-counties, water-power in the mid-state valleys, farmland in Connecticut, cranberry bogs in New Jersey, enter the picture. To think of all these acres as merely tributary to New York, to strengthen the lines of the web in which the spider-city sits unchallenged, is again to miss the clue. But to think of the region as a whole and the city as merely one of its parts – that may hold promise.

Lewis Mumford, Regions – to Live In[1]

In 1961, the same year that McHarg explored the ecology of urbanization on the television program *The House We Live In*, the French geographer Jean Gottmann declared the traditional distinctions that separated the city and countryside no longer held true. The mid-20th-century city was not a "tightly settled and organized unit in which people, activities, and riches are crowded in a very small area clearly separated from its nonurban surroundings." An exhaustive compilation of over 800 pages of information on demographics, manufacturing, trade, retail, banking, politics, land-use, and transportation that he recently completed indicated that urbanized areas contained houses, factories, farms, forests, and recreation areas. The resulting commingling of use made it all but impossible to discern metropolitan boundaries and signaled the need for a "profound revision of many old concepts such as the distinctions between city and country."[2] The geographical extent of Gottmann's research included a swath of land along the northeastern seaboard of the United States that stretched from Washington, D.C. to Boston, Massachusetts (Figure 8.1).

The Twentieth Century Fund (today, The Century Foundation) supported Gottmann's research. The Fund, which began operation in 1919, previously played an important role in the New Deal policies of the Franklin Roosevelt Administration. Following the conclusion of World War II,

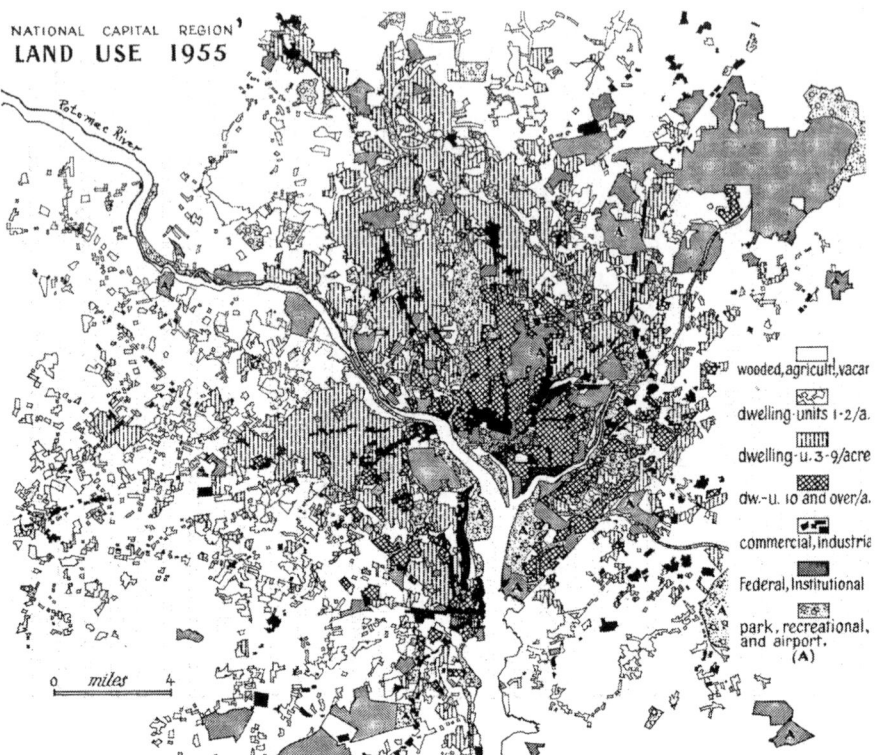

Figure 8.1 Land use map of Washington, D. C. © 1961 by The Century Foundation, Inc. Reproduced by permission from Jean Gottmann, *Megalopolis: The Urbanized Northeastern Seaboard of the United States.*

The Fund renewed its commitment to centralized government planning, which it considered the best means to advance industrial productivity and social prosperity, and thus provide the highest standard of living for the largest percentage of the population.[3]

In 1947, the Fund outlined its post-war agenda in an 817-page study titled *America's Needs and Resources.* The study patriotically presented a framework for action that transformed war production into an "arsenal of democracy" to ensure "the American consumer and his family" were "by far the best-fed, best-housed, and best-clothed civilians in the world." As this Cold War language of "guns and butter" indicated, the Fund would work to maintain the economic and social capital generated by World War II military spending as the country transitioned back to a civilian economy.[4] In support of these policies, the Fund became a major benefactor of the Marshal Plan, and it financed efforts to export this ideology of prosperity, along with goods and services manufactured in the United States, to the rest of the world.

America's Needs and Resources championed the notion that it was necessary to anticipate, plan, and to some extent, control production policies to promote the common good. Factors considered essential to the health and welfare of the country, and by extension the American vision of global order, were gainful employment, infrastructure spending, readily available natural resources, prime agricultural land, abundant water supplies, new housing, and ample consumer goods and services. The study assumed the prewar migration of the population from the countryside to the city would resume, but in the post-war period this movement would be matched by an outward flow from the city centers – urban cores – as people sought newer housing with water and sewer connections, central heating, and the electrical capacity to run the work-saving appliances that war-time innovation and mass-production had made available and affordable. Data indicated that 9.4 million dwellings would be built in the next decade, and this would produce a coincident demand for furniture, appliances, tableware, curtains, and upholstery.[5]

Gottmann's research updated *America's Needs and Resources* with the intent of elucidating the "processes and entanglements" that shaped the *"dynamics of urbanization."*[6] Which is to say, the mechanics of distribution and the flow of goods and services – the where, why, and how of development activities – was of more interest to Gottmann than population statistics or the number of highways and skyscrapers.

Gottmann returned to classical antiquity to select a title for his landmark study. He chose the ancient Greek city of Megalopolis, a vast showcase of metropolitan splendor with a population drawn from forty surrounding villages, to symbolize the grand aspirations and endeavors that he believed the northeastern region of the United States represented.[7] Besides suggesting that post-war prosperity had promulgated massive physical change, the reference to this ancient city set the philosophical foundation (and social tone) of his scholarship:

> The ancient philosopher of Alexandria, Philo Judaeus, taught there is a great *city of ideas* that predetermines and commands the material world in which we live, and this greater city of ideas Philo called Megalopolis. It seems then, especially fitting to apply the same name to this extraordinary region, the present shape and style of which arose from the beliefs and searching of those who settled there to bring a new order to their brethren on earth.[8]

An enthusiastic, and at times credulous, advocate for what he observed and recorded, Gottmann considered the urbanization of the northeastern seaboard of the United States to be an expression of historical progress that stood on the "threshold of a new way of life." But in a prudent display of pragmatism, he observed, "history records a long list of brilliant civilizations that have sunk under the pressure of internal decay, and external jealousy."[9] Decentralization and urban sprawl, distribution bottlenecks and highway congestion, and income inequality and environmental pollution were particularly problematic. At issue was how far and how fast this region would, or could, progress before it collapsed. For Gottmann, the resolution of these issues would determine the preeminence of the West, and at a more fundamental and alarming level, the ability of Western civilization to survive. Fully committed to the vision of productivity, prosperity, and national security advanced by his research sponsor, he insisted on the preservation of this way of life "to reassure the nation and the world about the kind of life modern urbanization trends presage for the future."[10]

Megalopolis was not a cautionary tale. Conversant in both the revolutionary rhetoric of liberal democratic society and the capitalist discourse of the business community, Gottmann endorsed social cooperation and economic competition. His analysis of the country's land use policies indicated the combined efforts of the American people and the American free market would overcome any obstacles that blocked the path to success. Challenges such as "urban crowding and the slums and mobs characteristic of it" were merely "growing pains in the endless process of civilization."[11] In his view, these physical conditions provided opportunities to advance science and technology, and this intellectual capital made cities innovative leaders in the search for security and freedom.

Planners and geographers generally acknowledged the importance of Gottmann's work even though they intensely debated the historical ramifications of his findings. A review of *Megalopolis* by William L. C. Wheaton, for instance, stated the book proposed "a new thesis on urbanization, one which will probably prove as important in the history of thought as *Democracy in America*."[12] In other words, Gottmann's treatise, in emulation of the work of the 17th-century French geographer Alexis de Tocqueville, revealed the complex transactions between individual liberty and government dictates, elitism and populism, and altruism and greed that defined American society.

Although it differed in tone, Gottmann's study of material progress and regional development also captured the complexity of the "unfettered suburban landscape" that Lewis Mumford described that same year in *The City in History*. Writing from the quiet comfort of his suburban home in Armenia, New York, Mumford castigated the stultifying physical uniformity and alienating social conformity of suburbia. In his estimation, this "low-grade environment" was a "revolt against order" that was indicative of out-of-date planning policies and romantic aspirations incapable of meeting the demands of modern society. He called for constraints that would limit the:

> . . . multitude of uniform, unidentifiable houses, lined up inflexibly, at uniform distances, on uniform roads, in treeless communal waste, inhabited by people of the same class, the same income, the same age group, witnessing the same television performances, eating the same pre-fabricated foods, from the same freezers.[13]

Despite his less than stellar opinion of suburbia, Mumford did, however, consider this landscape the perfect venue to study the processes currently shaping urbanization. If this terrain was properly appraised, selectively adapted, and subsequently improved, he believed it would make "a positive contribution to the emerging concept of the city as a mixed environment, interwoven in texture with the country"[14] Gottmann came to a similar conclusion regarding the analytic potential of suburbia, but differed in his evaluation of its physical and social morphology. In his assessment, the forces shaping the suburban landscape represented "one interwoven system bound to progress," and he opposed management policies that artificially constrained growth.[15]

A firm believer in exceptionalism, Gottmann placed the responsibility for overcoming the challenges threatening the region's energy and prosperity on the shoulders of an elite cadre of citizens crowded within the Megalopolis. These distinguished individuals were "*on the average*" wealthier, better educated, better housed, better serviced, and more technologically advanced than any other group of people in the world. "They ought to be able," he observed in a remark that mirrored the sentiments expressed by the historian Arnold Toynbee in his twelve-volume *A Study of History*, "to find ways of avoiding decline of the area." One of their first tasks involved the conservation of "the natural beauty of the landscape" and the development of policies that would ensure the "health, prosperity, and freedom of the people."[16]

Gottmann presciently realized that the patchwork terrain of suburbia also supported a rich and productive ecology. He was particularly intrigued by the way economic development fostered the abandonment of farmland and the subsequent regrowth of woodlands that were now more extensive than the greenbelt cities envisioned by Ebenezer Howard in *Garden Cities of To-morrow*. The resource potential of this spontaneous secondary growth, he observed, indicated that vacant land was not an "abandoned space" or a "leftover landscape," but instead a valuable source of wood that would sustain home building and economic growth.[17]

Gottmann depicted the natural history of suburbia using a 1958 cover of *The New Yorker*. The cover illustration, by the artist Garrett Price, depicts two sportsmen walking through a suburban woodlot in pursuit of country pleasures. They are empty handed, without bagged prey. A deer, squirrels, rabbits, raccoons, pheasants, quail, and chipmunks sit quietly in the foreground and watch the hunters as they leisurely stroll toward their parked car. In this arcadian pastoral, people co-exist, albeit ambiguously with nature, suggesting that the maintenance of ecological diversity in the wake of urban development was not an issue for Gottmann. Indeed, as an early, and un-witting advocate of the concept of urban re-wilding, he matter-of-factly claimed that the modern urbanite loves to hunt and "he finds plenty of game in the Megalopolis." Having thus passed judgment, he further observed that suburbia was not the purest form of nature in the sense advocated by Thoreau and Emerson, but it was:

> . . . on the whole 'natural enough' to provide city folks, who spend most of their time in an environment of bricks, cement, glass, and steel, with an environment where flora and fauna develop according to the basic laws of Nature, even though they are under the influence and supervision of civilization.[18]

Megalopolis caught the attention of McHarg, who, after the successful conclusion of *The House We Live In,* began to explore the ecology of suburbia. The numerous maps and charts produced by Gottmann provided a road map that he could follow and visually complement as he appraised, selectively adapted, and in his mind, improved the land use patterns of this peri-urban region.

To fully understand what McHarg sought to achieve in his explorations of suburbia, it is necessary to think of his maps and ideas as parallels of the work and ambitions of Gottmann. Both men studied climate, topography, hydrology, soils, vegetation, transportation, land use, and settlement patterns, but their work differed in emphasis. Gottmann mapped the figure ground of built form and McHarg mapped the figure ground of undeveloped land. In combination, their studies illustrated how these twin landscapes, and the flow of goods and services between them, overlapped in space and time. Yet, in contrast to the laissez-faire capitalism that permeated *Megalopolis,* McHarg produced data sets and maps to curb excesses in the system, including policies that privileged the production of consumer goods and ignored the protection of the natural resources that made the production of these goods and services possible. Closer to Mumford than to Gottmann in this regard, he created "negative development maps" to redirect the focus of urban planning away from built form and toward an in-depth discussion of critical ecological processes, resource conservation, and landscape preservation.[19] To advance this agenda, McHarg embarked on a series of land use studies that reconceived regional development and the Garden City within the context of post-war suburban sprawl. His objective, in emulation of Ebenezer Howard, was a unifying synthesis of the city and countryside that incorporated the best attributes of both. Ecology – variously conceived as a socio-political stance, moral philosophy, economic engine, utilitarian process, procedural method, and creative experiment – unified his ideas and efforts. Students and colleagues served as co-investigators.

The Delaware River watershed

McHarg's exploration of the suburban landscape began with a survey of land use patterns along the Delaware River that he conducted at the request of William L. C. Wheaton, his academic colleague and Director of the

Institute of Urban Studies at the University of Pennsylvania. The impetus for the study came from a court-ordered mandate that was issued following an attempt by New York City to impound the water of the river for its sole use. Signatories to the agreement – the states of New York, Pennsylvania, and New Jersey – had to ensure dependable water supplies for all communities within the Delaware River watershed. The court order required these parties to document the impact of development along the river and its tributaries. The resulting data on water quantity and quality would support land preservation, reduce flood risk, ameliorate pollution, and provide increased opportunities for recreation. Similar to government initiatives such as the Tennessee Valley Authority (1934–1936) and the Rural Electrification Administration (1942–1945), the work established the framework of a hydrologic infrastructure that included dams and energy production.[20]

McHarg's survey of the Delaware River watershed culminated in the fall of 1960 with a design studio at the University of Pennsylvania. A project summary, written by McHarg in the spring of 1961 titled "New Approaches to the Reservation of Open Space in the Delaware River Basin," explained how a land use survey that documented the "real value" of the land, as opposed to its "speculative value," would stop the "unprecedented assault" of urbanization that had developed in the absence of planning and planning tools. To protect the region's forests, farmland, and water resources from development, ten students under his direction sub-divided the landscape into five zoning categories designed to serve as alternatives to "socio-economic" determinism. The zoning categories differentiated what superficially appeared to be a uniform "sea of green" in normative depictions of undeveloped land into distinct functions and roles: (1) Flood Plain and Riparian Zoning, (2) Aquifer and Aquifer Recharge Zoning, (3) Forest and Agricultural Zoning, (4) Fish, Wildlife, Ecological and Geological Zoning, and (5) Historical Area Zoning. The students mapped each category. Areas that remained clear – those with no zoning categories – were suitable for development. Following the completion of the land use survey, the students designed greenway systems for the cities of the Delaware River watershed – Scranton, Trenton, and Philadelphia. These "fingers of green" preserved stream corridors and woodlands, enhanced water quality, and provided opportunities for recreation (Figure 8.2). Innovative and yet conventional, this was another example of the way McHarg adaptively repurposed existing practice to serve his planning agenda. In the project discussion, McHarg claimed the land use and greenbelt studies were "rational and useable but, at the same time, also inspirational to the people of the watershed."[21]

Metropolitan open space from natural resources

In 1963, shortly after McHarg and David Wallace founded the firm Wallace-McHarg Associates, they embarked on a new planning project that extended the findings of the 1960 Delaware River studio to include

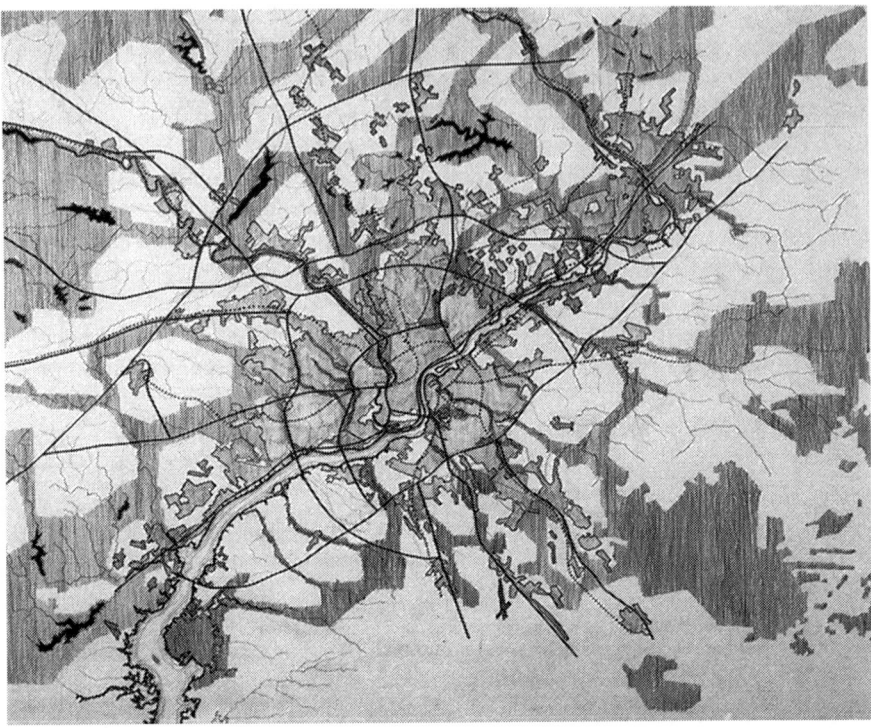

Figure 8.2 Urban Settlement and Open Space System from the 1960 Delaware River design studio. Philadelphia City Planning Collection, The Architectural Archives of the University of Pennsylvania.

the Schuylkill River and the metropolitan region of Philadelphia. William McLaughlin, a member of the federal government's Urban Renewal Administration and the states of Pennsylvania and New Jersey, supported the work (Figure 8.3). The findings appeared in *Design with Nature* and in *Metropolitan Open Space and Natural Processes.*[22]

The Wallace-McHarg study, following procedures developed in the Delaware River design studio, divided the land into zoning categories related to water quality: (1) surface water, (2) marshes, (3) flood plains, (4) aquifer and aquifer recharge areas, (5) steep lands, (6) forests and woodlands, (7) and prime agriculture land (Figure 8.4). The categories were individually mapped on vellum and a photographic negative of each map was produced. To determine the total area of each category, the negatives were placed on a light table and the amount of light that passed through them was subtracted from the total amount of light emitted by the light table. A proportional formula mathematically compared the amount of light and determined the overall extent of a particular land use. In his description of the project, McHarg noted

Figure 8.3 Photograph of the Schuylkill River from *Design with Nature*. Grant Heilman Photography. Reprinted by permission.

the formula avoided "time consuming and expensive" planimeter measurements.[23] The negatives were then layered to determine overlapping categories of land use. An accompanying matrix listed eighteen possible combinations of the seven land use categories. These combinations were mapped in multiple shades of blue, green, and gray. The resultant mosaic of woodlands, water, and steep slopes turned areas that were often blank (white) in nominal development maps into important and highly visible landscape attributes (Figure. 8.5).

A bird's-eye perspective, following a representation method used by Patrick Abercrombie in the Plan for London, visualized the conservation strategy. This mode of representation, which Abercrombie adapted from the Valley Section of Patrick Geddes, provided the client, and the public at large, an image of what the color-coded markings in land use maps represented. In this particular instance, the perspective illustrated how the project preserved the forests, farmlands, scenic quality, and water resources of the metropolitan region of Philadelphia. Also in keeping with the traditions of British surveying, the team created a block diagram of the landscape that tied the region's hydrology – rivers, marshland, floodplains, and aquifer recharge areas – to the underlying geology[24] (Figure 8.6).

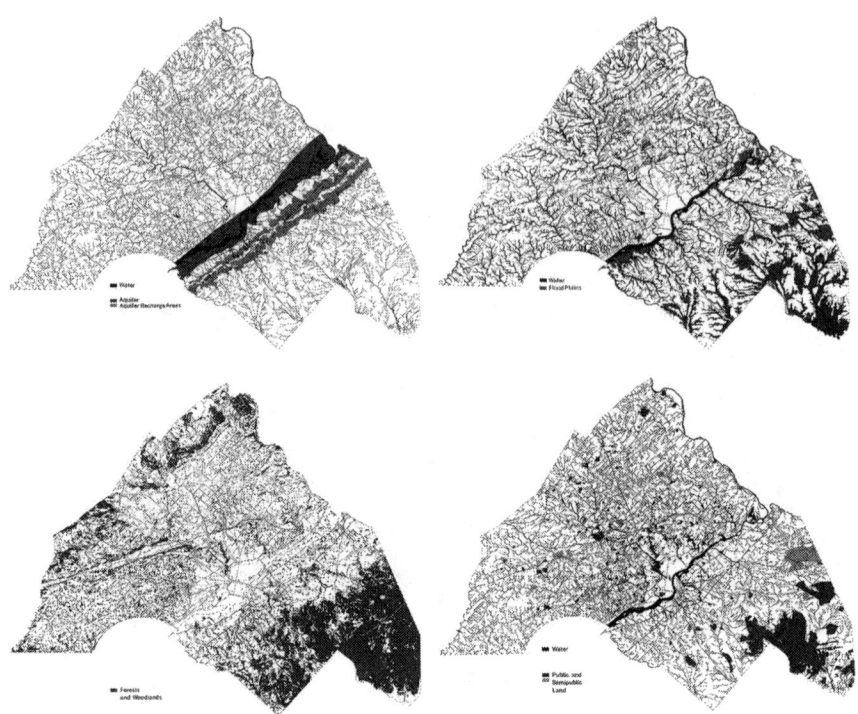

Figure 8.4 Land use maps from *Metropolitan Open Space and Natural Processes.*
UPP Publication: *Metropolitan Open Space and Natural Process* (1971)
Edited by David A. Wallace. Reprinted by permission The University of
Pennsylvania Press.

A draft project report, titled *Metropolitan Open Space from Natural Resources* (Metropolitan Open Space), written by McHarg and edited
by his wife Pauline, explained the conservation strategy presented in the
drawings.[25] The essay began with a preamble that established what was a
stake – nature receding under the onslaught of uncontrolled development.
The resulting "low-grade urban tissue," McHarg observed, in a statement
that reprised Mumford's less than stellar assessment of suburbia, was devoid
of beauty and made the American Dream of a home in the country an un-
attainable fantasy. The preamble's excoriating denunciation of wanton development and resource destruction established the need for ecological land
planning, as well as the philosophical foundation (and social tone) of the
argument:

> Today a million acres of land each year are transformed from farm-
> land and forest to hotdog stand and diner, gas station, rancher, split
> level, concrete, asphalt, billboards and sagging wire, shopping centers,
> parking lots and car cemeteries. There is little metropolitan planning;

Figure 8.5 Metropolitan Open Space land use survey summary map from *Design with Nature*. © 1992 by John Wiley & Sons, Inc. Reprinted by permission of John Wiley & Sons, Inc.

development occurs without reference to natural phenomena, on flood plains, marshes and steep slopes. Woods, forests and farmland are destroyed without remorse, streams culverted, ground, surface water and atmosphere polluted, floods and droughts exacerbated, beauty superseded by vulgarity and ugliness.[26]

To counter what he referred to as poorly planned "growth by accretion," McHarg divided the landscape into a "web of urban land" that occupied "areas best suited for development," and a "web of open space" that served "specific functions in natural process and urban recreational space needs."[27] He borrowed density numbers from Gottmann to demonstrate that urban sprawl was not due to the scarcity of land, but to planning policies that failed to address resource distribution.

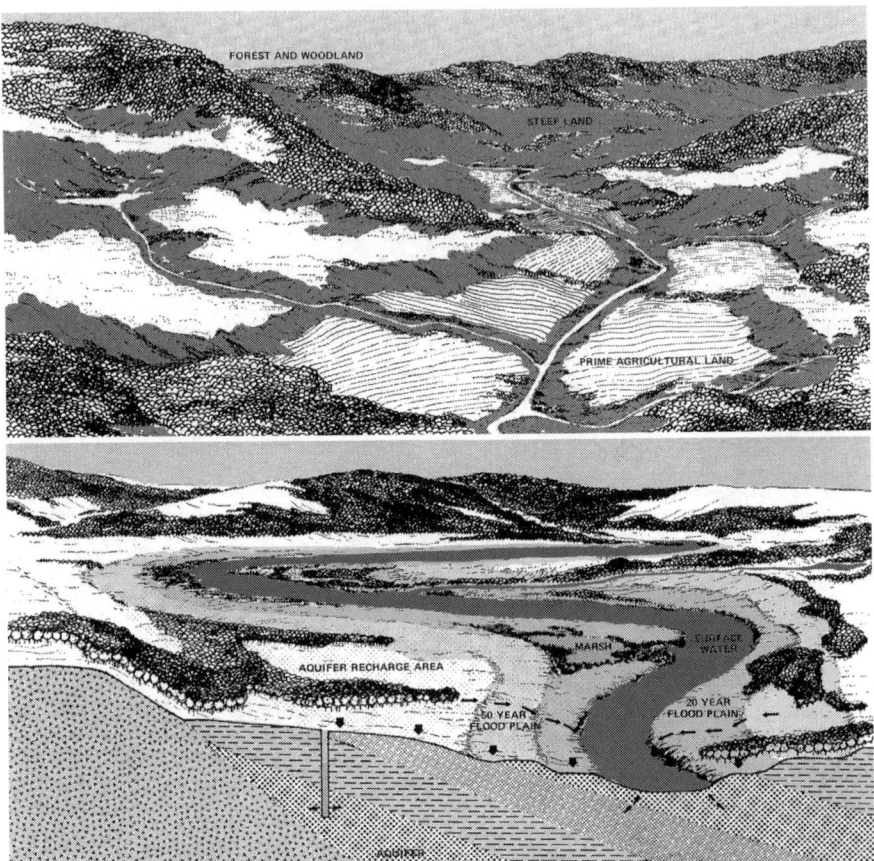

Figure 8.6 Metropolitan Open Space study aerial perspective and aquifer recharge diagram. Aerial perspective from The Potomac River Basin study, UPP Publication: *Metropolitan Open Space and Natural Process* (1971) Edited by David A. Wallace. Reprinted by permission The University of Pennsylvania Press. Aerial perspective of aquifer recharge from *Design with Nature*. © 1992 by John Wiley & Sons, Inc. Reprinted by permission of John Wiley & Sons, Inc.

The study's prioritization of the watershed and water resources clearly reflected the mandate of The Urban Renewal Administration, but it also allowed McHarg to display his new understanding of ecosystem dynamics and to think strategically in terms of environmental relationships and natural patterns of organization. Using ideas garnered from the ecologists Robert MacArthur, Paul Sears, and Aldo Leopold as points of reference, he proceeded to outline a design strategy that linked the hydrologic cycle to physiography, physiography to nutrient recycling, nutrient recycling to solar energy, solar energy to climate, climate to plant and animal diversity, and plant and animal diversity to land use and environmental health.

After McHarg enumerated the benefits of the ecosystem approach, he emphasized the fragility of the natural landscape and the destruction that occurred when it was disturbed. Using timber harvesting on steep slopes as an example, he explained how this action denuded the land, increased soil erosion, caused the siltation of lowland marshes, and heightened the risk of flooding. To avoid time-consuming and costly remediation, it simply made more sense to preserve forested hillsides. To ensure this outcome, he emulated Leopold, and called for regulations that permitted certain uses and prohibited others.

For the proposed ecosystem approach to be considered a viable alternative to normative planning practice, undeveloped land had to be valued at or above its market price. This ran counter to the current system, which McHarg noted, was geared toward a "commodity consumer relationship" that stressed the "short term financial advantages of prospective development."[28] Change necessitated a shift in attitude, and this, as he had learned from Mumford on *The House We Live In*, required the proper information and an education campaign. To entice people to willingly accept his land valuation system, he explained how they would profit, both privately and publicly, using evidence that he believed to be indisputable. First, the preservation of undeveloped land would personally benefit suburban homeowners because it ensured that their houses remained close to nature, which, after all, was why they had moved to suburbia in the first place. Second, the preservation of areas of critical ecological importance benefited the general public because this land provided valuable environmental services – microclimate amelioration, water purification, flood control, soil stabilization, pollution mitigation, forest resources, and healthy wildlife populations – for everyone, free of charge. Third, if the land worked as intended, this would reduce the risk of property damage, which again was in everyone's interest because it lowered the physical and social costs of urbanization.

To justify his valuation system for undeveloped land, McHarg crafted a deft synthesis of ecosystem theory, landscape stewardship, and capitalist economics that explained how the natural landscape provided services that were sustainably powered by the energy of sunlight as it moved through the soil, water, plants, animals, people, families, and communities of the metropolitan region. The inputs and outputs of this natural process balanced the inputs and outputs of industrial manufacturing. The "net result" was a "complex process of energy utilization and production" that naturally augmented the work of people.[29] Compliance demanded minimal personal sacrifice. All that was required for this vision to become a reality was careful study and documentation by experts, followed by the codification of these efforts in government policies. Even Gottmann, who argued against stringent development regulations, noted the positive outcome of the efforts initiated by the Delaware River Basin Commission to regulate the use of resources that supported growth.[30]

McHarg's conflation of economics, development, ecology, and stewardship mirrored an approach ingrained in the policies and practices of American land conservation. As explained by the environmental historian Donald Worster, this overtly managerial approach to land stewardship reflected a practical compromise that combined two foundational principles of the United States – the mercantile virtues of free enterprise and the agrarian virtues of land husbandry.[31] In this worldview, everything had to earn its way in life and produce goods for consumption, including the land. The operative phrase was maximum organic utility through minimum organic cost. Worster further observed that the nascent science of ecology, which came of age in the 19th century during a period of substantial industrialization and capitalist expansion, adopted a complementary rationale and language. In sympathy with the free-market fundamentalism of these historical precedents, McHarg conceived of natural processes as integral components of the country's production system, and he used the technocratic terms optimization, efficiency, energy utilization, and productivity to describe his objectives.[32] Again, it is not surprising that he would assume this posture, as MacArthur had used similar language to describe species distribution and resource utilization when he lectured in Man and Environment.

McHarg's argument also channeled the moral strains of Leopold's land ethic and in a similar manner challenged privileged, self-centered codes of conduct that made the land nothing more than a metaphorical slave tasked with dripping milk and honey into the mouth of the patriarch Abraham.[33] In other words, his resource categorizations, vertical organization strategies, and economic valuations were decidedly technocratic, but they also envisioned the land to be, as he had imaginatively fantasized to the theologian Paul Tillich on *The House We Live In*, a truly metropolitan place where people and the "micro-organisms of the soil, sun, star, wind, reptiles, amphibians, mammals exist in one system, and no one organism is unique."[34]

The natural beauty of the landscape provided additional inspiration, and McHarg deployed aesthetics to argue that people had the right to use the land, but not to abuse it. With this objective in mind, he argued for conservation practices that did not harm, and perhaps even enhanced the function and beauty of "marsh and forest, rivers and clouds, ground and surface water."[35] Watershed protection, sustainable forestry, and soil husbandry were the codes of conduct that advanced his aspirations and calmed the fears associated with technological hubris, resource exploitation, and environmental pollution. This approach humanized industrial progress, but it did not exonerate bad practices. Instead, and as argued by the historian David Lowenthal in his study of George Perkins Marsh, it ensured a world in which bad practices no longer governed.[36] The resulting beauty served as an index of social health and economic prosperity.

The resource inventory map produced for the project reflected all of these ideas. The color-coded pattern of this document not only indicated what to conserve and where to conserve, it also expanded the concept of scenic beauty to include the ethics of stewardship, the economics of production, and the aesthetics of ecology. In a similar project completed for Staten Island, McHarg described his intent as follows:

> Normally land use maps, and even planning proposals, show broad categories of uses. The maps in this [Staten Island] study are more like mosaics than posters – for good reason. They result from asking the land to display discrete attributes which, when superimposed, reveal great complexity. But this is the real complexity of opportunity and constraint. Yet it may appear anarchic, but only because we have become accustomed to the dreary consistency of zoning, because we are unused to perceiving the real variables in the environment and responding to this in our plans.[37]

The question of limits pervades this discussion of land use: How much abuse can the land take and how much alteration in behavior can people accept before things fall apart? The solution depicted in the resource inventory map deliberately inverted the operative assumptions and spatial hierarchies of development in order to elevate the importance of natural resources in this scenario. Essentially, a critical reflection of American society and its values, the map's visually powerful, but spatially ambiguous mosaic of ecologically sensitive areas called attention to the disruptive impact of spatial fragmentation and the benefits of balance and connectivity. As a counterpart to the work produced by Gottmann in *Megalopolis*, this map dissolved the boundaries that separate the city and countryside. But McHarg went further and claimed his landscape surveys turned an irreconcilable choice – land development or land preservation – into a liberating act of environmental equality. In *Design with Nature* he explained the issue this way: "It is not a choice of the either the city or countryside: both are essential, but today it is nature, beleaguered in the country, too scarce in the city which has become precious."[38]

During this process of discovery, McHarg concluded that metropolitan Philadelphia was simply too big and too complex to be completely organized and controlled. However, with planning policies that took into account the opportunities and constraints afforded by the land, he believed it was still possible to achieve some type of balance between the competing interests of economic expansion and the conservation of the resources that made expansion possible.[39] In this, he simply enlarged the design thesis that governed his earlier advocacy of the urban townhouse with an interior garden.

The draft essay that McHarg wrote for the project contained an assortment of cartoons cut from the pages of popular magazines. These vignettes, inserted by McHarg's wife Pauline, created a humorous autobiographical

"*Know what this road needs? Billboards.*"

Figure 8.7 Cartoon by Alan Dunn from the *New Yorker* that appears in the draft
report of Metropolitan Open Space. The Mary Petty and Alan Dunn
Estate at Syracuse University. Reproduced by permission.

sub-text. One cartoon, for instance, referenced the negative impressions of
the urban fringe that they had jointly recorded during a cross-country road
trip in the summer of 1949. The cartoon depicts a man and a woman in the
front seat of a convertible driving through what appears to be an endless
housing development. The caption reads: "Know what this road needs? Billboards"[40] (Figure 8.7).

Two case studies completed by McHarg while he was working on Metropolitan Open Space provide further insight into his ideas and methods.
One project, done with students at the University of Pennsylvania, enriched
his understanding of ecology. The other, a professional project for a client
with specific planning needs, nostalgically played to a picturesque vision of
land stewardship. Together, they reveal the complexity and contradictions
in McHarg's thinking, and the compromises he was willing to make as he
struggled to reconcile his ecological idealism with the realities of practice.

Sea and survival

In March of 1962, the Ash Wednesday Nor'easter pounded the Atlantic
coastline of the United States. For three days, gale force winds and thirty-five
foot waves buffeted the barrier islands of New Jersey. Dunes were breached.

Homes, roads, and utilities were destroyed. Several people were killed and many more were injured.[41]

The catastrophic loss of lives and property prompted McHarg and his students to suspend a studio project on housing and devote the rest of the semester to the study of littoral currents, sand flow, and barrier island formation. The resulting case study of the New Jersey shore, which appeared in a chapter of *Design with Nature* titled "Sea and Survival," called attention to the dynamic nature of the shifting sand and the relative stability provided by the native plants that colonize the sand dunes and the coastal wetlands.[42] The study is noteworthy for its innovative use of an ecological transect of a barrier island in the design decision-making process. The transect extended from the ocean to the bay and subdivided the barrier island into geomorphologic zones – ocean, beach, primary dune, trough, secondary dune, back dune, bayshore, and bay. A complementary cross-section illustrated how the interaction of wind, sand, and salt spray had extended the island profile over time and promoted the growth of plant communities uniquely adapted to the different microclimates of this landscape (Figure. 8.8).

McHarg completed a similar analysis of barrier island vegetation in a 1956 class studio that examined the landscape of Cape Hatteras.[43] In this study, which followed protocols outlined in a 1948 issue of *Architectural Forum* dedicated to "measure" and a 1951 article on temperature and wind movement by the architect Victor Olgyay that also appeared in *Architectural Forum*, students diagrammed the impact of vegetation on wind flow to ascertain the optimum location for recreation facilities and sunbathing. This analysis included the creation a prototypical cross-section of Cape Hatteras that divided the landscape into geomorphologic zones. Additional sections noted the plant associations in each zone, as well as their development (succession) over time[44] (Figure 8.9).

In their study of the New Jersey shore, McHarg and his students redeployed the Cape Hatteras analysis of geomorphology, wind flow, and vegetation to locate housing. The accompanying project narrative, in keeping with the concepts of territorial dynamics, species diversity, and succession detailed by MacArthur, described how the pioneering plants that colonized the dunes supported a dynamic community of organisms that waxed and waned in response to the shifting sand, salt, and water regimes, and the physical disruption of storms, but nevertheless worked to stabilize the land and shape the island profile. Photographs of the storm destruction indicated that it was foolhardy and dangerous to remove the dunes and their vegetation (Figure 8.10).

To preserve the dunes and wetlands, McHarg and his students limited development to the central spine of the barrier island and the back slope of the dunes facing the bay. The plan, which was all about habitat partitioning and the optimization of resources, balanced the needs of development against the needs of the island's plant and animal species. It positioned houses

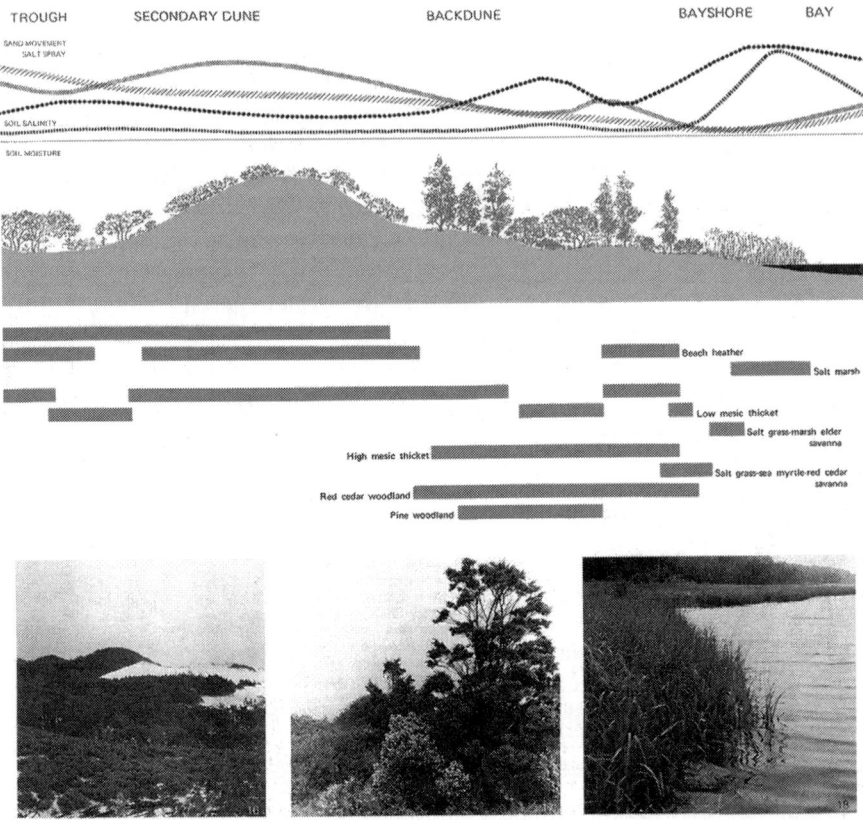

Figure 8.8 New Jersey barrier island transect and vegetation from *Design with Nature*. © 1992 by John Wiley & Sons, Inc. Reprinted by permission of John Wiley & Sons, Inc.

to maximize protection from storm damage, and to minimize injury to the sand dunes and plant communities that stabilized the dunes. A central road, elevated to function in a manner analogous to the flood control dikes in Holland, connected the proposed housing and anchored the scheme. Boardwalks provided beach access and protected the dunes and vegetation. Strict limits on the number of wells and septic systems maintained the fresh water table that supported the plant communities. The report noted that future reductions in storm damage would offset the economic penalties imposed by the restrictions.

The urban planner David Wallace attended the design review at the end of the semester. Impressed by the environmental analysis and the insightful arrangement of houses in conformance with the natural landscape, he asked McHarg to collaborate on a project he was about to commence

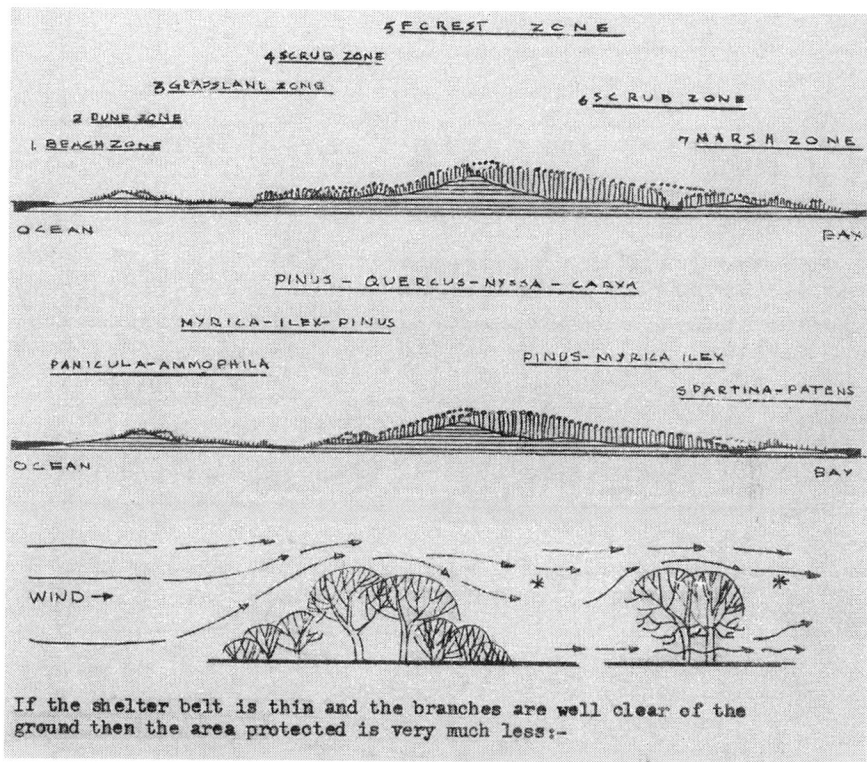

Figure 8.9 Barrier island wind movement, plant communities, and geomorphologic zones from Cape Hatteras National Seashore Recreation Area study. Ian L. McHarg Papers, The Architectural Archives of the University of Pennsylvania.

in Maryland. They formed Wallace-McHarg Associates and began work on an ecologically inspired development proposal, which they documented in a report titled the *Plan for the Valleys*.[45]

Plan for the valleys

The *Plan for the Valleys* was completed under the auspices of the Worthington Valley Planning Council, Inc. – a non-profit citizens group established to ensure "preservation of the highest level of amenity with optimum development." The study area, which included 70 square miles of land northeast of Baltimore that had recently come under development pressure due to the construction of an interstate highway, contained pastoral river valleys, wooded slopes, and upland plateaus. The primary development issue was not land scarcity, but the preservation of the rural landscape.[46]

Figure 8.10 Photograph of storm destruction from *Design with Nature*. The Architectural Archives of the University of Pennsylvania.

To control development and preserve the agrarian ambiance of the site "at densities consonant with housing and market preferences," the proposal controlled the location of sewers, which it clustered, along with houses, in suburban enclaves – town centers and surrounding villages and hamlets – on the upland plateaus. In addition to protecting ground water from contamination and preserving the scenic quality of the stream valleys, this strategy reduced the cost of construction, as the areas slated for development were flat and there were few trees to remove. Growth projections, density numbers, development acreage, lot sizes, construction costs, and appraised real estate values confirmed the economic viability of these aesthetic moves.[47]

The *Plan for the Valleys* also incorporated aspects of the New Town design that McHarg had proposed for the Firth of Clyde in Scotland, as seen, for instance, in the decision to place high-density residential towers on the plateaus and low-density, single-family houses on the hillsides.

But here, rather than slavishly repeating the orthogonal super block organization of his earlier scheme, the hillside houses threaded through the existing trees in conformance with the lot sizes and curvilinear lines of suburban development. In both projects, the strategy provided the best views of the landscape.

The project report stated the proposal used "wisdom, skill and taste" to enhance the beauty of the land and create an "ennobling physical environment" that brought this area "a step toward the American Dream."[48] Photographs, sketches, and a map of the project, color-coded in several shades of green to denote river valleys and forested slopes, demonstrated how the proposed design preserved ecologically critical areas and the landscape's picturesque river valleys (Figure 8.11). A build-out plan, ominously titled The Spectre of the Future, in conjunction with photographs of rows of identical suburban track houses, inspired fear and established what was at stake – nature receding under the onslaught of monotonous development (Figure 8.12). McHarg conceived of the design as series of environmentally friendly, pro-growth propositions:

> The area is beautiful and vulnerable; Development is inevitable and must be accommodated; Uncontrolled growth is inevitably destructive; Development must conform to regional goals; Observance of conservation principles can avert destruction and ensure enhancement; The area can absorb prospective growth without despoliation; Planned growth is more desirable and as profitable as uncontrolled growth; Public and private powers can be joined in partnership to realize the plan.[49]

To sell the ecological principles of the proposal to the client, the team decided, as later noted by William Roberts, to connect the scheme to traditional ideals of landscape beauty using evocative photographs and sketches that deployed the aesthetic conventions of Romantic treatises on the pastoral.[50] Playing upon the historical resonance and popularity of this type of imagery, the report adroitly painted a picture of benevolent development that indicated how the visionary plan in conjunction with the visionary foresight of the client had saved the area from destruction:

> Looking backward to today there were two extreme possibilities. Twenty years before the area could have been quickly despoiled. The other possibility lay in the realization of a Plan. It was only through private concern, reflected in private and governmental partnership and action, in a continuing process, that such a beneficent result ensured.[51]

The covert reference to the utopian writing of the socialist Edward Bellamy in the description of the plan is telling.[52] In keeping with Bellamy, whose novel *Looking Backward* inspired Ebenezer Howard to develop his treatise on the Garden City, the report suggests that the economic excesses of

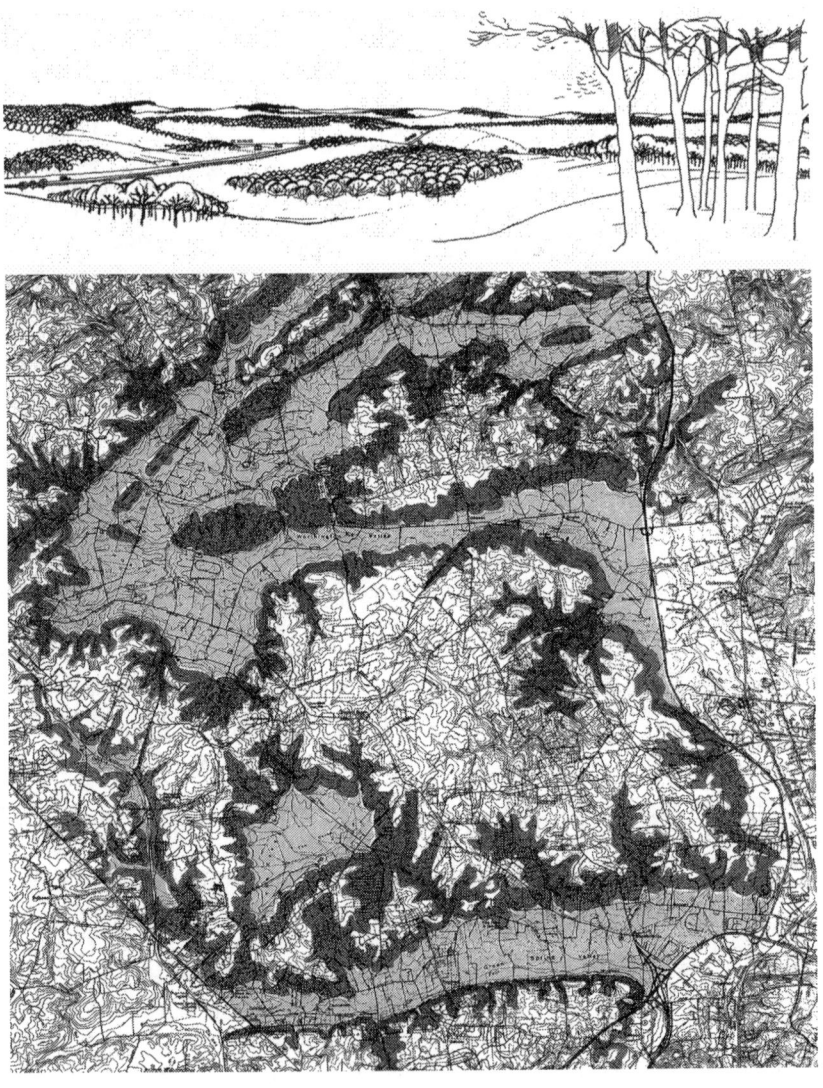

Figure 8.11 The Basic Amenity plan and accompanying sketch by William Roberts from *Plan for the Valleys*. Courtesy WRT Design. Reprinted by permission.

modernity could be held in abeyance through cooperative public-private agreements that protected and conserved natural resources. There were, however, inherent tensions between the existing economic and social realities and the project's nostalgic backward gaze. The proposed scheme did create communally owned land that protected the Worthington Valley from the unwanted intrusion of suburban development, but it also helped a

Figure 8.12 The Spectre of the Future and photograph of suburban housing from *Plan for the Valleys*. Courtesy WRT Design. Reprinted by permission.

syndicate of wealthy landowners preserve the visual quality of their land-holdings, and thus the economic value of their property. The project report touted the scheme's pastoral beauty and landscape conservation initiatives, but failed to acknowledge the less than egalitarian ramifications of the design or how easy it was to manipulate an ecological agenda to reflect a socially exclusive, class-conscious respectability. Nevertheless, as one of the first private development projects in the United States to base landscape conservation on a survey of natural resources, the *Plan for the Valleys* soon garnered critical attention.

In light of these projects, perhaps two other cartoons from the draft essay of Metropolitan Open Space more accurately capture McHarg's ecological agenda. The first cartoon depicts a middle aged couple enjoying a day at the beach. The man stands facing the ocean with his right arm extended, palm up, signaling the waves to stop, while the woman unpacks a picnic basket. The caption, typed by Pauline, states: "It would therefore be unwise to underestimate the immense importance of non-human systems, particularly the water related ones in this man-dominated region." The second cartoon depicts a man and women in a dilapidated convertible stopped by the side of the road under a tree. The caption reads: "Let's abandon it here, where the simple honesty of its rusting bodywork will provide an astringent contrast with the banal uniformity of the countryside's greenness."[53]

Notes

1 Lewis Mumford, "Regions – to Live In" *The Survey Graphic* LIV no.1 (May 1, 1925): 151–152.
2 Jean Gottmann, *Megalopolis: The Urbanized Northeastern Seaboard of the United States* (New York: The Twentieth Century Fund, 1961), 4–5.
3 See The Century Foundation, "About the Twentieth Century Fund," https://tcf.org/about/, and "Century Foundation Records, New York Public Library Archives and Manuscripts, Century Foundation Records, "Biographical/Historical Records," http://archives.nypl.org/mss/18811.
4 J. Frederick Dewhurst ed., *America's Needs and Resources* (New York: The Twentieth Century Fund, 1947), 3, 6 and 17.
5 Ibid., 40–43, 84, 142–145, 152, 165, 173–175, and 196.
6 Gottmann, *Megalopolis*, 10–12.
7 In 1956, the nuclear physicist Robert Oppenheimer, in his role as director of the Twentieth Century Fund, invited Gottmann to study the metropolitan region of New York. Oppenheimer suggested the book should be titled *Megalopolis*. See "Jean Gottmann 1915–1994," in The British Academy. *Proceeding of the British Academy, Biographical Memoirs of Fellows, 11*, ed. P. J. Marshall (Oxford: Oxford University Press Scholarship Online, 2012): 201–218. See also Elizabeth Baigent, "Patrick Geddes, Lewis Mumford and Jean Gottmann: divisions over megalopolis" *Progress in Human Geography* 28 no. 6 (2004): 687–700.
8 Gottmann, *Megalopolis*, 772.
9 Ibid., 12–13.
10 Ibid.
11 Ibid., 15.
12 William L. C. Wheaton, "Jean Gottmann. Megalopolis: The Urbanized Northeastern Seaboard of the United States" *The Annals of the American Academy of Political and Social Science* 341 no. 1 (May, 1962): 166–167.
13 Lewis Mumford, *The City in History*, 486.
14 Gottmann, *Megalopolis*, 496. Mumford maintained an ambiguous relationship with the city and urbanization. This conflicted stance led him to argue for the retention of metropolitan centers like New York, and the planned dispersal of industry and culture to smaller regional enclaves. For Mumford, the issue was not

the city versus the countryside, but an issue of maintaining a balance and scale that would preserve physical and social integrity. See also Lewis Mumford, *The Culture of Cities,* and *The City in History.*

15 Gottmann, *Megalopolis,* 737.
16 Ibid., 12 and 15. See also Toynbee, *A Study of History.*
17 Ibid., 236–237, 383.
18 Ibid., 237, 369.
19 McHarg, "Ecology of the City," 53–65. The term "negative" development is also a pun that refers to the photograph negatives used to create this type of map.
20 The regional planning study of the Delaware River originated in response to a 1954 Supreme Court decision that called for the equitable distribution of water between the four states that line the banks of the river and New York City. See The Delaware River Basin Commission," The Role of the Delaware River Master in Interstate Flow Management," http://nj.gov/drbc/programs/flow/river_master.html; Delaware River Basin Commission, *Delaware River Basin Compact* (1961, Reprinted 2009), http://www.nj.gov/drbc/library/documents/compact.pdf.
21 The Architectural Archives of the University of Pennsylvania, Ian L. McHarg Collection, "Landscape Architecture 500/600 Delaware River Basin Fall 1960," 109.I.E.4.14; and The Architectural Archives of the University of Pennsylvania, Ian L. McHarg Collection, "New Approaches to the Reservation of Open Space in the Delaware Valley," 109.II.C.67. This report documented an address given by McHarg on Wednesday, June 21, 1961 to the Museum of the College of Art, under the auspices of the Philadelphia Housing Association. This is the first mention of the use of the overlay method by McHarg, which he referred to as "transparencies."
22 See McHarg, *Design with Nature,* 55–65; David Wallace et al., *Metropolitan Open Space and Natural Processes* (Philadelphia, PA: University of Pennsylvania Press, 1970); and Ann L. Strong and George E. Thomas, *The Book of School,* 169. Funding for the study was provided by a grant awarded to the Institute of Urban Studies by the Department of Housing and Urban Development, under the provisions of Sect. 314 of the Housing Act of 1954. When the report was published in 1970 the name of the Institute had changed to the Institute for Environmental Studies of the Graduate School of Fine Arts of University of Pennsylvania. In addition to Wallace and McHarg, team members included William Roberts, Ann L. Strong, William Grisby, Narendra Juneja, and Carol Levy Franklin.
23 The Architectural Archives of the University of Pennsylvania, Ian L. McHarg Collection, "U.R.A. Open Space Research Project," 109.II.E.1.30.
24 See Geddes, *Cities in Evolution,* 164–165; Pauline K. M. van Roosmalen, "London 1944: Greater London Plan" in *Mastering the City: Northern European City Planning, 1900–2000,* eds. Koos Bosma and Helma Hellinga, (Rotterdam: NAi Publishers, 1997), 266–273; and C. C. Fagg and G. E. Hutchings, *An Introduction to Regional Surveying* (Cambridge: Cambridge University Press, 1930), 91–116. There are several other inspirations for this study. In his description of the project, McHarg noted the greenbelt plans for Boston developed by Frederick Law Olmsted and Charles Elliott, the metropolitan park system for Chicago developed by Jens Jensen, the post-World War II reconstruction plan for London by Sir Patrick Abercrombie, and the environmental corridor studies developed by Philip Lewis for the state of Wisconsin. The decision by Lewis to divide landscape resources into two categories – intrinsic resources created by nature and extrinsic resources created by people – is noteworthy here. Lewis argued that planning policies had to align with both resource systems if they were to resolve the conflicting needs of

landscape preservation. See also Philip H. Lewis, "The Landscape Resources of Wisconsin" in ed. H. Rupert Theobald *The Wisconsin Blue Book* (Wisconsin Department of Resource Development, 1964), 130–142, http://digicoll.library. wisc.edu/cgi-bin/WI/WI-idx?type=turn&id=WI.WIBlueBk1964&entity=WI. WIBlueBk1964.p0145&q1=Lewis; Jens Jensen *Shiftings* (Chicago. IL: Ralph Fletcher Seymour, Publisher, 1939); Charles William Elliot, *Charles Eliot Landscape Architect: A Lover of Nature and of his Kind who Trained for a New Profession* (New York: Houghton Mifflin, 1902); and Norman T. Newton, "Olmsted's Work in Boston" in *Design on the Land* (Cambridge, MA: The Belknap Press of Harvard University, 1971), 290–306.

25 The Architectural Archives of the University of Pennsylvania, Ian L. McHarg Collection, "Metropolitan Open Spaces from Natural Processes," 109.II.C.61.

26 Ibid., 3–4.

27 Ibid., 5 and 9.

28 Ibid., 12.

29 Ibid., 16 and 19.

30 Gottmann, *Megalopolis*, 732.

31 Donald Worster, *Nature's Economy: A History of Ecological Ideas* (New York: Cambridge University Press, 1997), 291–294. According to Worster, it was generally believed that the agrarian ideal of self-sufficiency would temper the worst excesses of urban consumerism and transform competition into cooperative production.

32 "Metropolitan Open Spaces from Natural Processes," 17–19.

33 Leopold, *A Sand County Almanac*, 203.

34 "Dr. Paul Tillich," 9; and *The House We Live In* Paul Tillich.

35 "Metropolitan Open Spaces from Natural Processes," 19.

36 David Lowenthal, "George Perkins Marsh and the American Geographical Tradition," *Geographic Review* 43 no. 2 (April, 1953): 201–213; and David Lowenthal, "Nature and Morality from George Perkins Marsh to the Millennium" *Journal of Historical Geography* 26 no. 1 (2000): 3–27.

37 McHarg, *Design with Nature*, 115.

38 Ibid., 5.

39 Gottmann came to a similar conclusion when he noted that regional development in the mid-20th century was too complex and fluid to be charted as orderly hierarchies of elements, or geometric structures: "There are too many relationships that link any given community or area of some size to several other areas, cities, or hubs. Perhaps the best comparison of its structure, at a time when astronomical comparisons are in fashion, would be with the structure of the nebula." See Gottmann, *Megalopolis*, 736.

40 "Metropolitan Open Spaces from Natural Processes."

41 Emil R. Salvini, "The Great Atlantic Storm of 1962," (March 6, 2012), www. njtvonline.org/news/uncategorized/the-great-atlantic-storm-of-1962/.

42 McHarg, *Design with Nature*, 4–17. See also Ian McHarg, "Ecology, for the Evolution of Planning and Design," in *VIA 1: Ecology in Design*, ed. Rolf Sauer, et al. (New York: Grossman Publishers, Inc., 1968): 47–49.

43 Ian McHarg, *Cape Hatteras National Seashore Recreation Area*" (Philadelphia, PA: Landscape Architecture Department, School of Fine Arts, University of Pennsylvania, 1956).

44 Robert Geddes noted that the November 1948 issue of *Architectural Forum* devoted to "measure" served as an important reference for the Providence, Rhode Island Collaborative Thesis project. Personal communication, Robert Geddes, August 18, 2015. See also *The Architectural Forum Magazine of Building*, "Measure" 89 no. 3 (November 1948); and Victor Olgyay, "The Temperate House" *Architectural Forum Magazine of Building* 94 no. 3 (March, 1951): 179–197.

45 Ann L. Strong and George E. Thomas, *The Book of School*, 116.
46 Wallace-McHarg Associates, *Plan for the Valleys* (Philadelphia, PA: Wallace-McHarg, Associates, 1964), 3. The analysis in *Design with Nature* was re-colored to visually enhance the critical ecology. Copy of the *Plan for the Valleys* courtesy of William Roberts.
47 Ibid., 1, 3, 14–17 and 25.
48 Ibid., 1.
49 See: Wallace McHarg Associates, "Plan for the Valleys," 2; and Ann L. Strong and George E. Thomas, *The Book of School*, 213.
50 Personal communication, William Roberts, July 2016.
51 Wallace-McHarg Associates, *Plan for the Valleys*, 5.
52 See Edward Bellamy, *Looking Backward* (New York: Oxford University Press,1888).
53 "Metropolitan Open Spaces from Natural Processes."

9   Natural beauty

As noted, the hallmark of conventional wisdom is acceptability. It has the approval of those to whom it is addressed. There are many reasons why people like to hear articulated that which they approve. It serves the ego: the individual has the satisfaction of knowing that other more famous people share his conclusions. To hear what he believes is also a source of reassurance. The individual knows that he is supported in his thoughts – that he has not been left behind and alone. Further, to hear what one approves serves the evangelizing instinct. It means that others are also hearing and are thereby in process of being persuaded.

John Kenneth Galbraith, The Affluent Society[1]

On February 8, 1965, President Lyndon Baines Johnson delivered a *Special Message to Congress on Conservation and Restoration of Natural Beauty* that so entwined cultural history with natural geography it was virtually impossible to separate the character of the people from the character of the land. He began with the following statement:

> For centuries Americans have drawn strength and inspiration from the beauty of our country. It would be a neglectful generation indeed, indifferent alike to the judgment of history and the command of principle, which failed to preserve and extend such a heritage for its descendants.[2]

To be appraised properly, it is necessary to consider Johnson's remarks within a certain cultural tradition – something tacitly understood by his audience and therefore unnecessary to state – and this was the belief America was exceptional. The hand of providence had blessed the country and created a landscape that surpassed the rest of the world in beauty, and in the quality and quantity of natural resources (Figure 9.1). It was, therefore, the God-given right of the people to put this abundance to good use and prosper from the opportunities that it afforded. This model of behavior made progress and America's pre-eminent global position a self-fulfilling prophecy. But it also contained another, less noble truth that revealed

Figure 9.1 Photographs from *The Potomac: The Report of the Potomac River Planning Commission* published by the American Institute of Architects (AIA). Reprinted by permission.

a deep-seated contradiction in the country's vision and understanding of itself. When Johnson idealized the beauty of the American landscape and its resources, he also made the land a handmaiden to power.

An important precedent for Johnson's statement, and its didactic moralism, can be found in an address delivered by the historian Frederick Jackson Turner at the World Columbian Exposition held in Chicago in 1893.[3] During this lecture, Turner discussed the significance of the frontier in shaping the American psyche, and he equated the country's prosperity to a pioneering spirit defined by freedom (which he conceived as the ability to compete unrestrictedly for the resources of a continent) and democracy (which he conceived as a government that willingly commandeered indigenous lands and handed over vast tracts of this territory to its non-native citizens). According to Turner, this opportunistic blend of laissez-faire capitalism and resolute colonialism enabled waves of immigrants to cross a continent, win a wilderness, and progress from the "primitive economic and political conditions of the frontier into the complexity of city life."[4] His argument prized rugged individualism, self-sufficiency, and mobility. Ambition, freedom, and vast tracks of land assured success.

In the summer of 1960, John F. Kennedy reanimated the American pioneering spirit in the acceptance speech that he delivered at the Democratic National Convention in Los Angeles, colloquially known as "The New Frontier Speech." The tone and content of his remarks evoked the rugged individualism of Turner's populous rhetoric:

> For I stand tonight facing west on what was once the last frontier. From the lands that stretch three thousand miles behind me, the pioneers of old gave up their safety, their comfort and sometimes their lives to build a new world here in the West. They were not the captives of their own doubts, the prisoners of their own price tags. Their motto was not 'every man for himself' – but 'all for the common cause.' They were determined to make the new world strong and free, to overcome its hazards and its hardships, to conquer the enemies that threatened from within and without.[5]

Kennedy dreamed of a new beginning, and he made the city and its over-crowded schools, cluttered suburbs, and squalid slums the new frontier. To address the challenges posed by this problematic and overlooked landscape, he reframed Turner's thesis from taking to giving and called for national service. This action, he assured his audience, would allow the county's values to triumph and endure.[6]

All three of these individuals, in their various pronouncements about the land and people, framed the great subject of American history around the notion of prosperity – its achievement, its maintenance, its equitable distribution, and its relationship to personal freedom. Although idealistic in tone, this discourse was all about defining the achievable rather than the perfect. According to the Harvard economist John Kenneth Galbraith, this meant supplying the requisite consumer goods and services to maintain the good life, while also ensuring health and well-being. In terms of the environment, it meant dealing with the physical damages of air and water pollution, as well as the "visual pollution from the intrusion of production and sales activity, particularly retail sales activity on the urban and rural landscape."[7] Galbraith, in his advice to both the Kennedy and Johnson administrations, argued that conventional wisdom indicated American affluence required public investment in infrastructure and education.[8]

When Johnson addressed Congress, he channeled conventional wisdom and claimed the country's prosperity was under threat. A storm was brewing. The prevailing winds of modernization advanced technology, bolstered industrial output, and improved life, but they also polluted the environment, poisoned wildlife, and littered the countryside with the skeletons of abandoned cars. If these conditions were left unchecked, they would "blight and diminish in a few decades what it [America] has cherished and protected for generations."[9]

For Johnson, this climate of uncertainty represented a political opportunity, and he vowed to preserve the way of life of the countryside and salvage "the beauty and charm of our cities."[10] He sought, in emulation of Kennedy, to ensure the equal distribution of wealth and opportunity among all segments of the population, and he included measures to deal with the physical and visual problems of environmental pollution. To protect health and welfare, he called for clean water free of pesticides, agricultural run-off, and sanitary and industrial waste, and for clean air free of particulate emissions from cars, manufacturing, and home heating. He also stressed that this was a bipartisan issue as pollutants showed no respect for political boundaries, and there "was certainly not a household in the whole country unaffected by one of these issues."[11]

In a brilliant justification of his proposed agenda, which required an unprecedented (even by New Deal standards) expansion of the federal government and its regulatory powers, Johnson framed his call to arms as the triumph of beauty over ugliness, and he asked his audience to consider what the appearance of the landscape said about America, its values, and its place in the world. As he continued to speak, Johnson painted the portrait of a great society. The landscape he conjured radiated health, and he adopted the language and logic of visual aesthetics to sketch the details of its appearance. Similar to the argument of the 18th-century politician and philosopher Edmund Burke in his thesis on the sublime and beautiful, Johnson claimed it was hard to quantify landscape beauty, but it's worth could be defined by its opposite.[12] It was easy, for example, to calculate the cost of removing ugly soot from historic buildings and the price of a new recreation area to replace a natural landscape that had become a suburb. But Johnson was careful not to place beauty on a pedestal. Instead, he claimed it was a right that all people should be able to enjoy in their daily lives. To achieve this democratic ideal he called for a "new conservation" dedicated to the dignity of the human spirit. "Nature," he stated in a remark delivered with a disarming air of objectivity that made his declaration of aesthetic equality a self-evident truth, "is nearly always beautiful."[13]

Johnson's subsequent commentary masterfully captured the politics of the situation and how he planned to achieve legislative success through a dual agenda of economic growth and landscape conservation. People were cut off from nature, crowded in cities, and deprived of the right to live in decent surroundings, and these circumstances called for bold action and increased spending. He sanctioned commerce to address the needs of a growing population, and he reassured his audience that he would secure areas of natural beauty to provide recreation and pleasure. For the city, he sanctioned funds to create small parks, squares, pedestrian malls, and playgrounds. For the countryside, he sanctioned funds to conserve the land and water, protect wilderness areas, create sanctuaries for migratory birds, and institute "cooperative programs to improve the beauty of privately owned lands."[14]

Highways were of paramount concern. Johnson sought funding to make the country's "automobile society," less onerous and more attractive for the millions who commuted to and from work. In a statement that acknowledged the universal right of car ownership, he noted this action would "greatly enrich the lives of nearly all our people in city and countryside alike." His highway beautification project included the control of billboard advertising, the elimination of beauty-destroying junkyards, and the planting of trees. "The roads themselves," the President argued, "must reflect, in location and design, increased respect for the natural and social integrity and unity of the landscape and communities through which they pass."[15]

At the end of the address, Johnson turned his attention to the country's waterways and issued a challenge. Through hard work, the American people tamed the nation's rivers, harnessed their power, and used their water to make whole regions prosper. Now was the time to preserve the rich physical and cultural heritage of this landscape before growth and development made it a memory. To advance this agenda, he called on Secretary of the Interior Stewart L. Udall to prepare a study of the Potomac River that would "serve as a model of scenic and recreation values for the entire country." The study would be done under the auspices of a special task force, and in line with the vision of health and prosperity that he had outlined, it would indicate how to cleanse the Potomac of pollution and make it safe for boating, swimming, and fishing. He also called for the completion of the already authorized George Washington Memorial Parkway to foster commerce and industry. His final request was a Conference on Natural Beauty that would "rescue our cities and countryside from blight with the same purpose and vigor with which, in other areas, we moved to save the forests and soil."[16]

The conference on natural beauty

The Conference on Natural Beauty was held in May of 1965 in the White House. For two days the invited participants, including McHarg and his University of Pennsylvania colleagues Loren Eiseley, William L. C. Wheaton, and Ann Louise Strong, discussed parks and townscapes, water and waterfronts, highways and scenic roads, rural landscapes and suburbia, landscape reclamation and automobile junkyards, and education and citizen action programs.[17] This agenda clearly addressed current needs, and yet in many ways it retained a strong allegiance to the civic design ideals expressed at the turn of the 20th century by advocates of the City Beautiful movement. This earlier conception of national beauty, which arose in response to the principles of harmony and order displayed in the design of the World Columbian Exposition, promoted an atavistic vision of classical order and denounced billboard advertising.[18]

The First Lady, Mrs. Lyndon B. Johnson, opened the proceedings with the observation that the "ugliness and the decay of our cities and countryside" were worrisome afflictions that plagued the nation's heritage of beauty. To explain what she meant by beauty and how it related to the work of the conference, she stated that Rome, Paris, Vienna, and Florence were great artistic achievements worthy of emulation. She also cautioned this level of excellence would be difficult to achieve in a society where there was freedom of action but virtually no artistic control. She then called on her audience to accept this challenge and work together to devise policies that promoted the inherent beauty of the county. In spirit and substance, she posed a collective civic mandate comparable to the one expressed by the poet Walt Whitman in the preface to the *Leaves of Grass*. The conference attendees would, as Whitman had figuratively imagined, "incarnate its [America's] geography and natural life and rivers and lakes," and by so doing create a national vision that romantically venerated the majesty of the land and the riotous energy of the people.[19] She optimistically concluded this goal was within reach and the conference would produce outstanding proposals.

During the proceedings, McHarg sat in the front of the room on a piano bench near the President, First Lady, and Stewart Udall, ostensibly because there was no other seat available.[20] When it was his turn to speak, he used the opportunity to further his ambitions for environmental design. In line with the overall theme of the event, he conceived of beauty as the harmonious co-existence of the people and the land, but he also declared, somewhat contrarily, that a realistic study of "the place of man in nature and the place of nature in the environment of man," must supplant aesthetics as the primary concern of the conference.[21] He then proposed the landscape survey that he developed for the Delaware River as the starting point for the new conservation and a national land use policy.

The Potomac River watershed

The Potomac Planning Task Force was already at work when the White House Conference on Natural Beauty convened. A press release issued by The Department of the Interior on April 27, 1965 announced the commencement of the project and the list of Task Force members.[22] Task Force participants came from the disciplines of geography, engineering, architecture, planning, and landscape architecture. McHarg and Grady Clay, the editor of *Landscape Architecture* magazine, represented the profession of landscape architecture. Following the completion of the work in 1967, the American Institute of Architects (AIA), the organization leading the effort, published the development proposal under the title *The Potomac: A Report on Its Imperiled Future and a Guide for Orderly Development*.[23] That same year McHarg received a sabbatical leave to write *Design with Nature*. The project occupies a place of prominence in the book.[24]

In his role as a member of The Potomac Planning Task Force, McHarg continued the explorations of ecological processes and land use that he initiated in his studies of the Delaware River and metropolitan Philadelphia. In characteristic fashion, he divided the work between his students and professional colleagues. Students gathered and analyzed data. Colleagues at Wallace, McHarg, Roberts and Todd (WMRT), the successor firm to Wallace-McHarg Associates, refined the information for public presentation.[25] Narendra Juneja, who taught with McHarg at the University of Pennsylvania and worked with him at WMRT, coordinated the effort and oversaw the production of the maps, diagrams, and sketches that appeared in the AIA report and in *Design with Nature*.[26]

From September 1965 through May of 1966, students under the direction of McHarg and Nicholas Muhlenberg, a junior faculty member who studied ecology, forestry, and resource management under Marston Bates and Paul Sears, investigated the Potomac River watershed.[27] The students collected social and economic data, and correlated this material to physical geography (physiography) and human settlement patterns. In keeping with precedents established by McHarg in previous case studies, the work involved the production of a balanced, science-based synthesis of multiple data sources. In keeping with the directives of the President and First Lady, it called attention to the innate strength and beauty of the land. In keeping with standard conservation practice, it paid close attention to the resources that contributed to economic prosperity. By the end of the year, the students produced over 303 charts, maps, and diagrams.[28]

In the fall semester, students cataloged and mapped the natural features of the watershed – geology, topography, climate, hydrology, soils, vegetation, and wildlife – and the region's socio-economic patterns – demographics, income, land use, appraised value, transportation, and mineral reserves. This fact-finding phase of the project also documented maximum and minimum water flow, surface and groundwater quality, flood plains and aquifer recharge areas, industrial pollution, and sedimentation volumes.

Following the completion of the natural features survey, the students analyzed the data to determine optimum land use. This exercise involved the creation of intrinsic suitability maps for agriculture, forestry, recreation, and urbanization. To complete this assignment, students matched specific natural features to particular land uses. For example, climate, geology, topography, and hydrology (combined in that order) revealed slope, sun exposure, and water availability, which, in turn, determined the areas best suited for agriculture, forestry, and recreation. The selection of lands for urban development likewise began with climate, geology, and hydrology, but this assessment included highways as an important determining factor. To minimize the impact of development on water resources, the urbanization assessment excluded aquifer recharge areas and steep slopes. A summary map combined the data from the intrinsic suitability maps, and indicated the areas of the watershed best suited for agriculture, forestry, recreation, urbanization, coal mining, and water management (Figure 9.2).

An accompanying matrix resolved discrepancies and determined the appropriate land uses when several occupied the same location in the summary map (Figure 9.3). The vertical axis of the matrix organized the natural features into categories of use, and the horizontal axis tested these uses against each other to determine compatibility, or incompatibility with the physiography and critical ecology. The matrix presented the decision-making

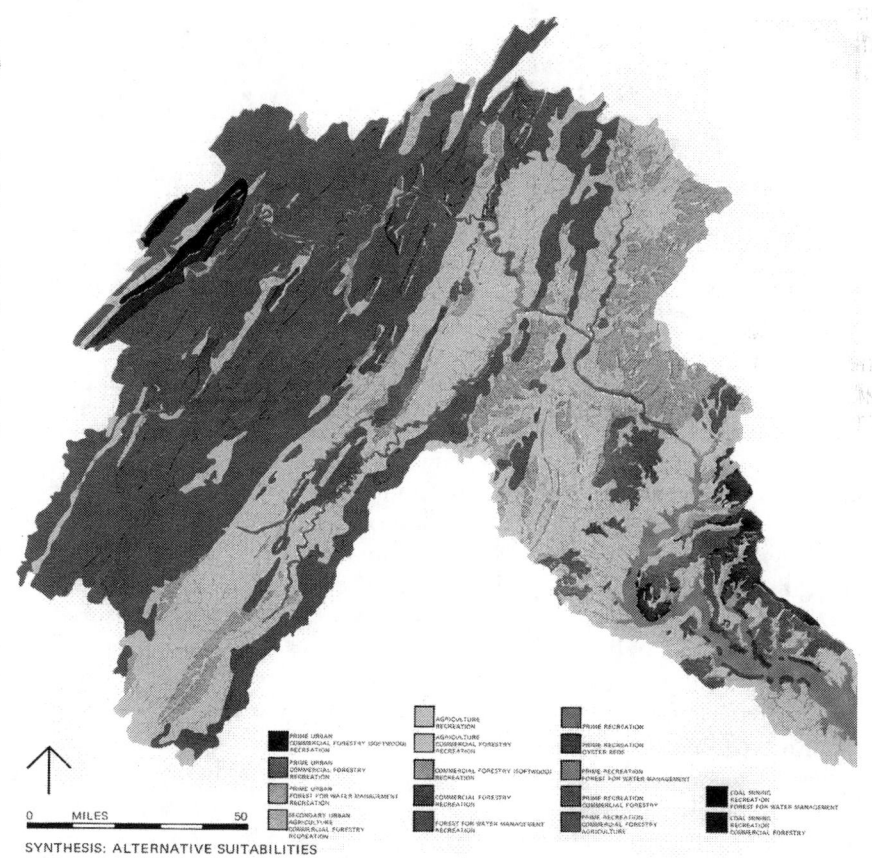

Figure 9.2 Composite Suitability map the Potomac River watershed study. © 1992 by John Wiley & Sons, Inc. Reprinted by permission of John Wiley & Sons, Inc.

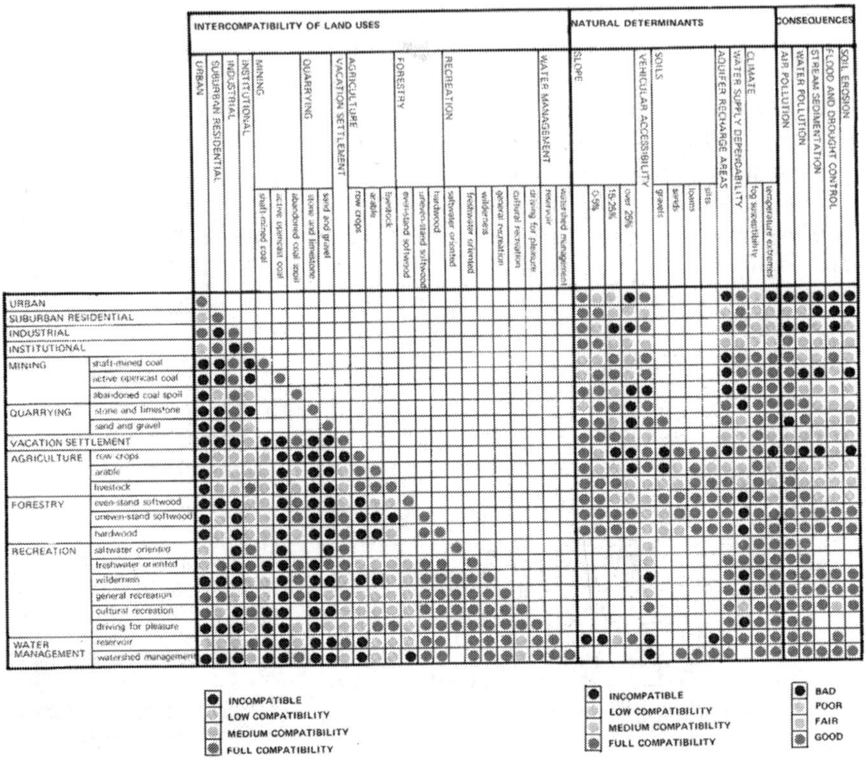

Figure 9.3 Land use matrix The Potomac River watershed study. © 1992 by John Wiley & Sons, Inc. Reprinted by permission of John Wiley & Sons, Inc.

process as a matter-of-fact determination of options and consequences. But the value-laden terminology used to describe the options and consequences displayed an inherent bias. Urban land uses were "bad" in terms of soil erosion, and water and air pollution; forestry, recreation, and water management were "good" in terms of these same categories; agriculture, quarrying, and mining fell between these extremes.

Muhlenberg played a key role in the development of the project's location-based analysis and the related application of geographic distance in the land use determinations. The areas deemed suitable for agriculture, for example, incorporated the physical availability of raw materials – soil, water, and sunlight – as well as economic factors – travel distance, accessibility to markets, and net profit – as key indicators of success. The multiple and potentially conflicting land uses detailed in the matrix also reflected McHarg's recent participation in a March 1964 conference on The Conservation of Natural Resources organized by the Ford Foundation to discuss pressing environmental issues and potential ways to address these problems.

Several of the speakers, including E. M. Nicholson, the Director of the Nature Conservancy of Britain, argued for multi-disciplinary studies that "weigh each alternative land use, or combination of land uses, against others." When McHarg spoke at the conference, he likewise called for expansive, multidisciplinary action, and, in a statement that further clarified his intent, he referred to land use planning as:

> . . . an attitude of mind, a predisposition in favor of nature; it is neither a science or religion although it draws from both sources. The work implies protectionism although clearly an operative attitude toward nature must also accept change and indeed must offer a constructive conception of positive human interaction.

Also aware that the Ford Foundation convened the conference to explore potential funding initiatives, he indicated the need for a trained group of professionals, who could knowledgably "speak for geology, hydrology, precipitation, vegetative cover, coefficient of runoff, percolation, surface water, aquifers," and back their comments with facts.[29]

During the spring semester, students studied the six physiographic regions that comprised the Potomac watershed in greater detail. To illustrate the ideal pattern of land use for each physiographic region – Piedmont Plateau, Blue Ridge, Great Valley, Ridge and Valley, Allegheny Plateau, and Coastal Plain – they created prototypical cross-sections of the river valley and corresponding aerial perspectives (Figure 9.4). The cross-sections located the proposed land uses and their overlaps. The perspectives, following the traditions of British surveying, provided a bird's-eye-view of the watershed's hamlets, villages, towns, and cities, and their relationship to physiography, water, woodlands, agriculture, transportation, and industry.[30] These representational strategies, as previously mentioned in the discussion of the Metropolitan Open Space project, were the intellectual offspring of the plans, sections, and perspectives of the prototypical river valley that Patrick Geddes utilized to diagram regional urbanization.[31] The Valley Section, as explained by Geddes in *Cities in Evolution*, united physical and social geography into an alliance of city (human culture) and countryside (natural landscape) that so entwined the character of the people (folk) and their occupations (work) with the character of the land (place) that it was virtually impossible to separate these traits. His interest, as it was for McHarg, was environmental reciprocity. He sought to understand how the particularities of the land advanced or inhibited the development of people, cities, and civilizations.[32] These ideas are clearly at work in the Potomac River watershed study and its situation of people, their actions, and where they live as both a natural outgrowth of land's morphology and a statement of national purpose. As the Potomac River flowed between the narrow ravines and the forested slopes of its rocky headwaters, past the rolling terrain, agricultural fields and towns of the piedmont, and through the low-low-lying farmland

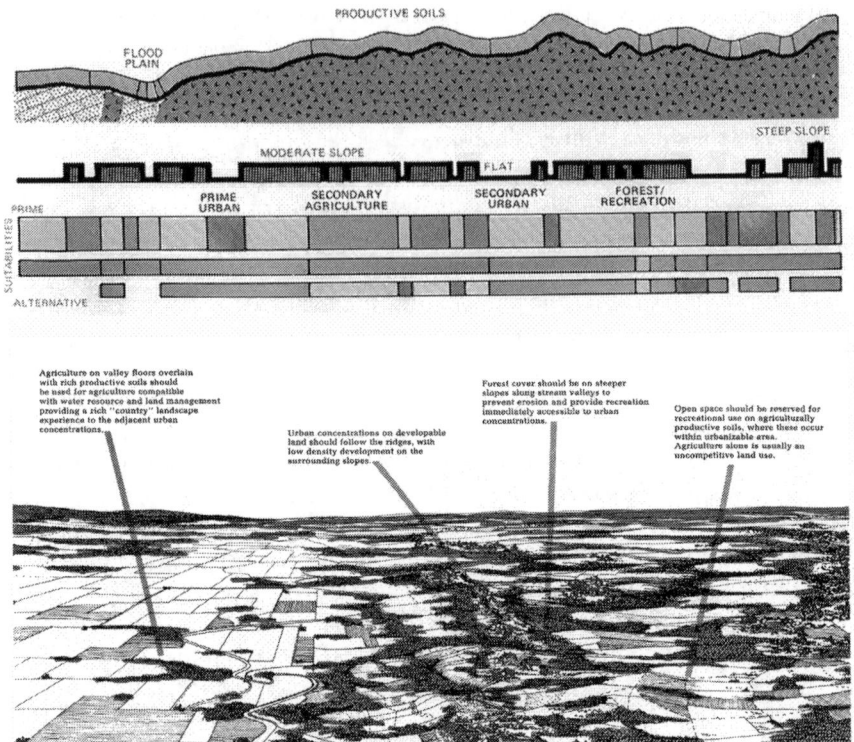

Figure 9.4 Section and aerial perspective The Potomac River watershed study. Section © 1992 by John Wiley & Sons, Inc. Reprinted by permission of John Wiley & Sons, Inc. Aerial perspective from *The Potomac: The Report of the Potomac River Planning Commission* published by the American Institute of Architects (AIA). Reprinted by permission.

and cities of the coastal plain before it dispersed into the sea, it narrated a journey of discovery that recapitulated the natural history of continent and the historical development of its pioneering settlements following European colonization.

A summary report of the student work, titled *Natural Resources and Wise Land Management*, outlined the objectives of the project and the challenges it addressed. In tone and substance, it echoed the words and politics of President Johnson and the First Lady:

> The Potomac is not a clean river. It does not flow through a landscape affording satisfaction to the eye that by nature it should afford. Its setting is marred by needless ugliness. The lands it drains are less

productive than they should be, and in many instances, once were. The human communities on its banks have left largely unrealized the contributions the river could make to the wellbeing of their citizens.[33]

The narrative's litany of ills did not stop at the riverbank:

There is poverty in Appalachia, uncontrolled urban sprawl where communities are prosperous and decay where they are not, a disintegrating agricultural economy on the coastal plain. In general wildlife is in retreat; we have even managed to extirpate the bald eagle, the national symbol, whose lordly form a few years past was a feature around the capital city.[34]

But the real issue, stated in deliberately exaggerated language and repetitive hyperbole that betrayed McHarg's hand in its formulation, was the prevalence of an aggressive mercantilism that caused people to act as marauding invaders rather than pioneering settlers:

We have felled the forests and left the hillsides to their fate, sacked the soil of its nutrients, stripped the coal from its seams and left the mines untopped to add their acid poisons to the mud in the streams from the slopes we have ravaged. Even when we approach nature as suitors we usually turn out to be despoilers. In our pursuit of pastoral idyll we bring with us the land hungry highway and rivers of traffic and treeless subdivisions enriched with drive-in hot-dog stands, used–car lots and shopping centers, their plains of roofs topped by naked poles between which skeins of wires disconsolately sag.[35]

Nature was the "long-suffering" partner in this relationship, but she was not "a weak and yielding handmaiden." If people overstepped their bounds, she responded with a force that turned "the homelands of her self-imagined conquerors into deserts." Her revenge was both overt and psychological and subtle, as manifest by the social ills of "alienation, dislocation and neuroticism."[36]

To bring order to this unruly relationship, the project narrative proposed practical restraints that included the retention of forested hillsides and the creation of riparian buffers to prevent soil erosion, water pollution, and floods. The report also set standards for population density, and industrial and commercial activity to limit the impact of development on scenic quality. But the real beauty of the proposal, at least for McHarg, was the way it allowed natural processes to work freely, without constraint, to purify water, impede erosion, create topsoil, and enhance comfort and health.

To convince the general public to accept restrictions on their freedom of action, the project narrative tallied the cost of the behaviors deemed ugly. Acid-leakage from coal mines, for example, spoiled irrigation water and decreased the productivity of farms located miles away, and a hillside bull-dozed for development caused the river to flow red with mud and diminished opportunities for recreation. The marketability of a property with a "desirable view of adjoining lake or beach" provided positive enticement for behavioral change.[37]

Natural beauty was not the only thing that occupied McHarg's thoughts when he added his commentary to the project statement; he was also troubled by the countervailing agencies of destruction that tear society apart:

> That, while 'aesthetic' and 'spiritual' values may be hard to pin down, it is hardly disputed that human beings take color from their environment and that conflict, discord and ugliness in what they see and hear will breed conflict, discord and ugliness within themselves.[38]

To explain his thoughts, McHarg acquiesced to his messianic impulses, channeled the religious piety ingrained in the American narrative of exceptionalism, and provided a God's-eye-view of his concerns:

> That, whatever the nature of the Creator, about which it is not necessary for us to be of one mind, indisputably He has expressed Himself in the natural world, which is of His creation, and that when we exploit His handiwork to unworthy ends or defile His masterpieces we commit not merely a crime against the human community but a sacrilege.[39]

In this providential vision of manifest destiny, the land was a divine gift, Nature the national religion, and the preservation of natural processes the cornerstone of a holistic system of social rights and obligations that vanquished discord and elevated society to a higher ethical plane. Indeed, as the operative creed for a great society, the ecological relationship of an organism to its surroundings was here conceived as a just way to see the world, a rational means to alter actions that were fundamentally flawed, and a method to guide development toward democratic ideals. By binding personal freedom (the rights and duties of the individual) to the processes of nature (the ability of the land to work freely without constraint), McHarg, like President Johnson, indelibly linked the beauty of the people to the beauty of land. In a similar manner, and despite his avowed attempts to do otherwise, he inadvertently made the land a handmaiden to power.

Interstate I-95 and Richmond Parkway

McHarg's participation in the White House Conference on Natural Beauty and The Potomac Planning Task Force advanced his career and provided new opportunities for WMRT. He made the most of the situation and actively positioned himself, and his firm, within the vortex of high-level politics.

As one might expect, the vision of order that guided this phase of his career aligned with the progressive policies and liberal dictates of the great and just society outlined by President Johnson in his 1965 Message to Congress. Fully embracing this communiqué's pioneering ethos, McHarg now produced work that conflated freedom and prosperity with mobility. In a productive society concerned with standards of health, people were free, within reason, to choose where they lived and worked. In the United States in the 1960s, this meant the right to own a new car and park the vehicle in the driveway of a new suburban home surrounded by a green rectangle of lawn. Moreover, as President Johnson argued, it was a national imperative to make the new homeowner's journey to and from work as pleasurable as possible.

Two highway studies conducted by WMRT are of particular note in this regard – one for Interstate I-95 between the Delaware and Raritan Rivers in central New Jersey completed in 1966 for The Stony-Brook Millstone Watershed Association, and one for the Richmond Parkway in Staten Island, New York completed in 1968 for August Heckscher, the Parks Commissioner of New York City.[40] McHarg, with the assistance of Narendra Juneja, supervised both studies. The I-95 corridor work was subsequently published in *Landscape Architecture* under the title "Where Should Highways Go?"[41]

The cover of the I-95 project report prominently displayed an excerpt from the Message to Congress delivered by President Johnson. This reference, in addition to being a shrewd marketing ploy, positioned the project as a consummate expression of participatory democracy in action:

> In almost every part of the country citizens are rallying to save landmarks of beauty and history. The government must do its share to assist these local efforts which have an important National purpose.
>
> . . . I hope that at all levels of government our planners and builders will remember that highway beautification is more than a matter of planting trees or setting aside scenic areas. The roads themselves must reflect in location and design increased respect for the natural and social integrity and unity of the landscape and communities through which they pass. . .[42]

In line with the President's executive policies, the report affirmed the economic importance of highways. It also noted, in a statement that laid the groundwork for the subsequent analysis, that highways were engineered for speed and economic efficiency, but they ignored "important aesthetic and social values" that weaken the bonds of community. Even worse, this omission turned an essential infrastructure and its designers and patrons into remorseless, anti-democratic "tyrants" who carved through "the hearts of cities," destroyed urban parks and natural resources, and inflamed the citizenry to oppose their construction.[43]

To distinguish this study from normative engineering solutions, and to indicate why it was indeed superior, the report explained how the WMRT alignment promoted commerce, safety, convenience, and pleasurable driving, while also protecting the land, water, air, and biotic resources of the communities that it served. This meant the shortest distance between two points was not necessarily the best or cheapest option. "**The best route is that which provides the maximum social benefit at the least social cost**," the report stated in bold face type.[44]

To determine the route that would satisfy these project objectives, WMRT analyzed ten planning parameters – topography, land values, urbanization, residential quality, historic value, agricultural value, recreation value, wildlife value, water value, and susceptibility to erosion – and individually mapped them on transparent acetate overlays in shades of gray. In contrast to former studies, they divided each parameter into categories of economic value that reflected: (1) the land's existing residential quality and historic importance, (2) its ability to support agriculture, hunting, fishing, recreation, and (3) the critical ecological services that it provided. The darkest sections of these maps indicated where road construction produced the greatest social and physical disturbance and the greatest economic cost. White indicated minimal social impact and the least economic cost. Intermediate shades of gray indicated areas where the impact of road construction lay somewhere between these extremes. Soil protection was paramount, as evidenced by the fact that the mapped parameters were bracketed on one end by topography and on the other end by soils susceptible to erosion. The valuation procedure was not based on strict quantification, but instead incorporated a qualitative and somewhat traditional vision of landscape beauty, bolstered, in turn, by the belief that development and the bulldozer's blade should not touch water resources, agriculture, historic sites, and recreation areas.

To produce the final alignment, the ten acetate maps were laid on top of each other in the order in which they were presented in the report. The compilation produced a composite map comprised of multiple shades of gray. The darkest areas of the map included ecologically valuable marshes and stream corridors, and economically valuable communities distinguished by high real estate assessments and sites of historical importance. The highway avoided these areas. Undeveloped land that lacked high real estate valuation, historical significance, recreation amenities, prime agricultural land, or critical water resources remained white and had the lowest assigned value. These areas indicated the optimum location for the highway even if the route was less direct and more expensive than the original proposal, which followed the selection criteria of the Bureau of Public Roads.

The project report stated the proposed highway alignment improved the lives of everyone in the region. And yet, similar to the *Plan for the Valleys*, the client consisted of wealthy and powerful stakeholders who lived in Princeton, New Jersey and did not want a highway carved through the middle of their community.[45] In all fairness, the proposed plan, in keeping with the firm's commitment (and President Johnson's directive) to protect the welfare of the general public, also avoided the less wealthy and more racially diverse neighborhoods of Trenton, New Jersey. The carefully selected (and pre-determined) design parameters indicated that in this particular instance it was less socially disruptive, and more cost effective to by-pass established communities and sacrifice undeveloped farmland. In terms of design precedent, this decision echoed the concept of "The Townless Highway" devised by Benton MacKaye, which meant that it allowed the residents of Princeton to maintain the ambiance of their community and easy access to a highway that conveniently connected their lives and livelihood to New York and Philadelphia.[46] But perhaps most important, the proposed road alignment, in keeping with one of McHarg's cherished design principles, preserved the natural beauty of the land, and thus the best attributes of the city and countryside. This action made it all but certain, as noted in the last paragraph of the report, that the plan and its method "received the approval of Mrs. Lyndon B. Johnson."[47]

Left unsaid in the I-95 report, because it was something tacitly understood by the political establishment and their cadre of designers, was the fact that there was a well-established canon of taste that guided highway beautification in the 1960s. Peter Blake, the editor of *Architectural Forum* and one of the select members of this like-minded society, distilled these principles in *God's Own Junkyard: The Planned Deterioration of America's Landscape*.[48] The argument in this text, which Blake claimed he wrote in anger, stated that design and planning should not be driven by the attainment of wealth, but should instead promote civil society.[49] To illustrate his argument, he created a scenic tour of the United States that consisted of contrasting photographs – the pastoral scenery of a professionally designed parkway versus the visual clutter of a suburban commercial strip, the beauty of a tree-lined urban courtyard versus a treeless suburban housing development, junkyards versus agricultural fields, and trees versus utility wires. It was Blake's contention that the material culture of the country had to achieve a certain level of taste and refinement or it would disfigure the land and debase social discourse. Commercial establishments hawking cheap goods, hot dog stands, billboards, and tract housing developments were major villains. Symbols of prosperity that included the display of luxurious merchandise in fashionable stores, gracious urban brownstones, productive farm fields, picturesque country lanes, stately redwood trees, and majestic mountains were, however, sanctioned.

McHarg signaled his allegiance to the class-conscious, albeit politically correct, visual philosophy presented in *God's Own Junkyard* when he inserted a photograph from this text into the pages of *Design with Nature* (Figure 9.5).

Figure 9.5 Photograph of a suburban commercial corridor in Miami Beach, Florida by Wallace Litwin from *God's Own Junkyard*. © Stanley B. Burns, MD & The Burns Archive. Reprinted by permission.

The photograph captured the visual cacophony of a suburban commercial strip lined by advertising signs for the Tangiers Motel. The signs offered weary travelers free parking, air conditioning, a private pool and beach, the culinary convenience of a coffee shop, and the convivial atmosphere of a cocktail lounge. Seals of approval from the Automobile Association of America, Diner's Club, and Duncan Hines assured the quality of the experience and its affordability. "It is all but impossible," McHarg dismissively wrote, "to avoid the highway out of town, for there arrayed in all its glory, is the quintessence of vulgarity, bedecked to give the maximum visibility to the least of our accomplishments."[50] His adherence to

Blake's reactionary conception of visual culture is also apparent in the following statement:

> It matters not if you choose to proceed to the next city or return to the first. You can confirm an urban destination from the increased shrillness of the neon shills, the diminished horizon, the loss of nature's companions until you are alone, with men, in the heart of the city, God's Junkyard – or should it be called Bedlam, for the cacophony that lives here. It is the expression of the inalienable right to create ugliness and disorder for private greed, the maximum expression of man's inhumanity to man. And our cities grow, coalescing into a continental necklace of megapoles, dead grey tissue encircling the nation.[51]

The photograph from *God's Own Junkyard* that appeared in *Design with Nature* was part of a larger visual argument that paralleled McHarg's written commentary. Similar to the national tour devised by Blake, this argument positioned idyllic scenery – forested hillsides rising from misty valleys, productive farmland, redwood trees, mountains, and children at play – against dystopian landscapes – skyscrapers enveloped in smog, traffic congestion, polluted rivers, suburban tract housing, and a small child crawling on a dirty city stoop (Figure 9.6). By carefully omitting as much as he included in this visual journey, McHarg not only transformed the terrain of highway design into a landscape of progressive ideals, pastoral beauty, and environmental action, he also suggested that ecological design was an act of discriminating selection and exclusionary categorization that vanquished pollution, championed children, and combated rampant development. This, after all, was the politically correct thing to do, and, as previously noted, he was not above pandering to the powers that be, particularly when they aligned with his personal sensibility.

In 1968, McHarg and his colleagues at WMRT followed a similar mapping, valuation, and color-coding procedure to produce an alignment for the Richmond Parkway in Staten Island. The proposed route likewise preserved established communities and landscapes, and in contrast to an existing Tri-State Transportation Commission proposal that bisected Historic Richmond Town and paved over the Olmsted Trail, it bypassed these areas and also preserved Willowbrook Park and the William T. Davis Wildlife Refuge (Figure 9.7).

To counter the proprietary claims of engineers, the report positioned the study within the pioneering contributions made by landscape architecture in highway design, as exemplified by the Bronx River Parkway.[52] The proposed route, in conformance with this earlier project, would "rehabilitate a dirty river, an unkempt landscape, and provide a continuous scenic experience and create a new public resource and new public values." A pronouncement in bold faced type stated: **"Where resources are scarce and valuable it is preferable to use the public investment in highways to rehabilitate landscapes rather than destroy existing resources."**[53]

Figure 9.6 New York City and Pennsylvania farmland from *Design with Nature.*
New York photograph © 1992 by John Wiley & Sons, Inc. Reprinted
by permission of John Wiley & Sons, Inc.; Farmland by Grant Heilman
Photography. Reproduced by permission.

The Richmond Parkway study persuaded the Tri-State Commission that
the alternative route devised by WMRT would accrue savings in the long
run even though it initially cost more money to construct. To stave-off any
additional objections, the project summary strategically aligned the proposal's environmental agenda with the Commission's pro-growth policies and
mission of safety. It also noted that consultants, rather than government

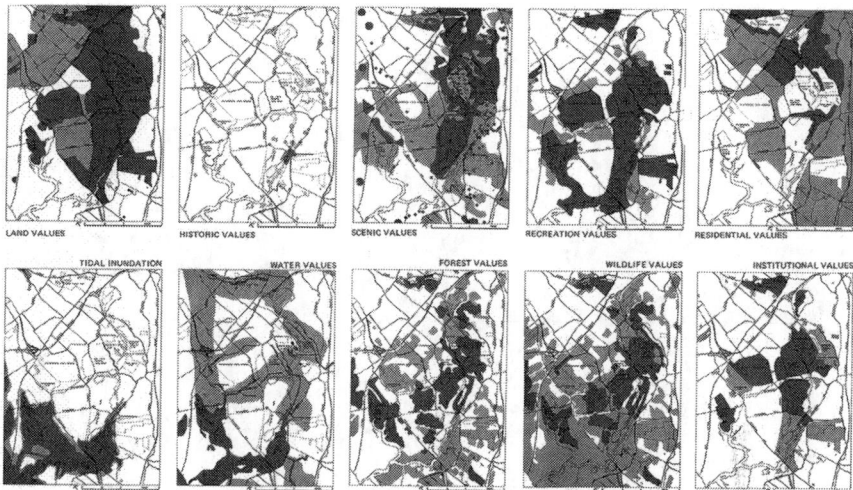

Figure 9.7 Land use valuation Richmond Parkway from *Design with Nature.* © 1992 by John Wiley & Sons, Inc. Reprinted by permission of John Wiley & Sons, Inc.

bureaucracies, best served the general public because they had the expertise to protect the land, accommodate development, and promote community. In this sense, The Richmond Parkway proposal was much more than a route selection that allowed people to live in suburban communities while enjoying a pleasant and convenient commute to work: it was also a politically calculating demonstration of the ability of landscape architecture to synthesize complex data into practical and imaginative solutions that addressed multiple problems in an equitable and entrepreneurial manner. The proposed alignment not only preserved historic Richmond Town, Willowbrook Park, and the Olmsted Trail, it also heralded the future transformation of an adjacent sanitary landfill into the 2,200-acre Freshkills Park.[54]

The most distinguishing feature of the Richmond Parkway project, however, was its explicit relationship to health. This can be seen, for instance, in the way the project report compared the proposed alignment to the surgical removal of an "embolism" from a major artery, and the subsequent claim that the proposed action was less invasive than standard engineering procedures where "the remedy often appears, for large sectors of the public to be more deadly than the malady."[55]

The overlay technique deployed in the study, with its x-ray like appearance, reinforced the reading of the project in terms of a medical prognosis (Figure 9.8). Given McHarg's personal history, the project's conflation of land planning and health is not surprising. Indeed, the summary of the Richmond Parkway project presented in *Design with Nature* reads as a

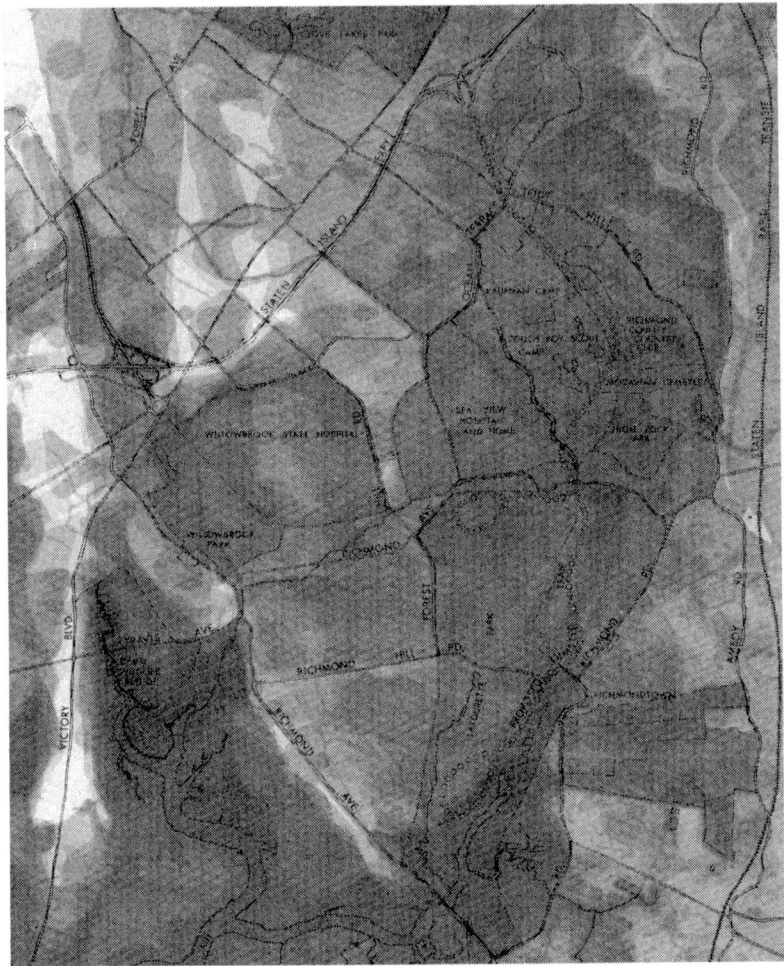

Figure 9.8 Composite land use valuation map Richmond Parkway from *Design with Nature*. © 1992 by John Wiley & Sons, Inc. Reprinted by permission of John Wiley & Sons, Inc.

thinly veiled reference to the medical diagnosis he received as a tuberculosis patient:

> The method was known but the evidence was not. It was necessary to await its compilation, make the transparent maps, superimpose them over a light table and scrutinize them for their conclusion. One after another they were laid down, layer after layer of social values, an elaborate representation of the island [Staten Island], like a complex x-ray photograph with dark and light tones. Yet in the increasing opacity there were always lighter areas and we can see their conclusion.[56]

Lewis Mumford would attest to the health-based ideology that permeated McHarg's planning principles in the remarks he composed for the introduction to *Design with Nature*, which he called:

> . . . a notable addition to the handful of important texts that begin, at least in Western tradition, with Hippocrates' famous medical work on Airs, Waters and Places; the first public recognition that man's life, in sickness and health, is bound up with the forces of nature, and that nature, so far from being opposed and conquered, must rather be treated as an ally and friend, whose ways must be understood, and whose counsel must be respected.[57]

In addition to praising his protégé for demonstrating how to work with natural processes to manage development and advance health, Mumford proclaimed, with his usual paranoid eloquence, that this approach would replace "the polluted, bulldozed, machine-dominated, dehumanized, explosion threatened world" with a natural intelligence that promoted the common good.[58] In light of the recent exposés of chemical toxins by Rachel Carson in *Silent Spring* and Lewis Herber (the *nom de plume* of Murray Bookchin) in *Our Synthetic Environment*, there was ample reason for concern.[59]

Energized by the critical and professional success of his high-profile government projects, McHarg expanded his efforts to include the restoration of American society, and he began to deploy the word "nature" in a noticeably different way. The vision of design and society that subsequently emerged, with its comparative analysis of health and pathology, is at once strangely disturbing and myopically hopeful.

Notes

1 John Kenneth Galbraith, *The Affluent Society* (Boston, MA: Houghton Mifflin Company, 1958), 11. McHarg met Galbraith at Harvard. Galbraith served as the teaching assistant in a course McHarg took titled "The Economics of Agriculture." See McHarg, *A Quest for Life*, 74.
2 White House Conference on Natural Beauty, *Beauty for America: Proceedings of the White House Conference on Natural Beauty, Washington, D.C., May 24–25, 1965* (Washington, DC: US Government Printing Office, 1965), 1. See also, The LBJ Presidential Library, President Lyndon Baines Johnson, "Special Message to the Congress on Conservation and Restoration of Natural Beauty, February 8, 1965, http://www.lbjlibrary.net/collections/selected-speeches/1965/02-08-1965.html.
3 Jackson presented his remarks to the American Historical Association in Chicago, on July 12, 1893. See Frederick Jackson Turner, *The Frontier in American History* (New York: Henry Holt and Company, 1921), 321. Project Gutenberg eBook #22994. http://www.gutenberg.org/files/22994/22994-h/22994-h.htm. See also "Frederick Jackson Turner: The Significance of the Frontier in American History 1993, Excerpts" http://nationalhumanitiescenter.org/pds/gilded/empire/text1/turner.pdf.
4 Ibid., 1.

5 John F. Kennedy, "Address of Senator John F. Kennedy Accepting the Democratic Party Nomination for the Presidency of the United States – Memorial Coliseum, Los Angeles, July 15, 1960." *The American Presidency Project*, www.presidency.ucsb.edu/node/274679.

6 In this address Kennedy presented American values through the lens of Cold War politics, opportunity, and freedom. The space race and who controlled this new frontier was a major concern. McHarg included an image of an Apollo spacecraft in *Design with Nature*.

7 John Kenneth Galbraith, *The Good Society: The Humane Agenda* (New York: Houghton Mifflin Company, 1996), 1–13, 82–88.

8 See Galbraith, *The Affluent Society*.

9 *Beauty for America*, 1.

10 Ibid., 2.

11 Ibid., 1.

12 See Edmund Burke, *A Philosophical Inquiry into the Origin of Our Ideas of the Sublime and Beautiful* (New York: Oxford University Press, 1757, 2015).

13 *Beauty for America*, 1–2. See also The LBJ Presidential Library, http://www.lbjlibrary.net/collections/selected-speeches/1965/02-08-1965.html.

14 *Beauty for America*, 3–6.

15 Ibid., 6–7.

16 Ibid.,7, 8, 15.

17 Landscape architects attending the White House Conference on Natural Beauty included Grady Clay, Garrett Eckbo, Paul M. Friedberg, Lawrence Halprin, Philip Lewis, Kevin Lynch, Hideo Sasaki, and William Whyte. See *Beauty for America*, 693–756.

18 See Jon A. Peterson, "The Birth of Organized City Planning in the United State, 1909–1910" *Journal of the American Planning Association* 75 no. 2 (Spring 2009): 123–133.

19 Walt Whitman, eds. William Everson and James David Hart, *American Bard the Original Preface to Leaves of Grass* (New York: Viking, 1982). See also Cecelia Tichi, *Embodiment of a Nation: Human Form in American Places* (Cambridge, MA: Harvard University Press, 2001).

20 McHarg, *A Quest for Life*, 188.

21 *Beauty for America*, 481–484.

22 The Architectural Archives of the University of Pennsylvania, Ian L. McHarg Collection, "Potomac River Basin Study 1965," 109.II.E.4.28.1.

23 Potomac Planning Task Force, *The Potomac: A Report on Its Imperiled Future and a Guide for Orderly Development* (Washington, DC: United States Government Printing Office, 1997).

24 McHarg, *A Quest for Life*, 127–151. See also McHarg, *Design with Nature*, 126–151.

25 The new partners included William Roberts and Thomas Todd.

26 The AIA report included work from the Plans for the Valleys. See Potomac Planning Task Force, 1967, 86–89. Narendra Juneja was born in Lahore, India. Prior to emigrating to the United States in 1963 to study landscape architecture at the University of Pennsylvania, he served as a planner in Delhi and London. As noted by McHarg, order fascinated Juneja. He made important contributions to the maps and matrices that appear in McHarg's studies, including the innovative and evocative use of color. See *The Book of the School*, 230.

27 Nicholas Muhlenberg joined the faculty in 1963. He received a Bachelor of Science (BS) and Master of Science (MS) degree in forestry from the University of Michigan, where he studied under the noted ecologist and co-chairman of the conference *Man's Role in Changing the Face of the Earth*, Marston Bates. Following a Fulbright research scholarship in New Zealand, he continued his academic

studies in ecological land management at Yale University, where he received a master's degree in conservation and Ph.D. in Resource Economics. See University of Pennsylvania, "Dr. Muhlenberg, Penn Design," http://www.upenn.edu/emeritus/memoriam/Muhlenberg.html; and 'Nicholas Muhlenberg,' www.legacy.com/obituaries/delawareonline/obituary.aspx?n=nicholas-muhlenberg&pid=145521555.

28 The Architectural Archives of the University of Pennsylvania, Ian L. McHarg Collection, "Natural Processes and Wise Land Management," 109.II.C.64.

29 See Rockefeller Archive Center, Ford Foundation Records, Education and Public Policy Program (EEP), Resources and Environment, Program Staff Files (FA641), Series I: Subject Files, Box 1, Folder "Conference on Natural Resources – Discussion Papers (1/2), (February 27–March 1, 1964), E. M. Nicholson "Ecological Research: As Approach to the Understanding and Management of Man's Natural Resources," Ian L. McHarg "Natural Sciences and the Planning Process," and Nicholas Muhlenberg, "Annex, Suggestions Relating to Academic Training in Regional Land Planning."

30 See C. C. Fagg and G. E. Hutchings, *An Introduction to Regional Surveying*, 91–116; and K. M. van Roosmalen, "London 1944: Greater London Plan."

31 Geddes, *Cities in Evolution*, 163–167.

32 Geddes derived his work, folk, and place diagrams from the work of the engineer and social scientist Frédéric Le Play and his triad of *lieu*, *travail*, and *famille*. As noted by Volker Welter, Geddes conceived of this triad as environment, function, and organism, and used it to argue that humankind distinguished itself from animals through its ability to alter the environment to meet daily needs. See Volker Welter, *Biopolis* (Cambridge, MA: The M.I.T. Press, 2002), 11–12.

33 McHarg and Muhlenberg, "Natural Processes and Wise Land Management," 3.

34 Ibid.

35 Ibid., 4.

36 Ibid., 5.

37 Ibid., 6, 8.

38 Ibid., 8–9.

39 Ibid.

40 Wallace, McHarg, Roberts, and Todd, *A Comprehensive Highway Selection Method Applied to I-95 Between Delaware and Raritan Rivers* (Philadelphia, PA: WMRT, 1965), and Wallace, McHarg, Roberts, and Todd, *The Least Social Cost Corridor for Richmond Parkway* (Philadelphia, PA: WMRT, 1968). See also McHarg, *Design with Nature*, 31–41. McHarg studied Highway Engineering in 1947 while a student at the GSD, and, based on his military training and experience, he excelled in the course. See The Architectural Archives of the University of Pennsylvania, Ian and Carol McHarg Collection, "Harvard Course Notes Highway Engineering, 1947," 365.I.11. The Staten Island highway alignment project was completed in tandem with a land use survey and open space study of Staten Island. See McHarg, *Design with Nature*, 102–115.

41 Ian L. McHarg, "Where Should Highways Go?' *Landscape Architecture* 57 no. 3 (April 1967): 179–181.

42 Wallace, McHarg, Roberts, and Todd, *A Comprehensive Highway Selection Method Applied to I-95 Between Delaware and Raritan Rivers*, cover.

43 Ibid., 2.

44 Ibid.

45 One of the citizen advocates for the project included the writer John McPhee. McPhee later noted that he actively supported this project because he did not believe there was a need for another north-south interstate highway through the center of the New Jersey. Email exchange, John McPhee, March 5, 2018.

46 Benton MacKaye, "The Townless Highway" *The New Republic* 62 (March 12, 1930): 93–95. See also Benton MacKaye and Lewis Mumford, "Townless Highways for the Motorist: A Proposal for the Automobile Age" *Harpers Magazine* (August 1, 1931): 347–356.

47 Wallace, McHarg, Roberts, and Todd, *A Comprehensive Highway Selection Method Applied to I-95 Between Delaware and Raritan Rivers*, 8. McHarg's personal files contained a hand-written copy of this quote, as well as a letter from the First Lady that thanked him for his remarks at the White House Conference on Natural Beauty and his efforts to combine social values with highway aesthetics. The First Lady, as the force behind Johnson's promotion of natural beauty, observed in her diary that the beautification of the country "knotted together" all of the President's objectives for The Great Society – recreation, pollution, mental health, crime reduction, rapid transit, highway construction, and the war on poverty. She also wrote that beautification encouraged shared values that dissipated economic, social, and racial tension. In 1965 the President passed The Highway Beautification Act, which limited billboard advertising and junkyards. When he signed the Act into law, Johnson stated: "Beauty belongs to all the people. And so long as I am President, what has been divinely given to nature will not be taken recklessly away by man." See "Lady Bird Johnson Beautification Campaign," Ian L. McHarg Collection, The Architectural Archives of the University of Pennsylvania, 365.I.6. See also, PBS, "Lady Bird Johnson, The Beautification Campaign," http://www.pbs.org/ladybird/shattereddreams/shattereddreams_report.html, accessed March 4, 2019: and The American Presidency Project, "Lyndon B. Johnson 576-Remarks at the Signing of the Highway Beautification Act, October 22, 1965," www.presidency.ucsb.edu/node/241177, accessed March 4, 2019.

48 Peter Blake, *God's Own Junkyard: The Planned Deterioration of America's Landscape* (New York: Holt, Rinehart and Winston, 1964).

49 Ibid., 7.

50 McHarg, *Design with Nature*, 20.

51 Ibid., 23.

52 See Norman T. Newton, *Design on the Land: The Development of Landscape Architecture* (Cambridge, MA: The Belknap Press of Harvard University Press, 1971), 596–603.

53 Wallace, McHarg, Roberts, and Todd, *The Least Social Cost Corridor for Richmond Parkway*, 4. Frederick Law Olmsted advanced similar ideas in a report on street improvements prepared in conjunction with the development of Prospect Park in Brooklyn. Olmsted observed that it was imperative to promote plans, no matter the initial cost, that supported orderly progress and social prosperity. This would return savings in the long run and reduce "the frequency of certain crimes, the prevalence of certain diseases." See Frederick Law Olmsted. *Civilizing American Cities: Writings on Landscape*, ed. S. B. Sutton (New York: Da Capo Press, 1997), 38, 36, and 94. The Olmsted essay was originally published in "Observations on the Progress of Improvements in Street Plans, with Special Reference to the Parkway Proposed to Be Laid Out in Brooklyn," Olmsted, Vaux and Company, 1868, 7–21.

54 See New York City Department of Parks and Recreation, "Freshkills Park," www.nycgovparks.org/park-features/freshkills-park.

55 Wallace, McHarg, Roberts, and Todd, "The Least Social Cost Corridor for Richmond Parkway," 2.

56 McHarg, *Design with Nature*, 35.

57 Ibid., vii.

58 Ibid., vii, viii.

59 See Carson, *Silent Spring*; and Lewis Herber, *Our Synthetic Environment* (New York: Alfred A. Knopf, 1962).

10 Health and pathology

Man stands aberrant to what he was or may become. Looking backward he shares a heritage with other mammals extending many millions of years into the past. Out of this legacy stem man's needs, capacities and potentialities for further evolution. Environmental design emerges as the means for promoting the fulfillment of this progression towards greater diversity of function, increased capacity to cope with complexity and change, heightened ability to profit from experience, and expanding facility to be creative. Such is the direction of man's heritage and his future.

John B. Calhoun, Design for Mammalian Living[1]

In 1966, while studying how to re-route an interstate highway around the wealthy enclave of Princeton, New Jersey, McHarg returned to a topic that he first proposed to explore in the summer of 1962 with the ecologist Robert MacArthur, and he asked his students to map the ecology of the city of Philadelphia. In contrast to the open space studies carried out in former studios, which conserved the natural beauty of countryside, or the highway studies carried out at WMRT, which advanced the pleasures of mobility, the Philadelphia study was about the lives of those left behind. It is also safe to assume that McHarg considered any action that made the core of the city more inviting and livable, also made suburban expansion and the consumption of undeveloped land less urgent and necessary. The study, which appeared in *Design with Nature* in a chapter titled "The City: Health and Pathology," reflected a new emphasis on human ecology in his work enabled by funding from the Resources and Environment division of The Ford Foundation[2] (Figure 10.1).

The proposed objectives of the ecological inventory also returned to issues that McHarg earlier explored in his essays on housing. Here, however, rather than focusing his attention on the favorable attributes of cities and their citizens, he instead documented the deleterious impacts of urban life, and why trees, sunlight, fresh air, and moments of quiet relaxation were so desperately needed. People could adjust to difficult living conditions, he observed, but this left indelible physical and psychological scars that adversely impacted health. Justification for this stance came from the

Figure 10.1 Philadelphia circa 1968 from *Design with Nature*. © 1992 by John
    Wiley & Sons, Inc. Reprinted by permission of John Wiley & Sons, Inc.

concept of General Adaptation Syndrome that the physiologist Hans Selye
described to McHarg on *The House We Live In*. Selye's research indicated
that prolonged exposure to stressful conditions caused disruptive chemical
changes in the body, which in conjunction with a depletion of energy reserves, weakened immune functions and the biological ability to mount an
effective response. Over time this bodily wear and tear led to "diseases of
adaptation" such as hypertension, cardio-vascular disease, stroke, ulcers,
and cancer that decreased longevity.[3] As the first step in an exercise of preventive medicine, the proposed study would locate the areas of the city that
fostered stress-induced illnesses. He explained his intent as follows:

> We need to know where are the environments of health for there the
> environment is fit; the adaptations are creative. There is a creative-
> fit-healthy environment. What are its components? All this we must
> know to create the humane city.[4]

Justification for the study's inquiry into environmental fitness came from the biochemist Lawrence Henderson and his argument that cooperative molecular adaptations, defined as the ability of chemical compounds to creatively interact, rather than competitive struggle, propelled evolutionary change. Henderson derived this concept of reciprocity from empirical observations of the chemical couplings of organic compounds. His research led him to postulate that the physical properties of matter and the biological phenomena of life co-evolved, and it was virtually impossible to separate the two. According to Henderson, the resulting physiochemical tautology selectively fitted the earth to life and life to the earth.[5] In his description of the ecological inventory, McHarg used this notion of cooperative evolution to claim that normative development policies not only ignored the adaptive transactions that had worked over time to make the environment a naturally favorable habitat: the scale and pace of change in modern life was now so vast and precipitate, particularly in urban environments, that it was all but impossible for Henderson's thesis of reciprocity to remain operable.

The Philadelphia study would advance Henderson's concept of environmental fitness by finally realizing, as previously mentioned, an urban ecosystem workshop that McHarg originally planned to conduct with a group of eminent ecologists. In keeping with the objectives of this earlier proposal, the students would act as urban ecologists and compile data on the impact of air and water pollution on people, plants, and animals. They would then use the gathered data to reconfigure the "distorted" morphology of the city to more closely conform to the creative biochemistry of life.[6] As McHarg observed in the essay "The Ecology of the City," it was possible to address air and water quality, uncontrolled growth, and overcrowding through urban designs that emulated the cooperative dynamics of natural systems.[7]

McHarg was not alone in his critique of urban America. Others shared his urban anxiety and his concern that planning policies that privileged economics over health and welfare either neglected or negated the positive appeal of the city and its simulating diversity and opportunities for intellectual growth. At the national level, Stewart Udall, the Secretary of the Interior during the Kennedy and Johnson administrations, captured the general spirit of this argument when he stated in *The Quiet Crisis* that cities "have grown too fast to grow well."[8] For Udall, the most egregious impact of rapid growth was "the corrosive" impact of "unrelieved tension, overcrowding, and confusion" on the human psyche. But he also argued "it was not too late to repair some of the mistakes of the past, and to make America a green and pleasant – and productive – land." To repair these mistakes and make cities green and pleasant, he called attention to the work of Frederick Law Olmsted, the elder statesman of landscape architecture in the United States. Udall highlighted Olmsted's efforts to achieve "a healthy balance between the works of man and the works of nature in an urban setting" through the creation of "diverse, and continuous enclaves of open space, green gardens, and public playgrounds." President Kennedy explained the politics of the situation in the introduction to *The Quiet Crisis* when he stated: "We must develop new instruments of foresight and protection and nurture in order to recover the relationship between man and nature and to make sure the national estate we pass on to our multiplying descendants is green and nourishing."[9] To be pro-landscape in the urban renewal debate in the mid-20th century, similar to being anti-billboard in the highway debate, was the politically correct thing to do. This stance required an operative attitude, compromise, and trade-offs. Planning in the United States in the 1960s, as noted by Sam Bass Warner in *The Urban Wilderness: A History of the American City*, had to provide "sufficient diet, decent housing, adequate medical care, and a safe environment for the lowest third of our citizenry," and it had to meet the "high standard of living" expectations of the upper two-thirds of society.[10]

The Philadelphia study

Beginning with the vague proposition that it was possible to distinguish healthy, or fit environments, from diseased, or unfit environments, McHarg designed a series of studio exercises to assess the impact of the urban landscape on physical and mental health. The initial exercise involved site-specific observation, and it sent students out into the neighborhoods of Philadelphia to record where "children laughed, or did not, the demeanor of policemen, the presence of garbage in the streets, broken glass or overturned automobiles, street trees, playgrounds, parks, defiant scribblings on walls, care, pride, or despair."[11] To verify the accuracy of these observations, McHarg instructed his students to create an ancillary inventory of urban pathologies using statistical data sets encoded by government agencies. In conformance with the survey method he developed in previous design studios, the data sets were geospatially located, individually mapped, color-coded in shades of gray, and overlaid to create a summary map. The lightest areas of the summary map indicated a healthy urban environment. Conversely, the darkest areas indicated the presence of pathological conditions. The center of each map corresponded to the intersection of Market and Broad Streets, the geographical heart of William Penn's colonial plan for Philadelphia. Penn's utopian scheme initially called for a park at this location, but it later became the site of City Hall. By making this location the origin point of discovery, the project reinforced the role of the government and political mandates in the organization of city life.[12]

Essentially a study of poverty, the maps produced by the students documented incidences of disease and crime, and they correlated this information to race, ethnicity, income, and air quality. The results indicated that the non-white residents who lived immediately north and south of the city center had lower incomes, less access to education, deficient housing, and greater illiteracy and unemployment than the city's white ethnic groups. The maps also indicated that these residents had lower life expectancies and they suffered from greater incidences of crime, hospitalization, infant mortality, tuberculosis, diabetes, and sexually transmitted diseases (Figure 10.2 and 10.3). "The pattern is very clear," McHarg stated in his diagnosis of the situation, "the heart of the city is the heart of pathology and there is a great concentration of all types of pathology encircling it."[13]

In addition to the maps, the project discussion contained two photographs that captured the emptiness and isolation that McHarg believed shaped the experience of mid-20th-century cities. The first photograph documents a destitute man in the skid row section of Philadelphia collapsed against a wall with his head bent, eyes closed, and arms protectively clutching his huddled body.[14] The second photograph shows a crowded sidewalk teaming with well-dressed but anonymous people. These melodramatic and

Figure 10.2 Philadelphia disease maps from *Design with Nature.* © 1992 by John Wiley & Sons, Inc. Reprinted by permission of John Wiley & Sons, Inc.

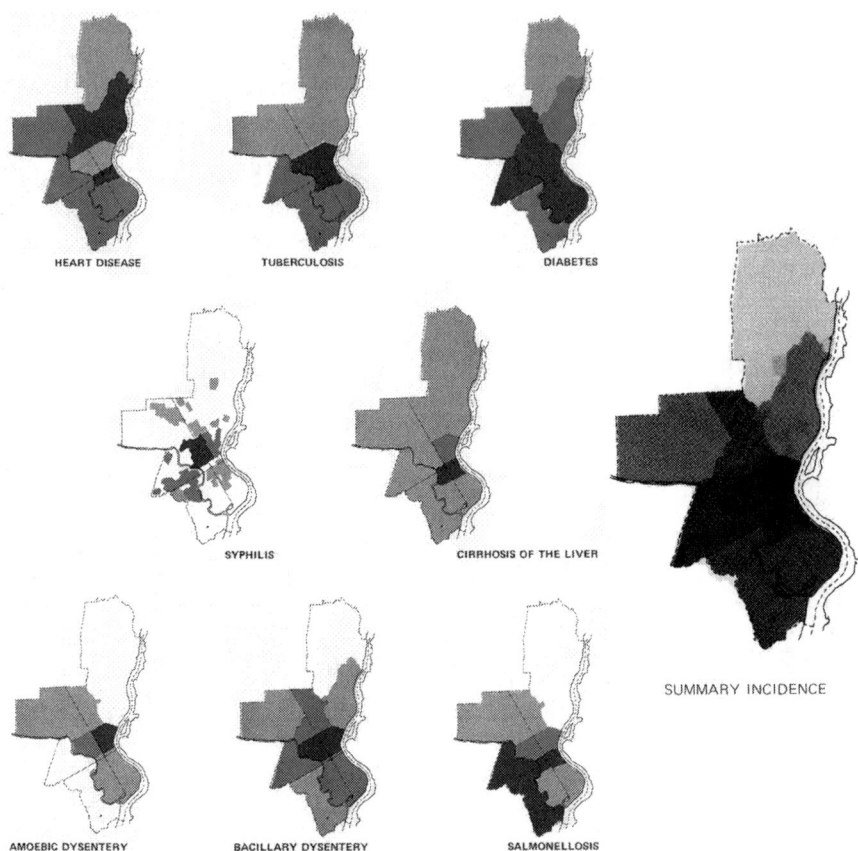

HEART DISEASE   TUBERCULOSIS   DIABETES

SYPHILIS   CIRRHOSIS OF THE LIVER

SUMMARY INCIDENCE

AMOEBIC DYSENTERY   BACILLARY DYSENTERY   SALMONELLOSIS

Figure 10.3 Philadelphia social pathology maps from *Design with Nature.*
© 1992 by John Wiley & Sons, Inc. Reprinted by permission of
John Wiley & Sons, Inc.

intentionally alienating images, in contrast to the maps, remove distance,
and force confrontation. As indices of health, they signal the study's inventory of inflictions threatened the destitute and the affluent, change was
desperately needed, and it was time for society to face its self-destructive
inclinations. As character studies, they objectify, and exploit, the predica-
ments of the people he sought to help and transform these individuals – a
social outcast ignored by those around him and a coerced labor force too
programmed to understand its plight – into the disposable resources and
waste products of modernity.

   The narrative McHarg created to explain the project was as complex,
multifaceted, discordant, and ultimately as prejudicial as the maps and
photographs of the pathologies that he sought to expose. He briefly discussed perception, physical organization, patterns of class and race, and

the biological basis of behavior. The resulting intellectual journey took him far away from the normative messages of landscape architecture and revealed a side of the profession that few willingly acknowledge. In the process, he discovered that human action, including his own advocacy of ecology, was more complicated, harder to explain, and difficult to justify than he had anticipated (or would admit).

McHarg began his project narrative with a description of The Pioneer Health Centre (1926–1950), a community facility established by the medical doctors G. Scott Williamson and Innes Hope Pearse in a poor neighborhood in southwest London.[15] As mentioned in a previous chapter, the luminaries of architectural modernity touted the facility as an exemplary model of progressive design. Gropius, for example, in *Rebuilding Our Cities* praised Williamson and Pearse for gathering, under one roof, the amenities and services that should be available to people in modern society – medical consultation rooms, playrooms, workshops, offices, and a nursery, theater, billiard room, swimming pool, and gymnasium. He also maintained the building's "club-like" facilities cultivated "the social soil," and thus promoted the organic growth of a community with the ability "to determine its own anatomy and physiology according to biological law."[16] Echoing this earlier accolade, McHarg claimed the facility and its services were a model of progressive urbanization that addressed the physical and social needs of the family, neighborhood, and community.

In his discussion of The Pioneer Health Centre, McHarg linked the urban pathologies that his students documented to the medical examinations conducted by Williamson and Pearse. As explained by Williamson in "The Individual and Community," an essay that appeared in the CIAM publication *The Heart of City*, these health appraisals were empirical and they recorded the small details that patients failed to report or failed to notice.[17] Following the construction of The Pioneer Health Centre, Williamson expanded these observations to include the daily lives and home environments of his patients. The information he compiled indicated these individuals, through a combination of poverty and social conditioning, lacked the ability to make informed choices regarding their health and welfare – they either made decisions that negatively impacted their lives or they passively accepted the choices they were given. These actions led to social fragmentation, isolation, stress, and illness. To address these conditions, he programmed the facility with activities designed to help his patients regain the ability to make informed choices about their lives and their surroundings. The behavioral changes he sought to instill required time and patience. When the program began chaos reigned and everything was broken and nothing was respected. Soon, however, "a sudden and spontaneous order emerged out of the previous disorder."[18] He attributed the change in behavior to altruism. Once people realized their choices impacted the lives of others, they refrained from selfish action. From this, he concluded that genetic inheritance determined what an individual could do, but the environment and socialization determined how they did it. It was impossible, he observed, to neglect one without denigrating the other. Williamson further surmised that genetic inheritance largely determines human nature, but it was possible to redirect human behavior through strategic initiatives focused on health. In other words, the services and facilities that comprised The Pioneer Health Centre enabled the two halves of human development – biology and behavior – to creatively "fit" in a self-affirming and self-sustaining manner. In his opinion, properly administered social modification advanced a holistic biological gestalt that enabled the organism-as-a-whole to evolve into a "Family-in-a-Home."[19]

Williamson was keenly aware of the ethical implications of his corrective procedures and selection protocols, and he cautioned the "power of architecture to fix the conditions in which life and living has to take place, is tremendous – almost frightening."[20] McHarg, who earlier used this statement in the essay "Open Space and Housing" to reprimand his colleagues for their failure to consider social organization in their design projects, again borrowed the quote. But here, in keeping with the health objectives of the Philadelphia study, he used Williamson's warning to call attention to the detrimental impacts of stress. He also cited *Mental Health in the Metropolis: The Midtown Manhattan Study*, a recent psychiatric evaluation of stress and disease in New York City, to support his unfavorable assessment of urban health.[21] This 1962 report, written at the height of Freudian psychoanalysis in the United States, deployed data sets and statistical correlations to argue that the sources of mental anguish may reside deep in the psyche but they were on full display in the collective consciousness of city dwellers.

More problematic, however, was McHarg's decision to couple his assessment of the urban conditions mapped by his students with the work of the behavioral psychologist Dr. John Calhoun, a research scientist employed by the National Institute of Health.[22] Calhoun studied populations of rats in cages in order to elucidate the role of overcrowding in the physical and social development of rodent communities. The manifold symptoms he recorded included almost every conceivable organic and social dysfunction. Among these behaviors were extreme passivity, territorial aggression, high infant mortality, physical violence, filicide, fratricide, cannibalism, and increased incidences of hyper-sexuality, asexuality, and homosexuality. His findings, which were published in *Scientific American*, *Ekistics*, and as a sensationalized synopsis in *Newsweek*, soon permeated the scientific literature and the public imagination.[23] Prominent figures in fields that ranged from architecture to zoology referenced Calhoun. As noted by the cultural historians Edmund Ramsden and Jon Adams, he was a particular favorite of researchers who sought to present "an unduly pessimistic and cataclysmic vision of man's future in a crowded world."[24] McHarg learned of Calhoun and his work through his interactions with the epidemiologist Leonard Duhl, who framed the inquiry of *The Urban Condition: People and Policy in the Metropolis* on this behavioral scientist's studies of overcrowding, stress, and disease.

The provocative language that Calhoun used to describe his experimental results contributed to his scientific notoriety. "Pathological togetherness," for example, denoted stress syndromes precipitated by a hierarchy of dominance in which only the strong and devious triumphed, and the term "behavioral sink" referred to ill-fated assemblages of vicious conduct.[25] Calhoun's unconventional vocabulary made it easy to create analogies between his experimental findings and the social discord that marked society in the 1960s as the country grappled with the repercussions of environmental pollution, urban race riots, an unpopular war, the women's liberation movement, counter cultural foment, and news reports of third world famine. Perhaps most disturbing by today's standards, however, was the way Calhoun's descriptions of demented rats in overcrowded cages became associated with urban slums. The rodent enclosures that he constructed facilitated this association. The concrete walls, narrow stairwells, and multiple levels of rooms that constituted these animal pens were remarkably similar to the low-income housing blocks – towers in the park – championed by the proponents of mid-20th-century urban renewal. Calhoun did not intend this association, but as Ramsden and Adams also observed, it was implicit in the physical structure of his experiments[26] (Figure 10.4). McHarg began his description of Calhoun's work with the

Figure 10.4 John Calhoun rat enclosure. Image reprinted by permission from the estate of Bunji Tagawa.

claim that it was hard to draw a direct analogy between the deviancies detailed in these studies and the social ecologies of dense metropolitan centers, and then (as many others had done) preceded to do so.

Less questionable than McHarg's Freudian flirtation with the racial and behavioral implications of Malthusian population dynamics was his refutation of slum clearance as a means to redress substandard living conditions and social inequality. Herbert Gans, McHarg's colleague at the University of Pennsylvania, provided support for this critical assessment of a standard urban renewal practice in a study that documented the deterioration of social cohesion and mental health in a low-income neighborhood of Boston, Massachusetts that occurred following slum clearance. Gans's analysis questioned the validity of the physical criteria, photographic documentation, and disciplinary terminology that many planners and government agencies used to justify *tabula rasa* urban renewal.[27] The highway alignment studies recently completed by McHarg similarly indicated that there was no need to destroy communities and displace residents to make room for infrastructure improvements.

Nevertheless, during this discussion of slum clearance McHarg again revealed his struggle to align his study of the city with empirical objectivity. Why, he asked, should the poor "live on the most expensive of central urban land, which requires the highest density?"[28] Was he indicating, as earlier expressed in "The Humane City," that the urban core must reflect the interests of wealth and power? Was he refuting *tabula rasa* urban renewal and the construction of low-income housing blocks? Was he searching for a compromise that provided all urban dwellers, not just the wealthy, access to fresh air and sunlight? Whatever his intentions, the implicit social stratification in this statement is disturbing.

The idea that the physical conformation of the city could either advance or hinder the lives of its citizens was, of course, not new, and following the 1958 Rockefeller Conference on Urban Criticism that McHarg's colleagues at the University of Pennsylvania organized it was a major topic in design discourse. Jane Jacobs, one of the conference attendees, notably detailed the elaborate bonds of community that united the residents of her street in *The Death and Life of Great American Cities*.[29] A similar interest structured the work of the urban planner Kevin Lynch, another conference attendee. Lynch, intrigued by the imaginative way people perceive and navigate their surroundings, coined the term "legibility" to describe "the ease with which its [the city's] parts can be organized into a coherent pattern."[30] To this end, he asked people to create maps of their communities using mental images drawn from memory. Lynch compiled this information into a list of the noteworthy physical features that people used to orient themselves as they traversed their neighborhoods. In *The Image of the City*, he argued that these paths, edges, districts, nodes, and landmarks explained how the residents of a neighborhood collectively saw, valued, and organized their surroundings. According to Lynch, most designers failed to appreciate "what a setting can mean in terms of daily delight," or how the seemingly ordinary served as "a continuous anchor" that added meaning and richness to life. Lynch also derided outside observers who were quite vocal about the dirt, the smoke, the heat, the congestion, and the chaos they perceived, but lacked the real intimacy that residents of a community have with their surroundings.[31] McHarg was similarly intrigued by the structure of the landscape and its imaginative potential, but in contrast to Lynch he used his findings to estab- lish the validity of a preconceived order. To put it another way, Lynch com- piled data to create an urban pattern book that contained myriad images and voices. McHarg (as one of those outside observers who was quite vocal about the dirt, smoke, heat, congestion, and chaos of the city) gathered data to establish the pattern of a singular truth. If, as he believed, people were happier and healthier in environments hygienically cleansed with sunlight and trees, then the most important thing about the appearance of cities was not that they were dirty and diseased, but how the absence of sunlight and trees made them dirty and diseased. The logical outcome of this argument was an urban environment that reflected the salutary values of his ecological agenda. As he later observed in the essay "Ecological Determinism," why should people live where it is impossible for trees to live?[32]

The Philadelphia study's portrayal of the city as loci of disease and disor- der also reflected the increasingly strong imprint of Lewis Mumford upon McHarg, and in particular his cautionary warning that modern society, co-opted by capitalism and in thrall to machines, could no longer meet the physical and social challenges posed by industrial production and advanced technology. Mumford, who also attended the Rockefeller Conference, would later argue, in both *The City in History* and to McHarg personally on the television program *The House We Live In*, that it was possible to address these challenges through communal acts of self-empowering coop- eration.[33] In his assessment, people could meekly acquiesce to the discord, disorder, and filth of the capitalist system, or they could take control of their lives and re-instill the peace, order, and cleanliness essential to their health and wellbeing. To do the latter, however, did not mean the rejec- tion of science and technology, but rather the redefinition of the objectives of science and technology to align with the organic order that Lawrence Henderson detailed in *The Fitness of the Environment*.[34] If this was not done, he warned, the resultant physical disarray would inevitably incite social disarray. Death was the unavoidable outcome of Mumford's natural history of the city, particularly when physical conditions passed their biological limits, but it was also the prelude for a new beginning.

McHarg, in an attempt to eradicate the diseased urban tissue docu- mented in the ecological inventory, aligned his prescription for the city with the organic order and the health-inducing ambiance of green leaves and sunlight envisioned by Mumford. Using the maps produced by his students as critical evidence, he reasoned that verdant cities, in conjunction with the strategic removal of excess population to new satellite communities in the countryside, would relieve stress, promote wellbeing, and increase property values. The resulting accumulation of physical, social, and economic capitol could then be re-invested in the long-term care of the city and its citizens.

Yet, as just discussed, McHarg's data collection and project discussion contained assumptions about class, race, social values, and biology that negated his objectivity and called into question his desire to generate a truly pluralistic solution. In spite of these glaring flaws, the project did link living conditions and social economic opportunities to health and longevity, and it illustrated that not everyone shared the same advantages in this regard. Nevertheless, little or no mention was made of the class divisions and discriminatory practices that abetted the documented pathologies, or the fact that planned dispersal could promote a more extreme segregation of individuals and communities. Indeed, there are more than a few moments when it is hard to read McHarg's discussion of the work and the color-coded meanings that he ascribed to the dark areas of the maps, particularly in light of the social changes that have occurred in the intervening years. What is perhaps most disconcerting is the implicit acceptance of the idea that design could change society through sanctioned social engineering, and the explicitly white, middle class, and gendered presumptions of how this society should function, and how it should look.

The validity of this type of study was not questioned at the time. The same thinking can be found in the political calculations of Steward Udall, architectural modernity's hygienic discipline of sunlight and fresh air, and the Garden City inspired dispersal plans proposed by Lewis Mumford and his colleagues at the Regional Planning Association of America. Even earlier, in 1910, the Second National Conference on City Planning formalized similar anti-congestion policies to combat sub-standard living conditions. According to Benjamin Marsh, in his executive summary of the conference, strategic initiatives that limited economic exploitation, promoted transit systems that enable dispersion, and advanced fair wages for work could overcome the challenges of excess population.[35] And if we return to the mid-19th century when the United States first began to grapple with rapid urban grow and physical congestion, this way of looking at cities and people aligns perfectly with the progressive traditions of Victorian social reform, as notably seen in the photographs of slum conditions and urban degeneracy documented by Jacob Riis, and the essays on civic design written by Frederick Law Olmsted.[36]

Notes

1 John B. Calhoun, "Design for Mammalian Living" *The Architectural Association Quarterly* 1 no. 3 (1969): 24–35.

2 McHarg, *Design with Nature*, 187–195. The Department of Landscape architecture received two grants from the Ford Foundation. In 1965, The Foundation provided $134, 879 to hire faculty and develop a curriculum in regional planning. In 1967, The Foundation provided an additional $200,000 to advance the

regional planning program. McHarg used the funds to hire faculty in meteorology, geology, hydrology, soils, plant ecology, wildlife ecology, and to buy computers. The funding followed McHarg's participation in the 1964 Conference on Natural Resource Conservation convened by the Ford Foundation. During this conference, McHarg submitted a paper prepared by Nicholas Muhlenberg that outlined the design curriculum that The Foundation subsequently funded. See Strong and Thomas, *The Book of School*, 148; *The Ford Foundation Annual Report 1965*, "Resources and Environment Grant Funding" (New York: Hillison & Etten Company, 1965), 117; and *The Ford Foundation Annual Report 1967*, "Resources and Environment Grant Funding" (New York: Hillison & Etten Company, 1967), 26; and "Nicholas Muhlenberg, "Conference on Natural Resources – Discussion Papers (1/2), (February 27–March 1, 1964).

3 Hans Selye, "The General Adaptation Syndrome and the Diseases of Adaptation" *The Journal of Allergy and Clinical Immunology* 17 no. 4 (July 1946): 358–398. See also "Program 11: Dr. Hans Selye."

4 McHarg, *Design with Nature*, 195.

5 Henderson, *The Fitness of the Environment*, 1958.

6 "Ecology of the City Participants," 6.

7 See McHarg, "Ecology of the City."

8 Stewart L. Udall, *The Quiet Crisis* (New York: Holt, Rinehart and Winston, 1963), 161.

9 Ibid., viii, xiii, 163.

10 Sam Bass Warner, Jr., *The Urban Wilderness: A History of the American City* (Berkeley, CA: University of California Press, 1972), 230.

11 McHarg, *Design with Nature*, 193.

12 John W. Reps, "William Penn and the Planning of Philadelphia" *The Town Planning Review* 27 no. 1 (April, 1956): 27–39. See also Pennsylvania Historical & Museum Commission, *Pennsylvania Architectural Field Guide*, "Second Empire/ Mansard Style 1860–1900," www.phmc.state.pa.us/portal/communities/ architecture/styles/second-empire.html.

13 McHarg, *Design with Nature*, 193.

14 Email exchange, Eileen Christelow Ahrenholtz, the photographer who took the photograph, March 30, 2019.

15 See G. Scott Williamson M.D. and I. H. Pearse M.D., *Biologists in Search of Material: An Interim Report on the Work of the Pioneer Health Center* (London: Faber & Faber Limited, 1938). The Center closed during World War II.

16 Walter Gropius, *Rebuilding Our Communities*, 53.

17 G. Scott Williamson, "The Individual and Community," 30–35.

18 Ibid., 33.

19 Ibid., 35.

20 Ibid., 33.

21 Leo Srole, Thomas S. Langner, Stanley T. Michael, Marvin K. Opler, and Thomas A.C. Rennie, *Mental Health in the Metropolis: The Midtown Manhattan Study* (New York: McGraw-Hill Book Company, Inc., 1962).

22 See John B. Calhoun, "Population Density and Social Pathology" in *The Urban Condition: People and Policy in the Metropolis*, ed. Leonard Duhl (New York: Basic Books, Inc., 1963), 33–43. The Calhoun essay essay appeared in the first section of the text immediately preceding the essay by McHarg.

23 John B. Calhoun, "Population Density and Social Pathology" *Scientific American* 206 no. 2 (February 1962): 139–146, 148; John B. Calhoun, "Space and the Strategy of the City" *Ekistics* 29 no. 175 City of the Future (July 1970): 425–437; and Stewart Alsop, "Dr. Calhoun's Horrible Mousery" *Newsweek* (August 17, 1970): 96.

24 Edmund Ramsden and Jon Adams, "Escaping the Laboratory: The Rodent Experiments of John B. Calhoun and Their Cultural Influence" *The Journal of Social History* 42 no. 3 (1 March 2009): 761–797.
25 Calhoun, "Population Density and Social Pathology."
26 Ramsden and Adams, "Escaping the Laboratory."
27 Herbert Gans, *Urban Villagers: Group and Class Life in the Italian-Americans* (New York: Simon & Schuster, 1962).
28 McHarg, *Design with Nature*, 195.
29 Jane Jacobs, *The Death and Life of Great American Cities* (New York: Random House, 1961).
30 Kevin Lynch, *The Image of the City* (Cambridge, MA: The M.I.T. Press, 1960), 2–3. For information on the relationship of Lynch's work to perception and visual order, see Orit Halpern, *Beautiful Data: A History of Vision and Reason Since 1945* (Durham, NC: Duke University Press, 2014), 110–122.
31 Ibid., 2, 47–48.
32 Ian L. McHarg, "Ecological Determinism."
33 See Mumford, *The City in History*; *The House We Live In*, Lewis Mumford, and "Lewis Mumford (edited)."
34 Henderson, *The Fitness of the Environment*, 1958.
35 See Benjamin C. Marsh, "Causes of Congestion" in *Proceedings of the Second National Conference on City Planning and the Problem of Congestion, Rochester, New York, May 2–4, 1910.* (Boston, MA: National Conference on City Planning): 35–39, http://urbanplanning.library.cornell.edu/DOCS/marshpop.htm.
36 See Jacob A. Riis, *How the Other Half Lives* (New York: Charles Scribner's Sons, 1890); and Olmsted, *Civilizing American Cities*, 23–42.

11 Fit, fitting, and most fit

Amid all the revolutions of the globe the economy of Nature has been uniform, and her laws are the only things that have resisted the general movement. The rivers and the rocks, the seas and the continents have been changed in all their parts; but the laws which direct those changes, and the rules to which they are subject, have remained invariably the same.

Charles Lyell, Principles of Geology[1]

McHarg was not, as he would have liked his readers to believe, the first environmental planner to advocate water as a unifying element, a watershed as a regional planning unit, physiography as the basis of analysis, or rivers as emblems of civilization. His methods and ideas continued a well-established exploration of the land and life that posed critical questions about perception, the nature of design, and the perfectibility of society (Figure 11.1).

The cross-sections and perspectives produced for the Potomac River basin study, for example, recapitulated the Valley Section – a multipartite diagram of human settlement patterns developed by the Scottish planner Patrick Geddes.[2] In plan, the Valley Section traced the course of a prototypical river as it flowed from the mountains to the sea. In section, it deployed archetypal occupations – Miner, Woodman, Hunter, Shepard, Peasant, Farmer, and Fisher – to illustrate the patterns of work that occurred along the course of the river. In perspective, it nestled the hamlets, villages, towns, cities, forests, farms, and factories associated with these occupations within the physical terrain. Each of these representations entwined the character of the people (folk), and their occupations (work), with the character of the land (place). Geddes often used an image of the Acropolis in Athens to symbolize the transcendent beauty of this diagram.

The entangled narrative of water, topography, rivers, cities, and people that Geddes diagrammed in the Valley Section, recapitulated a vision of natural order elucidated in the 19th century by the biologist T. H. Huxley. Huxley, who served briefly as Geddes's teacher and mentor, believed that empirical observation coupled with scientific inquiry demonstrated "the existence of immutable moral and physical laws" that revealed humankind's true place in nature.[3]

Figure 11.1 The Grand Canyon from *Design with Nature*. © 1992 by John Wiley & Sons, Inc. Reprinted by permission of John Wiley & Sons, Inc.

Huxley presented his study of natural order in *Physiography: An Introduction to the Study of Nature*.[4] He began the text by asking his readers to imagine themselves in the center of London, standing on the London Bridge looking down at the water as it ebbed and flowed with the daily rhythms of the tide. From this well-known vantage point he embarked on a tour de force of scientific discovery that charted the river's course, mapped its geologic terrain, evaluated its climate and rainfall, discussed the relationship of the moon and the tides, and explained how the sun provided the energy for all of these operations. A watershed map of the Thames River, which served as the frontispiece of the text, situated this developmental history within a geographic framework that his readers could understand, particularly as it related to their lives and livelihood. He concluded his journey as follows:

> And thus we reach, at last, the goal of our inquiry. At the furthest point to which we have pushed our analysis of the causes of the phenomena presented to us, the sun is revealed as the prime mover in all

the circulation of matter that goes on, and has gone on for untold ages, within the basin of the Thames; and the spectacles of the ebb and flow of the tide, under London Bridge, from which we started, proves to be a symbol of the working forces which extend from planet to planet, and from star to star throughout the universe.[5]

To fully understand Huxley's interest in physiography, and why he selected the flow of water and a watershed to explain the patterns and processes of physical and cultural development, it is necessary to digress deeper into the terrain of natural history and explore earlier descriptions of time, water, land, and life. Others have examined this landscape in detail, but it is worth repeating here because of the way it influenced, both directly and indirectly, McHarg's belief that designers must understand how organisms and the environment are naturally fit and fitted to each other.[6]

In this particular instance, the historical meander into deep-time begins in the Scottish Highlands with the geologist James Hutton as he walked along the banks of the River Tilt and pondered how long it had taken the flow of water to carve a narrow stream valley "as steep as is possible for earth and stones to lie." The river valleys of the Highlands, he postulated, acquired their "present form by the operation of water running upon the surface of the earth," and he speculated this process of erosion was not new, but instead existed for an indefinite period of time.[7] From this conjecture he envisioned a world in which water traveling from the mountains to the sea, dissolved rocks, eroded mountains, formed soils, created lowland plains, and made the land fit for plants and animals. In 1788, he assembled his observations into a dynamic, instrumental, and orderly depiction of time, energy, and material processes titled *The Theory of the Earth*:

> This globe of the earth is a habitable world; and on its fitness for this purpose, our sense of wisdom in its formation must depend. To judge of this point, we must keep in view, not only the end, but the means also by which that end is obtained. These are, the form of the whole, the materials of which it is composed, and the several powers which concur, counteract, or balance one another, in procuring the general result.[8]

In 1830, the British geologist Charles Lyell followed a similar path of panoramic discovery, and presented his empirical observations and imaginative deductions in the *Principles of Geology*.[9] In a carefully reasoned argument, Lyell detailed how erosion, deposition, and sedimentation operated over time in a slow and steady manner to engender geologic change. Observations in the present indicated what happened in the past, he claimed.[10] From the outset of his argument, Lyell established a strong link between living and non-living matter, as seen, for instance, in his definition of geology as an investigation into "the successive changes that have taken place in the organic and inorganic kingdoms of nature."[11] Not content, however, to limit his discussion to biophysical processes, Lyell equated the laws that govern change in the natural world to the laws that govern reason in an ideal society. In this analogy, geologic strata served as an index, or sign, that operated, in sense of Auguste Comte, to trace progress as humankind, in a slow and incremental manner, refined scientific knowledge, political action, and moral behavior to conform to the dictates of Nature's laws.[12]

One year later, the HMS Beagle sailed from London on a journey of discovery at the behest of the British Navy to gather information on the natural history of the earth. As the ship circumnavigated the globe, the crew charted ocean shorelines, mapped geologic terrain, and studied the distribution of plants and animals. The Beagle contained an impressive library of scientific treatises that included the *Principles of Geology*.[13]

Charles Darwin, a former divinity student immersed in the natural theology of God's plan in Nature served as the expedition's naturalist. As he traveled the world, reading, observing, collecting samples, examining fossils, and investigating the formation of coral reefs, Darwin began to ponder how groups of animals with common characteristics came into existence. Intrigued by Lyell's dictum of uniform change, he applied the principle to biology and speculated that a similar process had caused life to branch and diversify.[14]

In 1859, after years of painstaking refinement, Darwin published a geologically inspired theory of speciation titled *On the Origin of Species by Natural Selection, or The Preservation of Favored Races in the Struggle for Life*.[15] His argument outlined the processes in which a new species (a group of living organisms with the ability to reproduce with each other) could arise over time through incremental changes that occurred in response to local context (the environment). This descent from an ancestral species with modification (evolution) reflected the capacity of organisms to adapt to their surroundings and successfully reproduce (natural selection). Over many generations, these environmental adaptations resulted in a cumulative change in biological characteristics. To situate his ideas within a context that his readers could understand, Darwin equated natural selection to the actions of animal breeders when they culled the weak and sick, and allowed only the strong and healthy to survive.

In the Darwinian worldview only the fittest, as the philosopher Herbert Spencer argued, successfully adapted to change and lived to produce offspring.[16] But adaptive success, as Spencer also observed, was contingent and related to the physical characteristics of place:

> From the meanest zoophyte, up to the most highly organized of the vertebrata, one and all have fixed principles of existence. Each has it varied bodily wants to be satisfied – food to be provided for its proper nourishment – a habitation to be constructed for shelter from the cold,

or for defense against enemies – new arrangements to be made for bringing up a brood of young, nests to be built, little ones to be fed and fostered – then a store of provisions to be laid against the winter; and so on, with a variety of other natural desires to be gratified. For the performance of all these operations, every creature has its appropriate organs and instincts – external apparatus and internal facilities; and the health and happiness of each being, are bound up with the perfection and activity of these powers. They, in their turn, are dependent upon the position in which the creature is placed. Surround it with circumstances which preclude the necessity for any one of its faculties, and the faculty will become gradually impaired. Nature provides nothing in vain.[17]

Natural selection, put simply, was opportunistic and it provided no guarantees. Species, and by extension, cities, societies, and civilizations could arise like continents, and the streams of time could wash them away. These ideas challenged normative thoughts and perceptions – chance and competition supplanted the absolutism of divine will, cause preceded effect, and the empirical study of the Laws of Nature evolved into the science of ecology and its pursuit of the relations, both organic and inorganic, between organisms and their surroundings.

It was, therefore, only natural that Huxley, who garnered the appellation "Darwin's bulldog" for his staunch defense of natural selection and descent with modification, would use *Physiography* to explain, in a step-by-step manner, how the operations of the natural world shaped human imagination and the terrain of daily life.[18] Extremely skeptical of religious authority, he sought "for truth not among words but among things," and he developed a cosmology of reason where science explained "the order which pervades the multiform and endlessly shifting phenomena of nature."[19] He formulated his scientific exposé on the Thames basin to demonstrate to his readers that something as commonplace as the daily ebb and flow of the tides linked human action and the historical development of the city to the stars in the sky, the heat of the sun, the evaporation of water, the fall of rain, the formation of streams, the erosion of mountains, the creation of soil, and the growth of plants and animals. Overtly nationalist in tone, his empirical, and imperial, natural history of London indicated why it was fitting and all but inevitable that this great city, located at the mouth of a great river, had become the capital of a great nation, and the heart of a great global empire.[20]

But perhaps most important, at least in relationship to McHarg's vision of order, the natural history detailed in *Physiography* positioned the energy of the sun as the prime mover in the myriad relationships and adaptive struggles that constituted the surface of the earth. Whether intentional or not, McHarg signaled his allegiance to Huxley's argument when he placed starkly dramatic photographs of the sun and the earth on the front and back cover of *Design with Nature*. For McHarg, as for Huxley, the juxta-position of these cosmic bodies symbolized the intrinsic value of the nat-ural world and the triumph of reason over falsehood and dogma. When this thinking became commonplace, both men believed that people would acknowledge that they, too, were a part of nature and the earth's evolu-tionary history was their evolutionary history. In a similar manner, they reasoned this knowledge would prompt the pragmatic realization that it was advantageous to selectively organize the objects and events of daily life to enhance survival in the struggle for existence.

Huxley's student Patrick Geddes likewise believed in the value of empirical evidence, but in contrast to Huxley's scientific geography of what is, he stud-ied the social terrain of what ought to be. Geddes did not believe that ruth-less competition was the inevitable by-product of a Darwinian worldview. Intrigued by the generative possibilities of vital force and the cooperative dynamics of mutual dependence, he studied the flow of water to determine how the land shaped common values and patterns of thought, and thus in-digenous patterns of settlement and work.[21] It was his intent, as explained in *Cities in Evolution*, to shape the topography of everyday life to promote health and prosperity. For Geddes, the "geotechnics" of this manipula-tion began with the soil and cultivation, and with small demonstrations of cooperative action, such as the community gardens in the urban slums of Edinburgh, Scotland that he built with the assistance of children.[22] What he failed to mention, however, was once it was assumed that culture could beneficially shape human evolution, it was also assumed that there were nat-urally favored individuals who would shape the evolution of culture.[23]

Geddes's acolyte Lewis Mumford likewise explored the terrain of evolu-tionary potential and what ought to be, and, in a similar manner, he as-sociated urban planning with soil, cultivation, health, and survival. Less optimistic by nature than Geddes, Mumford believed with some justifica-tion, that even though modern science and technology miraculously pene-trated deep into the terrain of molecules and atoms and far into the cosmic reaches of space, these powerful insights corrosively spawned an illusion of power that abetted socially predatory, economically greedy, and environ-mentally destructive behavior. In his keynote address at the conference *Man's Role in Changing the Face of the Earth*, he claimed the selective pressures at play in contemporary society enabled the "blind forces of urbanization" to flow along the path of least resistance and disrupt the patterns of life that sustained physical, social, and economic prosperity. Refusing to accept this state of affairs, he argued for "far-sighted and provident" development that combined "the resources of modern science" with "the techniques of ecological balance."[24] At the end of the conference, when he returned to the podium to deliver his closing address it was only reasonable, considering his close relationship with Geddes, that he referenced the holistic ideals of 19th-century organicism. "You will recall," he stated, "that even before the great Darwin, Herbert Spencer had begun the modest work of synthesizing all knowledge on the basis of the evolutionary formula." As reprised by Mumford, Spencer's synthetic philosophy detailed an increasingly ordered natural history where life, through processes of never-ending change, progressed from "indefinite simple homogeneity to definite complex heterogeneity." In this narrative, Spencer's greatest contribution to evolutionary theory, and by extension to progressive modernity, was the understanding that human nature could be selectively cultivated to evolve, as intended, into an ecology of reason characterized by "ideal ends, imaginatively conceived, and rationally criticized."[25]

Mumford was also an incurable romantic, committed socialist, and utopian dreamer infatuated with noble sacrifices made in the name of progress. By nature, he gravitated toward small-town living.[26] If these rural traits were somehow implanted into the genetic make-up of the city, he reasoned they would propagate cooperative values and communal ideals that would challenge, and eventually out-compete the competitive behaviors that threatened his idyllic vision of the good life. To advance this outcome, he called for the study of natural history – geology, soils, climate, vegetation, and animal life – and cultural history – laws, manners, customs, and patterns of communal organization. As he observed in *The Culture of Cities*, an understanding of both histories was necessary as they defined an interactive field of forces that overlapped in space and time, and any changes made to one required compensatory changes in the other.[27]

Mumford discovered an evolutionary model of cooperative action in the science of the biochemist Lawrence Henderson. Intrigued by the natural order of "day and night, the changing but recurring seasons, the fertilizing sunshine and rain, the powers of the human hand, and all the beauties and mysteries of nature," Henderson developed an argument that explained why the earth provided a comfortable biological existence: oxygen, water, and a temperature between the freezing and boiling point of water. If the planet had been a little bit smaller, he observed, and the temperature a little bit different life would not exist. For Henderson, this circumstance alone was surprising; but even more amazing was the fact that living organisms, in their struggle for existence, helped maintain the earth within these comfortable tolerances.[28] From this, he concluded that organic and inorganic evolution were indivisible aspects of the same phenomena:

> Darwinian fitness is composed of a mutual relationship between the organism and the environment. Of this, fitness of the environment is quite as essential a component as the fitness which arises in the process of organic evolution; and in fundamental characteristics the actual environment is the fittest abode of life.[29]

When utopia arrived it would be ushered in, at least for Mumford, on a wave of chemically inspired altruism that allowed human action and natural processes to join forces and make the environment a fit expression of life.

In 1923, Patrick Geddes traveled to the United States at the behest of Mumford to discuss geotechnics and how to remodel the terrain to enhance health and prosperity. During the visit he met Benton MacKaye, Mumford's colleague at The Regional Planning Association of America.[30] Inspired by Geddes, MacKaye updated the principles of geotechnics to account for the impact of modern transportation on the movement of raw materials and finished goods between the countryside and the city. The explorers of the 19th century, he explained in "The New Exploration: Charting the Industrial Wilderness," had "unraveled the labyrinth of river and coastline" and discovered the great story of the earth.[31] In the 20th century, these same impulses would chart the frontier of the industrial wilderness and unravel the great story of urbanization. Once the coordinates of this urban wilderness were located and its movements and flows were mapped, it was his contention that the future would "naturally fall into place." He also claimed that physiography determined the most efficient pattern of movement. In a clear reference to the Valley Section, he plotted the logistics of this terrain – its rivers, railroads, and highways – to illustrate the ideal movement of raw materials and finished goods across the state of Massachusetts as they flowed from the Berkshire Mountains to Boston and the sea. Economic efficiency, orderly movement, and spatial optimization structured his diagram.

In 1928, MacKaye published *The New Exploration: A Philosophy of Regional Planning* to explain how to harness the processes of nature to advance the civilizing ambitions of capitalist democracy. He began the text with a reference to T. H. Huxley, and asked his readers to imaginatively stand on the London Bridge watching the River Thames as it ebbed and flowed with the tides. He then described how Huxley traced the river back through space and time until he revealed that the energy of the sun powered everything that people see and do, and many things that they take for granted:

> Here is a goal indeed – the revelation of the prime mover in our earth's material affairs! . . . for nobody knew better than Huxley that the full comprehension of those forces which extend throughout the universe is the subject of an everlasting quest, and the proximate understanding which we thus have attained is indeed but an "introduction" to the vast study of nature.[32]

Later, in "Regional Planning and Ecology," MacKaye reformulated Huxley's riparian journey into a paradigm of scientific management that united the natural landscape, and the resources and ecological services that it provides, with human action, and the social development and cultural enrichment that it provides.[33] To institute this "twin system," it was necessary to harness "the potential" of natural systems for the betterment of society. The flow of water illustrated the means and methods of his intent. Planning, he observed, in a statement that reprised the role assigned to it by Huxley and

Geddes, was comparable to charting the course of a river – it was best to know when to let the water flow naturally, and when it was appropriate to channel and control its movements. MacKaye imbued this proposition with democratic idealism. The regional connections that he envisioned were destined, by nature, to flow from their source and branch into a system of organic alliances that advanced the "good life" and fostered Thomas Jefferson's "pursuit of happiness." McHarg, as previously discussed, used the concept of a twin system to construct the logic for Metropolitan Open Space, and, following his participation in The Potomac Planning Task Force he was committed to the vision and values of a great society. He also explored MacKaye's ideas in his course Man and Environment.[34]

McHarg revealed his allegiance to this illustrious intellectual terrain in several chapters of *Design with Nature* devoted to natural history and creative evolution.[35] He explained his thoughts through an imaginative journey of empirical speculation that illustrated how an individual versed in natural history and the principles of ecology, but with no knowledge of cultural history, viewed the world and learned to understand it. The fictive explorer in this tale (a combination of rational space-age savant and intuitive noble savage – a.k.a. the ideal landscape architect) humbly discerned, through careful observation, the deep history of geologic time and the adaptive responses of organisms as they navigated this terrain of endless change and possibilities.

McHarg illustrated this parable of innocence and wonder with mountains and marshes, bird beaks and claws, coral polyps and frog embryos, and a drawing of a DNA helix appropriated from *Scientific American*.[36] Photographs of the Grand Canyon, Devil's Tower, and the Dakota Badlands – the iconographic landmarks of an earlier adventure with his wife Pauline – related these biological wonders to the grandeur of the American landscape. Snowflakes and electron micrographs of a platinum atom and virus particles established a link between living and non-living matter. A quick reference to the essay "The Modular Principle in Biological Form," by the molecular geneticist Conrad Waddington, established the proposition that the growth of living organisms, in contrast to the mechanical operations of architecture, reflected the rhythms of biology rather than "mere modular increase."[37]

To organize this panoply of ideas and images, McHarg reformulated his imaginative journey of discovery into an ersatz cosmology of adaptive fitness symbolized by the Naturalist:

> The Naturalists employ both conceptions of fitness, that propounded by Henderson and that by Darwin. Thus the environment is fit for life, for the forms which had preexisted, those which do now exist and those of the future . . . Evolution then consists of a tendency toward increasing fitness whereby the organism adapts the environment to make it more fitting and, through mutation and natural selection, adapts itself toward the same end.[38]

To prevent unfavorable traits from populating his vision, McHarg created a set of nominally scientific principles. The first principle honored the negentropic capture of sunlight by plants, as it cascaded up the food chain and energized the world. The second principle involved apperception, interpretation, and the transmutation of environmental "to whom it may concern messages" into meaning and purposive response. The third principle promoted symbiosis, which, in conjunction with apperception, led to cooperative arrangements that permitted increasing levels of order. The fourth principle embraced fitness, which he defined as the selection of an environment by an organism and the subsequent adaptive responses that accomplished a better fitting. The final principle discerned the presence of health or pathology and supplied the empirical evidence of a fit or unfit environment. Considering the critical role of sunlight in this theory of design it is not unexpected, but nevertheless ironic and fitting, that he framed his discussion around the algae experiment that he had witnessed early in his career when he assisted the architect Louis Kahn select the site of a research laboratory that specialized in space travel and military weaponry.

According to McHarg, his Naturalist cosmology fostered a democratic harmony notable for the way it allowed people to feel at home in the world and bask in the sunny joy of "the transcendent form of love." If the natural world was an "ordered place" where there was no central authority, anarchy, or tyranny, but instead a cooperative system of behavior governed by a natural "way of things," he reasoned that it should be possible to institute a system in which the laws that governed natural change in an ideal world also governed human action in an ideal society.[39] This was not a stringent command, but instead a moral obligation based on the deep ecology of rights and a presumption in favor of natural processes as proposed by the legal scholar Clarence Morris.[40] In consonance with McHarg's messianic predilections, this code of conduct also operated, in the biblical sense, as a natural theology in which good opposed evil.

He correspondingly championed freedom and made uniqueness an important component of his cosmology. Evolutionary processes, McHarg stated, guaranteed that every creature was "absolutely, not metaphorically, unique," and this singular distinction gave them the right to exist freely, find their own place in the scheme of things, and simply be. Uniqueness, he also explained, led to diversity and the creation of a vast web of life – the self-sustaining and self-regulating biosphere of Henderson – where everything was connected to everything else and there was no such thing as autonomous action.

Following the traditions established by Geddes, Mumford, and MacKaye, McHarg substantiated his cosmographic study of deep time and evolution through the agency of politically idealistic land use surveys that indicated how design could assist nature, in all of its guises, as it advanced in a slow but inexorable march toward physical complexity and social perfection:

> Consider a time when ecological inventories [of the United States] have been completed . . . Those interviewed are asked to enumerate the ideal

characteristics of climate, scenery, recreation, employment, residence that constitute their utopian preferences. The results are searched for consequences from which emerges a social program for a new urban America.[41]

McHarg returned to a barrier island, and water, wind, and sand to represent the processes of evolution, and, by analogy, to illustrate the natural revitalization of the urban ecosystem. Beginning with a barren sand dune, he explained how a "high in entropy and low in order" landscape, could, if left to its own devices, evolve into a "low in entropy and high in order" deciduous forest containing a diversity of species, numerous trophic levels, and many energy pathways.[42] As he earlier argued, a city was not a forest, but it was possible to emulate the processes that shaped the structure of a forest so that cities, too, progressed through successive stages of development until they obtained the same stratigraphic complexity and morphologic stability.[43] A wind swept sand dune and a forest suffused with sunlight illustrated his intent (Figures 11.2 and 11.3). A snow covered mountain peak, mystically shrouded in clouds, added a touch of the sublime.

A dichotomous chart explained the benefits that accrued when his design system was followed, and the consequences that occurred when it was ignored. One side of the chart, which nominally referenced the appearance of a sand dune, listed the physical characteristics of a "primitive state of

Figure 11.2 Photograph of a New Jersey barrier island sand dune. Courtesy Kathleen John-Alder.

Figure 11.3 Photograph of a mature forest. Grant Heilman Photography. Reprinted by permission.

existence" – simplicity, uniformity, instability, few species, no symbiosis, high entropy. The other side of the chart, which nominally referenced the appearance of a temperate forest, listed the physical characteristics of an "advanced state of existence" – complexity, diversity, many species, stability, symbiosis, low entropy. The resulting matrix of possibilities created a gestalt that was definitely more complex than the sum of the parts, but there was no moral ambiguity. Designers could select progressive, orderly traits that allowed society to evolve and flourish, or they could ignore his advice and allow civilization to regress into disorder, decay, and oblivion. The dichotomous structure of his approach made it easy to challenge the existing system, and it positioned design as the guardian of creative evolution.

Undeniable eugenic overtones permeate the chart's contrasting lists of landscape characteristics. McHarg's biologically inspired vision championed diversity, cooperation, and symbiosis, but these qualities required a fit land-scape populated by fit people. In light of the relationship of his ideas to the reform agendas of Spencer, Huxley, Geddes, and Mumford, this association was perhaps inevitable. He was aware of this conceptual tension, but found it difficult, if not impossible, to either renounce or resolve, as it was inher-ent to the sovereign concepts of selection and adaptation that governed his design paradigm; and, equally important, it coincided with his spiritual conviction that life was indeed evolving toward perfection as postulated by Teilhard de Chardin.[44] The following statement reveals his commitment to biological control, his instinctive recoil away from things that were marred or disabled, and the convoluted ambiguity of his reasoning:

> Is the house fit, or the street, the neighborhood or the city? What is unfit? There are dangers here, for we do not know what perfections are represented by the sloth we revile, but we can accept that the crip-pled animal whose grace earlier enchanted us is now no longer fit; our language conforms to this notion of unfit as the unhealthy, crippled, deformed, although there may be excellences that over come this.[45]

A statement, question, and resolution related this soliloquy on form, fitness, and perception to design:

> If the purpose of fitness is to ensure survival and evolutionary suc-cess for the organism, the species, the community and the biosphere, then adaptations are primarily directed toward enhancing life and evolution. Can we then avoid bringing concern with form into the realm of the enhancement or inhibition of life and evolution? When we link form to life we must retreat to a more basic, but united con-cern with adaptation as creative or destructive. Fitness is then by definition creative and will be revealed in the form of fitness that is life-enhancing.[46]

Things that developed organically and lived commensurately with their surroundings, such as a beehive, nautilus shell, coral, the Taos Pueblo, and Fallingwater, represented fit morphologies. Things once beautiful but now abused and no longer alluring or useful, such as a broken piano, de-faced painting, mutilated sculpture, junked auto in a wild landscape, and an anarchic city, represented ill-fit, misfit, and unfit morphologies. Here again, McHarg wanted his readers to realize that the everyday world, with its constant disorder and disturbance, did not reflect the aspirations of transcendent humanism. He also wanted them to understand that this less-than-ideal state of affairs was due in large part to thoughtless action and inattention. The term "fitting" (precisely because it denoted choice, selection, health, and perpetuation) articulated how to rectify the situation.

McHarg correspondingly stated that there was an ideal ecology of objects and actions that transcended the messy realities of everyday life, and when designers dispersed this ecology throughout the land, it would change the nature of reality. He offered the following truth in explanation: "The man who seeks to create metaphysical symbols – design with nature – is really concerned with idealizing." Another way to explain this presumptive imposition of signs upon the natural is to return to a lecture delivered by Morse Peckham in McHarg's seminar course Man and Environment. In a discussion of meaning in garden design, this scholar of 19th-century Romanticism observed that what people really wanted, whether they expressed it through science, religion, or design was a place perfectly adapted to their needs – a world of pure values and pure order where the divine was immediately accessible.[47] To design with nature, in this sense, enabled a prodigal return to paradise, as well as a refutation of those things that did not fit this ideal landscape. At its best, this worldview was an affirmative act of inclusion and acceptance, and at its worst, it was a domineering act of exclusion and refusal.

Washington, D.C.

In 1966, the National Capitol Planning Commission (NCPC) hired WMRT to prepare a landscape master plan for Washington, D.C. that would enhance and improve the function and appearance of the city. The decision to award the project to WMRT was no doubt due to McHarg's recent participation in the Potomac Planning Task Force, which, as previously noted, was convened by President Johnson to create a "new conservation" that would preserve and protect the natural resources and the national heritage of the Potomac River watershed, while simultaneously providing opportunities for development and recreation[48] (Figure 11.4).

McHarg utilized the opportunity to both promote his survey method and opportunistically plant the colonizing flag of ecology in the nation's capital. The project, which was managed by Narendra Juneja, illustrated the importance of broad-based inventories in design, planning, and policy decisions. In keeping with standard practice in master plan projects, the information was preliminary and designed to establish the foundation for future work. A 1967 report titled *Toward A Comprehensive Landscape Master Plan for Washington, D.C.* summarized the findings.[49] McHarg conducted the project while on sabbatical to write *Design with Nature*. The master plan appears prominently in the book, and, not surprisingly, it captures the democratic ideals, adaptive strategies, selective decision-making, instrumental reasoning, and symbolic promises of this ecological manifesto.[50]

The project report began with the claim that "memorable cities have distinctive characteristics," and these physical traits reflect the combined attributes of the natural landscape, referred to as "given" form, and the constructed landscape, referred to as "made" form. In terms of design,

Figure 11.4 Aerial view Washington, D. C. from *Design with Nature.*
© 1992 by John Wiley & Sons, Inc. Reprinted by permission of
John Wiley &Sons, Inc.

this meant that the natural landscape should be the priority in cities "built
upon beautiful, dramatic or rich sites," and, conversely, roads and buildings should be the priority when the land lacked physical drama. Places that
McHarg visited in his travels and memorable areas of Washington provided
examples of what he hoped to achieve:

> Mt. Fuji, the Bay of Naples, the Golden Gate, The Bois de Boulogne,
> the canals of Amsterdam and Venice enlarge the lives of all who live in
> their influence. So too do the Potomac and its Palisades, Rock Creek,
> The Mall, The Capitol, White House, the noble Monuments and
> Georgetown. These are public resources, important to Washington and
> the United States, enlarging to all who experience them. Their value
> and the value of many other aspects of the city should be incorporated
> into the planning process.[51]

The real challenge, however, was how to achieve these ambitions when the object of study – the landscape of natural (given) and built (made) form – constantly changed. Was it best to freeze everything in place and turn the city into a museum, or was it better to accept change? McHarg decided to do a little of both. He created a master plan that preserved memorable lands and buildings and he outlined conditions that allowed the less distinguished areas of the city to develop into memorable form. The section of the report titled "Search for Identity" articulated his intentions and the questions he sought to address: What differentiates the natural landscape and the built environment and what unites them? Can these twin systems peacefully co-exist? Can they adapt and change without losing their best attributes? Can they be deployed to alter the way people value the land and care for it?

McHarg structured his argument to unfold as follows: the given form of the land developed over time in response to processes determined by the biophysical laws of natural systems; the made form of the urban fabric developed over time in response to processes determined by the politics, economics, and social formations of the particular culture in which it exists. To understand the places created by these twin systems of production, and how they overlapped in time and space, it was necessary to return to origins and "begin at the beginning." It was assumed that once this history was revealed and documented, the future would fall into place and people would honor the land and the work it performed, the values it produced, and the opportunities and restraints that it proffered.[52]

The concept of genius loci, defined as "expressive" landscape features that evoke the "unique character" of the city, provided an ancillary framework for the discussion of memorable form. What could be perceived and remembered as orderly progress became the crucial object of this discussion, which again established time and change as key standards of assessment.

The survey of memorable form followed protocols that McHarg and his colleagues established in previous case studies, and it consisted of inventory maps of physiography (floodplains, marshes, and streams), plant communities, principal roads, railroads, existing open space, parks, and institutional land (Figure 11.5). Cross sections supplemented the maps and provided additional information on geology, topography, hydrology, vegetation, and land use. Historical images of the White House and the United States Capitol building dispersed among the geology, soil, and vegetation maps created an ancillary narrative that demonstrated the democratic co-existence of two different types of landscape production, or perhaps it is more accurate to say the arrangement of images dissolved the boundaries that separated "man from nature" by calling attention to the way given form served as the foundation of made form,

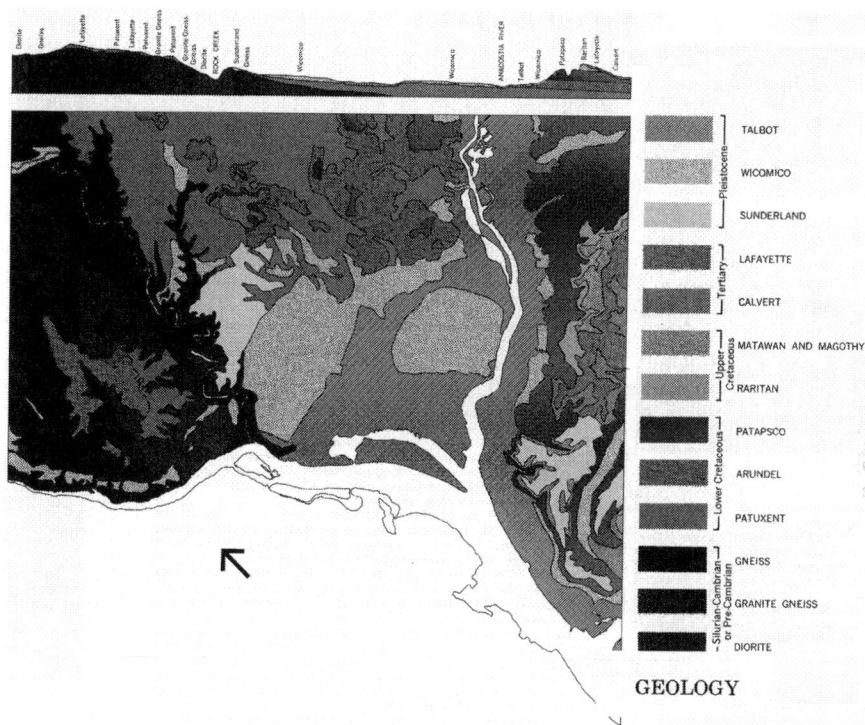

GEOLOGY

Figure 11.5 Geology analysis from *Toward a Comprehensive Landscape Plan for Washington, D.C.* Courtesy WRT Design. Reprinted by permission.

and, conversely, the way made form enhanced the natural attributes of given form (Figure 11.6).

The relationship of soil and vegetation was critical. The preservation of soil diversity preserved plant diversity, and together these physical conditions enhanced the memorable character of the city.[53] The report documented plant communities – defined as the plant associations that historically occurred in a particular place based on soil, water, and solar aspect – in plan to indicate location and extent; in section as ecological transects from low (moist) to high (dry); and through the creation of plant lists that indicated the species that populated the region prior to the founding of the city (Figure 11.7). Following a protocol first used by McHarg in a study of barrier island vegetation, photographs indicated the general appearance of the plant communities along each transect.

The documentation of made form began with the 1792 survey completed by the military surveyors Pierre Charles L'Enfant and Andrew Ellicott for

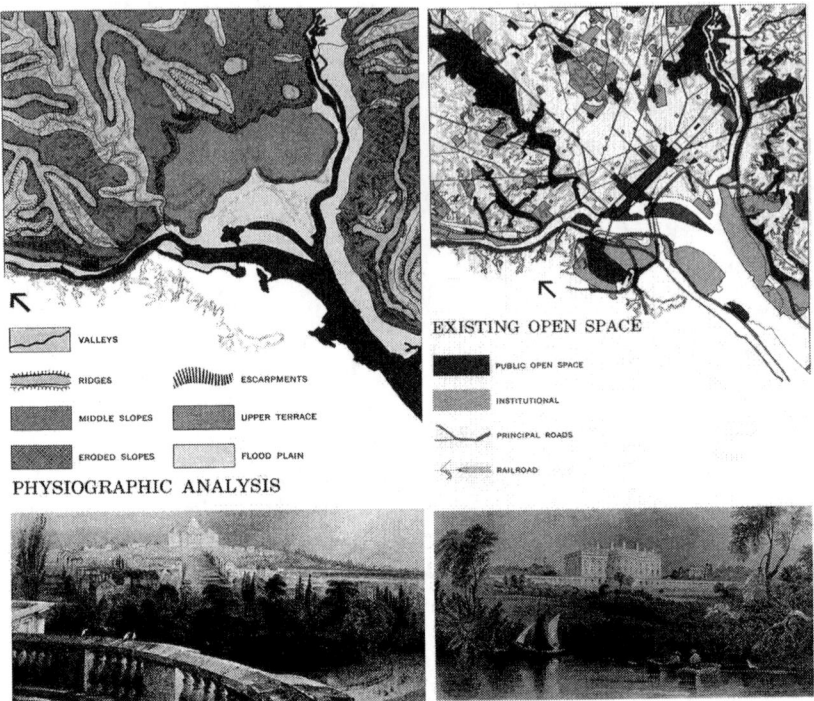

Figure 11.6 Physiography and Open Space analysis with images of the White House and The Capitol from *Toward a Comprehensive Landscape Plan for Washington, D.C.* Courtesy WRT Design. Reprinted by permission.

George Washington. The description was written in such a way that the praise bestowed on the earlier survey also applied to the comprehensive landscape master plan. The work of L'Enfant and Ellicott for example, displayed a keen "awareness of the natural characteristics of the site," and it recorded "all of the eminences, plains, commanding spots" with an eye toward creating a city that "would provide the best and most dramatic contrast between calculated artifice and nature." The report was less enthusiastic about L'Enfant's subsequent design for the city, and, in particular, the plan's "Renaissance" notions and vocabulary of forms. "One might not have selected the concept of the divine right of kings, the image of Louis XIV and Le Nôtre as the most propitious expression for the capital of American democracy," it was stated, "but such it is and it is irrevocable." This less than salutary assessment of historical precedent did not negate the fact that L'Enfant had bestowed "upon the natural site a new symbolism of government," which, in the true spirit of democracy, must be retained.[54] In the same admiring spirit, the 1901 plan for The National Mall prepared by the McMillan Commission received praise for the way it embraced the larger

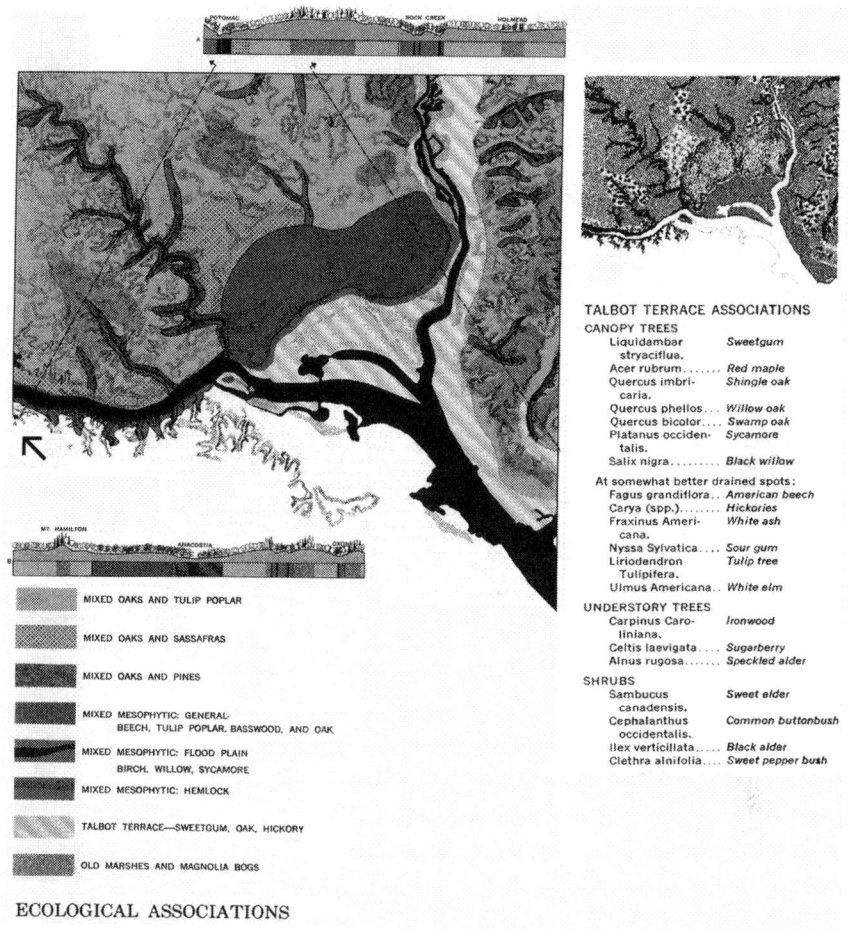

TALBOT TERRACE ASSOCIATIONS

CANOPY TREES

Liquidambar stryaciflua.	*Sweetgum*
Acer rubrum.......	*Red maple*
Quercus imbricaria.	*Shingle oak*
Quercus phellos...	*Willow oak*
Quercus bicolor....	*Swamp oak*
Platanus occidentalis.	*Sycamore*
Salix nigra.........	*Black willow*

At somewhat better drained spots:

Fagus grandiflora..	*American beech*
Carya (spp.)........	*Hickories*
Fraxinus Americana.	*White ash*
Nyssa Sylvatica....	*Sour gum*
Liriodendron Tulipifera.	*Tulip tree*
Ulmus Americana..	*White elm*

UNDERSTORY TREES

Carpinus Caroliniana.	*Ironwood*
Celtis laevigata....	*Sugarberry*
Alnus rugosa.......	*Speckled alder*

SHRUBS

Sambucus canadensis.	*Sweet elder*
Cephalanthus occidentalis.	*Common buttonbush*
Ilex verticillata....	*Black alder*
Clethra alnifolia....	*Sweet pepper bush*

MIXED OAKS AND TULIP POPLAR

MIXED OAKS AND SASSAFRAS

MIXED OAKS AND PINES

MIXED MESOPHYTIC: GENERAL-BEECH, TULIP POPLAR, BASSWOOD, AND OAK

MIXED MESOPHYTIC: FLOOD PLAIN BIRCH, WILLOW, SYCAMORE

MIXED MESOPHYTIC: HEMLOCK

TALBOT TERRACE—SWEETGUM, OAK, HICKORY

OLD MARSHES AND MAGNOLIA BOGS

ECOLOGICAL ASSOCIATIONS

Figure 11.7 Plant community analysis from *Toward a Comprehensive Landscape Plan for Washington, D.C.* Courtesy WRT Design. Reprinted by permission.

than life "neo-classical scale" and "heroic symbolism" of L'Enfant.[55] Here, noble architecture and landscape comprised the form of the city and stately avenues and vistas created linkages to the larger landscape; therefore, the appropriate response simply preserved and enhanced the memorable qualities of the original design. The recommendations for this area of the city, in a literal interpretation of the cultivation mandates of Geddes and Munford, also called for maintenance procedures that enhanced the soil and promoted species diversity to avoid stress and disease. Informal arrangements of plants, rather than rigid lines, were preferred as this conformation reflected the way native species naturally congregate in their indigenous habitats.

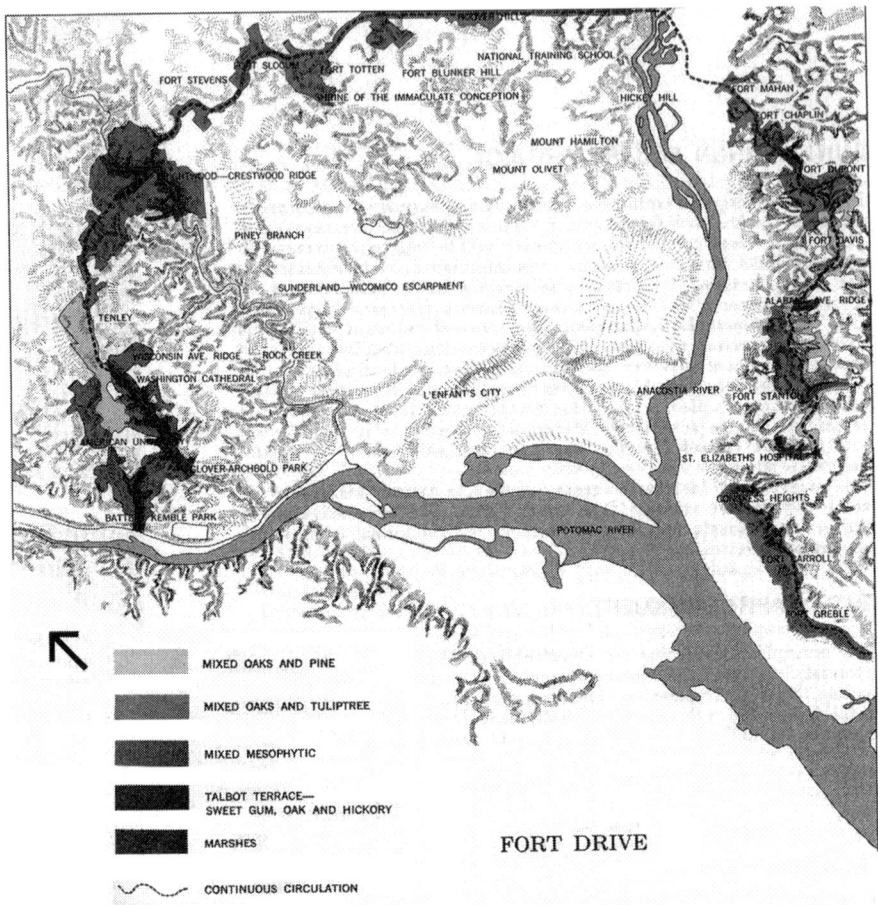

Figure 11.8 Fort River Drive from *Toward a Comprehensive Landscape Plan for Washington, D.C.* Courtesy WRT Design. Reprinted by permission.

The report also championed the MacMillan Commission's recommendation for Fort Drive, a proposed greenbelt parkway that encircled the periphery of the city and connected the Civil War fortifications[56] (Figure 11.8). It was argued that this project, which was located in an area that lacked buildings of distinction, would enhance the visual appearance of the city by connecting the memorable form of the historic Civil War fortifications with the memorable form of the hills, valleys, streams, and natural plant communities of the escarpment. The detailed visual and verbal description of the proposed route, in conjunction with the fact that the federal government supported the project, also suggests that McHarg considered Fort Drive a potential high-profile commission.

One cannot observe the maps, or read the following words from the report without feeling the presence of Hutton, Lyell, Darwin, and Huxley:

> What history of geology, mountain building and erosion, ancient seas and uplifting occurred upon this place? It is these which have given basic form. What is the nature of the underlying rock, its soils, micro-organisms, water, atmosphere, the plants and animals upon it?[57]

During this geologically inspired description of the urban terrain, the ecological relationship of an organism to its environment became an operation of historical recovery and textual analysis that revealed the meaning of the landscape's "to whom it may concern messages." The intent was not to trace the form of the city back to a singular source, but to demonstrate how its current appearance reflected adaptive change over time:

> Process has intrinsic form, form reveals past processes and present reality. They are indivisible. Thus the search for form must begin at the beginning. The place reveals its entire history, everything that has happened is written on the place-its geomorphology, its biomorphology, and its cultural history.[58]

This forensic reading of the land, in sympathy with McHarg's interest in evolutionary theory, quickly moved beyond the confines of the city and became a sweeping panorama of time and life:

> Processes alone explain physical and biological evolution, mountains and oceans, plains and plateaus, uplifting and sinking, erosion and sedimentation. When these are understood, then form and differentiation become comprehensible – ocean and land, Arctic and Equator, mountains and marshes, volcano and iceberg. . . Coastlines, deserts, ice sheets advance and recede, lakes fill while others form, mountains succumb to erosion and others rise. The lake becomes a marsh, the estuary a delta, the prairie a desert, scrub a forest, a volcano creates an island and continues to sink. Plants and animals respond to environmental variety, climate, water, soils, and occupy various habitats, revealing by their presence, their morphology, and their associations the character and diversity of environments.[59]

Nationalist overtones permeate the report's capsule history of biophysical evolution. Echoing Huxley's earlier description of London, Washington, D.C. was a "great city meeting a great river," and it was inevitable, and fitting, that it was the capital of a great nation endowed with physical attributes and social values the rest of the world emulated.[60] In a similar manner, geology, water, soil, plants, and sunlight grounded this empirical, and imperial, identity.

In this entwined paradigm of social progress and natural process, people cannot escape the influence of the world, and the world cannot escape the influence of people. This was the way of things, and the pragmatic approach accepted this condition and worked with it. And yet, the natural checks and balances that governed this vision of order were difficult, if not impossible, to maintain in a site of such symbolic importance. Washington, D. C. had to elevate and inspire, and this meant that design had to preserve, enhance, and exploit the best of what already exists. Fortunately for McHarg, this aspiration aligned with his concept of creative evolution. Healthy, life-affirming environments had to elevate and inspire and this meant that design had to preserve, enhance, and exploit the best of what already exists. Although they did not describe their work in this manner, McHarg and his colleagues were, in essence, functioning as selective breeders intent on making the capital city of the United States a fit expression of American exceptionalism.

Perhaps then it is only fitting that the description of memorable form detailed in this project reconciles the garden traditions that McHarg repeatedly claimed represented extremes of belief and action – the rigid pattern of Versailles and the desire to possess and control (in this case the L'Enfant urban plan), and the organic pattern of Blenheim Palace and a presumption toward natural freedom (in this case the city's wooded valleys and stream corridors). To explain the inherent beauty of this productive coupling, the report referred to L'Enfant's baroque framework of boulevards, vistas, and monuments as cultural gems that encrust the genius loci of bedrock geology, water, and trees with transcendent symbols of the country's highest aspirations. If we return for a moment to the existential dilemma posed by Paul Tillich on *The House We Live In,* this particular commingling of the "given" and "made" resolved the desire to have and the willingness to be, even if this required the acceptance of selective discrimination and more than a little bit of entitlement.

Notes

1 Charles Lyell, *Earth's Surface, by, Part II, Reference to Causes Now in Operation Vol. I* (London: William Clowes, 1830), 1.
2 Geddes, *Cities in Evolution,* 163–167. In his autobiography, McHarg noted that he met the curator of the Geddes papers when he worked in Scotland. He also stated that he studied Geddes and found him fascinating but hard to read. His personal files contain a folder with the Geddes diagrams. See McHarg, *A Quest for Life,* 112; and The Architectural Archives of The University of Pennsylvania, Ian L. McHarg Collection, "Geddes Diagrams," 109.I.B.1.10.
3 See Helen Meller, *Patrick Geddes Social Evolutionist and City Planner* (New York: Routledge, 1990), 38–39.
4 T. H. Huxley, *Physiography: An Introduction to the Study of Nature* (London: MacMillan and Co., 1884).
5 Ibid., 377.
6 See Loren Eiseley, *Darwin's Century: Evolution and the Men Who Discovered It* (New York: Barnes & Noble, 1958, 2009); Stephen Jay Gould, *Time's*

Arrow Time's Cycle: Myth and Metaphor in the Discovery of Geologic Time (Cambridge, MA: Harvard University Press, 1987); and Donald Worster, *Nature's Economy.*

7 Thomas Hutton, *The Theory of the Earth, with Proof and Illustrations Vol. III* (London: Geologic Society, Burlington House, 1788), 11 and 29–30. "The Theory of the Earth by James Hutton," www.gutenberg.org/ebooks/12861.

8 Ibid., 10.

9 Charles Lyell, *Principles of Geology: Being an Attempt to Explain the Former Changes of the Earth's Surface by, Reference to Causes Now in Operation Vol. I and II* (London: William Clowes, 1830).

10 Lyell's hypothesis of steady, incremental change in which the Earth's geologic processes operated in the same manner and same intensity in the past as they do in the present is known as uniformitarianism. This theory countered the notion that catastrophic events, such as the biblical flood caused geologic change. See *Encyclopaedia Britannica*, "Charles Lyell and uniformitarianism," www.britannica.com/science/Earth-sciences/William-Smith-and-faunal-succession#ref292518.

11 Lyell, *Principles of Geology*, 1.

12 Ibid. Lyell stated this idea as follows: "When we study history, we obtain a more profound insight into human nature, by instituting a comparison between the present and former states of society. We trace the long series of events which have gradually led to the actual posture of affairs; and by connecting effects with their causes, we are enabled to classify and retain in the memory a multitude of complicated relations – the various peculiarities of national character – the different degrees of moral and intellectual refinement, and numerous other circumstances, which, without historical associations, would be uninteresting or imperfectly understood." See also Michel Bourdeau, "Auguste Comte," *The Stanford Encyclopedia of Philosophy*, ed. Edward N. Zalta (Summer 2018 Edition), URL = https://plato.stanford.edu/archives/sum2018/entries/comte/.

13 John van Wyhe, ed., "Catalogue of the *Beagle* Library" in *The Complete Work of Charles Darwin Online*, 2000, Darwin Online http://darwin-online.org.uk/BeagleLibrary/Beagle_Library_Catalogue.htm.

14 See Kevin Padian, "From Geologizer to Theorist" *Bioscience* 68 no. 12 (December 2018), 1020–1021.

15 Charles Darwin, *The Origin of Species by Means of Natural Selection or The Preservation of favored Races in the Struggle for Life* (New York: Collier Books, 1970); and The National Academies of Sciences Engineering and Medicine Evolution Resources, "Definitions of Evolutionary Terms," www.nas.edu/evolution/Definitions.html.

16 See Herbert Spencer, *The Principles of Biology Vol. I* (London: William and Norgate, 1864).

17 Herbert Spencer, "The Proper Sphere of Government, Letter IX, Print of a Series of Letters," Originally Published in *The NonConformist* (London: John Green, 1843), 34–35, https://play.google.com/store/books/details?id=UKlXAAAAcAAJ&rdid=book-UKlXAAAAcAAJ&rdot=1.

18 Edinburgh University Library Special Collections, The Papers of Thomas Henry Huxley (1825–1895), https://archiveshub.jisc.ac.uk/data/gb237-coll-376. See also Cyril Bibby, "Huxley and the Reception of the 'Origin" *Victorian Studies* 3 no. 1 Darwin Anniversary Issue (September, 1959): 76–86.

19 T. H. Huxley, "Science and Culture" in *Science and Education*, ed. T. H. Huxley (New York: D. Appelton and Company, 1898), 150. In the context of this essay, it is important to note that Thomas Huxley was the grandfather of Julian Huxley.

20 Huxley, *Physiography*, vii.

21 See Meller, *Patrick Geddes Social Evolutionist and City Planner*, 38–39.

22 Geddes, *Cities in Evolution*, xviii, xxiii.

23 See William James, "Great Men Great Thoughts, and the Environment" *Atlantic Monthly* 66 (1880): 441–459; and Tim Lewens, "Cultural Evolution" in *The Stanford Encyclopedia of Philosophy* ed. Edward N. Zalta, (Summer 2018 Edition), https://plato.stanford.edu/archives/sum2018/entries/evolution-cultural/. Lewens notes that James sought to uncouple the notion of natural superiority from competitive struggle by arguing that the inherent quality of the environment could help those less well favored.

24 Lewis Mumford, "The Natural History of Urbanization" in *Man's Role in Changing the Face of the Earth*, ed. William L. Thomas, Jr. (Chicago, IL: University of Chicago Press, 1956), 387–398.

25 Lewis Mumford, "Summary Remarks: Prospect" in *Man's Role in Changing the Face of the Earth*, ed. William L. Thomas, Jr. (Chicago, IL: University of Chicago Press, 1956), 1143–1149.

26 See Lewis Mumford, "The Voice of the Village" and "Town Hall and Market Place" in *The City in History: Its Origins, its Transformations, and its Prospects* (New York: Harcourt Brace Jovanovich, Publishers, 1961), 126–133, 148–157; and Mumford, *The Golden Day: A Study in American Experience and Culture.*

27 Mumford, "The Natural History of Urbanization," 310–315.

28 Henderson, *The Fitness of the Environment*, 1.

29 Lawrence J. Henderson, "The Fitness of the Environment, an Inquiry into the Biological Significance of the Properties of Matter" *The American Naturalist* 47 no. 554 (February, 1913), 105–115. See also Lawrence J. Henderson, *The Order of Nature* (Cambridge, MA: Harvard University Press, 1917).

30 Benton MacKaye, *The New Exploration: A Philosophy of Regional Planning* (Urbana-Champaign: The University of Illinois Press, 1928, 1990), xviii. Mumford refers to the meeting in the introduction he wrote for this text. See also Thomas P. Hughes and Agatha C. Hughes, *Lewis Mumford: Public Intellectual* (New York: Oxford University Press, 1990).

31 Benton MacKaye, "The New Exploration: Charting the Industrial Wilderness," 155, 156.

32 MacKaye, *The New Exploration: A Philosophy of Regional Planning*, 4.

33 Benton MacKaye, "Regional Planning and Ecology" *Ecological Monographs* 10 no. 3 (July, 1940): 349–353.

34 The 1959 syllabus of "Man and Environment" lists MacKaye in a lecture titled "An Ethic for Man and Environment." There are no transcripts or notes of the lecture. See G. Holmes Perkins Papers, "1959 Syllabus Man and Environment."

35 See McHarg, *Design with Nature*, "The World is a Capsule," 94–101, "The Naturalist," 116–125, and "Form and Process," 162–173.

36 See McHarg, *Design with Nature*, 168.

37 C. H. Waddington, "The Modular Principle and Biologic Form," 20–37. Waddington based his discussion of biological form on the concept of epigenetics (a term derived from the combination of epigenesis and genetics) that he had formulated in 1957. His theory of genetic plasticity drew a distinction between the genotype (the genetic potential inherited from parents) and the phenotype (the appearance of an individual) to explain how the environment impacted the growth and physical characteristics of organisms. The terrain of his developmental topography consisted of alternative decision pathways, which he depicted as a valley with bifurcations. For Waddington, biological form was not preordained or written in stone, but was instead, as he observed in *The Strategy of Genes*, a dynamic process analogous to the flow of a river. In this particular essay, Waddington used the *Venus and Amor* by the Renaissance painter Lucas Cranach, a female figure from a Buddhist temple, the modular of Le Corbusier, a coral skeleton, seashells, bird beaks, an x-ray of the human hand,

the proportions of the human body as they changed from child to adulthood, and the evolution of a horse's hoof to document the characteristics of organic development. See also C. H. Waddington, *The Strategy of the Genes* (London: Allen and Unwin, 1957).

38 McHarg, *Design with Nature*, 120. The idea of the Naturalist may relate to an essay by Paul Sears that appeared in the *American Naturalist*. In this essay, Sears equated natural history to the observation of the heavens, the earth and living organisms, and he noted that Darwin was a "naturalist" who developed an integrated picture of living organisms by combining empirical observations of the environment with philosophical reflections on what he observed. Sears also stated that a similar synthetic approach was evident in the work of H. G. Wells, Patrick Geddes, Lewis Mumford, Henry Wallace, and Lawrence Henderson. Once these individuals were incorporated into the conventional lineage of ecology, and its study of material and energy cycles, he claimed this science would become "more luminous" as it moved from Alexander Humboldt to the plant geographers and taxonomists. McHarg's discussion of the naturalist also contained unacknowledged references to Robert MacArthur, Aldo Leopold, Clarence Morris, Lawrence K. Frank, and Hans Selye. See Paul B. Sears, "The Future of the Naturalist," *American Naturalist* 78 no. 774 (January-February 1944): 43–53.

39 Ibid., 123.

40 Morris discussed these ideas in a lecture delivered in the class Man and Environment. See The Architectural Archives of the University of Pennsylvania, Ian L. McHarg Collection, "Clarence Morris," 109.II.E.2.44; The Architectural Archives of the University of Pennsylvania, Ian L. McHarg Collection, "Clarence Morris Law and the Environment," 109.IIE.2.45; and G. Holmes Perkins Papers, "1959 Syllabus Man and Environment." See also Clarence Morris, "The Rights and Duties of Beasts and Trees: A Law Teacher's Essay for Landscape Architects" *Journal Legal Education* 17 no. 185 (1964), 185–192.

41 McHarg, *Design with Nature*, 197.

42 Ibid., 119–120.

43 McHarg, "Ecology of the City," 53–65. The photograph of an old growth forest in *Design with Nature* is attributed to Grant Heilman photography. The firm was unable to find the photograph used by McHarg, however, they did find this image in their files that closely corresponds to the one used by McHarg.

44 Teilhard de Chardin, *The Phenomenon of Man*, and Teilhard de Chardin, *Man's Place in Nature*.

45 McHarg, *Design with Nature*, 170.

46 Ibid., 173.

47 The Architectural Archives of the University of Pennsylvania, Ian L. McHarg Collection, "Morse Peckham, The Garden as Metaphysical Symbol," 109. II.E.2.55. This lecture occurred in 1960.

48 *Beauty for America*, 7.

49 See The Architectural Archives of the University of Pennsylvania, Ian L. McHarg Collection, Ian L. McHarg Collection, "Towards a Comprehensive Landscape Master Plan for Washington, D.C. (1966)," 109.III.C/22.1; and Wallace, McHarg, Roberts, and Todd, *Towards a Comprehensive Landscape Plan for Washington, D.C.* (Philadelphia, PA: WMRT, 1967).

50 McHarg, *Design with Nature*, 174–185; and The Architectural Archives of the University of Pennsylvania, Ian L. McHarg Collection, "McHarg Request for Sabbatical (1966–1967)," 109.II.E.1.19.

51 "Wallace, McHarg, Roberts, and Todd, *Towards a Comprehensive Landscape Plan for Washington, D.C.*, 3.

52 Ibid., 4.

53 The vegetation analysis reflected the input of the plant ecologist Jack McCormick. McCormick, who began his career at The American Museum of Natural History in New York prior to moving to The Franklin Institute in Philadelphia, specialized in plant inventories using aerial images, United States geologic quadrangle surveys, soil surveys, historical maps and reports, and field observations. McCormick also advanced McHarg's understanding of primary and secondary succession, and the concept of polyclimax. See: Jack McCormack, "Succession" in *Via 1, Ecology and Design*, eds. Sauer, *et al.* eds. (New York: Grossman Publishers, Inc., 1968), 22–24; and Jack McCormick, *The Pine Barrens: A Preliminary Ecological Inventory* (Trenton: New Jersey State Museum, 1971).

54 "Wallace, McHarg, Roberts, and Todd, *Towards a Comprehensive Landscape Plan for Washington, D.C.*, 19, 22.

55 For further discussion of the MacMillan Commission see Smithsonian Institution, "South Mall Campus Cultural Landscape Report (Washington, DC: Smithsonian Institution, June 2018).

56 See National Park Service: "Civil War Defenses of Washington" www.nps. gov/cwdw/planyourvisit/upload/Civial%20War%20defenses%20of%20 Washington_final.pdf; and Timothy Davis, "Rock Creek and Potomac Parkway, Washington, D.C.: The Evolution of a Contested Urban Landscape" *Studies in the History of Garden & Designed Landscapes* 19 no. 2 (1999), 123–237.

57 Wallace, McHarg, Roberts, and Todd, *Towards a Comprehensive Landscape Plan for Washington, D.C.*

58 Ibid., 2.

59 Ibid., 4.

60 Ibid., 25.

Part 4
The Patterns of paradise

12 Pardisan

Here in this carpet lives an ever-lovely spring;
Unscorched by summer's ardent flame,
Safe too from autumn's boisterous gales,
Mid-winter's cruel ice and snow
'Tis gaily blooming still.

The handsome wider border is the garden wall
Protecting, preserving the Park within
For refuge and renewal: a magic space
For concourse, music, and rejoicing,
For contemplation's lovely spell –
Conversations grave, or lover's shy disclosure.

Eyes hot – seared by desert glare find healing
In its velvet shade. Splashing fountains and rippling pools
In cool retreats sore-wearied limbs restore,
And tired hearts awake with joy once more.
The way was cruel.
Baffled by monotony and mocked by phantoms delirious,

Beset by stalking Death in guises manifold;
The dreaded jinns, the beasts ferocious,
The flaming heat and exploring storms;
From all these perils here at last set free:
In the garden all find security.

Here the long-laboring Earth at last gives birth.
From apparent death, a new and lovely world is born:
Below the desert's dusty floor, the jacinth imprisoned lies.
The stony wilderness so bleak and bare,
In ageless patience broods, aware of life within
The promise of fertility and abundance.
Ever longing for deliverance
In spangled meadows radiant
And valleys with thick verdure clad,

Sired by quiet waters flowing beneath,
Or pouring from the reservoirs of Heaven,
The world at last reveals its destiny.

Man but briefly in earthly gardens dwells;
His pilgrim way he henceforth must resume
Until at last if faithful to his Lord,
The fadeless Garden of Paradise will open wide to him.

　　　　　　Unknown Sufi poet (circa 1500), Ode to a Garden Carpet[1]

In June of 1972, the delegates attending the Stockholm Conference idealistically implemented a set of principles designed to "inspire and guide the peoples of the world in the preservation and enhancement of the human environment."[2] The meeting, sponsored by the United Nations and commonly known as the "Only One Earth" Conference (in reference to the iconic photograph of earth taken by the Apollo 11 astronauts), promoted international cooperation as a means to foster environmentally responsible urbanization.[3] Ecology and social justice, idealistically drove these discussions as the conference participants attempted to amicably resolve the "intractable conflict between environmental conservation and economic development, between global needs and national rights, between wealthy and poor countries."[4] To reconcile internal disagreements over the institution of these policies, the delegates pragmatically recognized the jurisdictions and the varying agendas of each country, and they concluded, in modest but optimistic language, that the global enhancement of the environment had to begin at the local level.

Eskandar Firouz, undersecretary for Iran's Ministry of Agriculture and vice chair of the Stockholm Conference, led the Iranian delegation.[5] Inspired by the forward-thinking tone of the conference and its unifying global perspective, Firouz envisioned Iran as a world leader in the science of sustainable development. His subsequent actions reveal that he drew particular inspiration from the themes of ecology, preservation, education, and informed decision-making enumerated in the conference recommendations.[6]

A member of a powerful and distinguished family with members strategically situated in prominent positions throughout the Iranian government, Firouz began his environmental activism in 1956 under the patronage of Mohammed Reza Pahlavi, the Shah of Iran, and his brother Prince Abdorreza when he became a member of the Game Council of Iran. This confederation of individuals from the country's elite families instituted and sold game licenses, established hunting seasons, and designated endangered species.[7] In 1967, when the government transferred the responsibilities of the Game Council to the newly formed Game and Fisheries Department of Iran, Firouz assumed leadership of the agency and ambitiously began to position his initiatives within a larger global perspective. One of his first actions involved the presentation of his country's conservation efforts at a scientific conference in Morges, Switzerland sponsored by the International Union for Conservation of Nature. In line with the concept of land management advanced by this organization, Firouz outlined a series of conservation policies that combined care for the rural poor with the protection of endangered animal species.[8] In many ways, this approach, which promoted the continuation of traditional hunting and farming, simply codified the practices that guided his efforts at the Game Council, as seen for instance, in the aristocratic obligation of responsibility toward the less fortunate, and the pragmatic belief that it was smart policy to protect the resources that advanced one's livelihood. Similar actions guided conservation efforts in Europe and the United States, however, for Firouz the approach was personal and tied to the historical legacy of his family.[9]

The Morges meeting set the stage for a conference held in the Iranian city of Ramsar in February of 1971. The meeting, which was co-chaired by Firouz, resulted in the Convention on Wetlands of International Importance (Ramsar) and the establishment of global standards for the conservation of wetlands and aquatic game birds. An address by Prince Abdorreza, in conjunction with the decision by the Iranian government to place in trust a wetland of global significance, added legitimacy to the proceedings. Firouz's skillful diplomacy with a recalcitrant Russian delegate is universally credited for the successful outcome of the conference.[10] One month later the Department of Game and Fisheries became the Department of Environmental Conservation.[11] The next year, as previously noted, Firouz led Iran's delegation to the Stockholm Conference.

Although it took years for Firouz to obtain the practical knowledge and international contacts that he needed to advance his environmental agenda, once he acquired this platform he acted quickly and decisively to protect the landscape of his rapidly modernizing country. His efforts included the establishment of national parks and land preserves.[12] His crowning vision – the act that would symbolize his ideals and instill a conservation ethic in his fellow countrymen – was the environmental park that he conceived during the Stockholm Conference. The park would embrace "all known aspects of the universe and the world. Ranging from its beginning; to the evolution of life; to man and his use and abuse of nature; to a future wherein he must find harmony with his world."[13] Venues would display the natural history of the earth and the cultural history of civilization, and they would present this narrative of "man and nature" as the ultimate synthesis of Eastern cultural tradition and Western technological acumen. To honor this conceptual unification, Firouz named the park Pardisan – an amalgamated concept of paradise derived from the cognate of the Greek *paradeisos* and the Arabic term for *many*.[14] In his mind, Pardisan would be a scientific and culturally resonant embodiment of the ancient Persian royal garden – a place to enjoy "all good things the earth provides."[15] The presence of wild and domesticated animals was both implicit to this concept, and assumed at the outset (Figure 12.1).

Figure 12.1 The Pardisan feasibility report cover (top) and the Pardisan master plan location map (bottom). Ian L. McHarg Papers and Ian and Carol McHarg Collection, The Architectural Archives of the University of Pennsylvania.

In the summer of 1972, shortly after the conclusion of the Stockholm Conference, Firouz approached the architect Jahangir Sedaghatfar, an expatriate Iranian living in the United States, to help implement his environmental park. Sedaghatfar was in Tehran to deliver talks on the urbanization of Iran to audiences from the Ministry of Science and Higher Education, the Ministry of Housing and Development, Tehran University, and the Shiraz (Pahlavi) University.[16] His remarks reflected material that he recently self-published in the book *Entopia*. As the title of this text suggests, he presented a buildable modernization plan for the city inspired by the work of the urban planner Constantinos Doxiadis.[17] Overcrowding and uncontrolled urban growth were major concerns, which he addressed through the construction of satellite communities that contained essential infrastructure, light industry, social facilities, single-family housing, and parks. The loss of Tehran's architectural heritage, symbolized by the demolition of the city's 19th-century walls and entry gates during the reign of Reza Shah Pahlavi provided historical background and indicated the need for equally decisive, albeit less destructive action.

Sedaghatfar's vision of a prosperous future combined with his respect for architectural history indicates why Firouz would seek his participation. An even more important aspect of their relationship, and this is something that gets to the heart of Pardisan and what it represents, is the fact that their complimentary visions of modernization reflected two important segments of Iranian society in the mid-20th century – a ruling elite, educated in the West, which, in allegiance with the Pahlavi regime had retained a tenacious grip on power, and a meritocratic middle class, also educated in the West, anxious to achieve the lifestyle made possible by industrial modernization. The sweeping scope of the environmental agenda that Firouz envisioned for Pardisan united these aspirations and differing conceptions of patrimony and progress.[18]

In his role as the deputized agent for Firouz, Sedaghatfar worked independently (and initially without pay) to select the lead design consultant for the project. He compiled a list of potential candidates that included Morris Ketchum, Van der Pool, and Skidmore Owings and Merrill. Each of these architecture firms recently designed and built major zoo facilities. The list also included Ian McHarg, who came to Sedaghatfar's attention through *Design with Nature*. He would later note that the logical intricacy of the text's planning method, the immaculate clarity of its environmental message, the hint of the fantastic in its futuristic vision, and the folk-hero status of its author were factors in his decision to place McHarg on the candidate list.[19]

At the conclusion of this extensive interview process Sedaghatfar presented his findings to Firouz. The meeting occurred in a hotel room in Chicago and McHarg was the top candidate on the list.[20] Firouz approved his recommendation and subsequently traveled to Philadelphia to meet McHarg. McHarg, in turn, traveled to Iran to learn more about the proposal.[21] Sedaghatfar, however, declined an offer to manage the project. He was convinced that Pardisan, although an intriguing idea, was a utopian dream outside the realm of possibility.[22]

McHarg, as subsequently reflected in his vigorous defense of the project, relished the opportunity to create an environmental park that was modern in its application of science, just in its actions toward people, and ethical in its behavior toward the land. He had already demonstrated the agency of this approach in a diverse array of projects in the United States, and he had received critical acclaim for his efforts. Pardisan situated this work within a global context, as a progressive emblem of post-colonial modernization and environmentalism. Equally important, the project provided the opportunity to create a contemplative garden retreat – something he explored in theory, but never materialized, particularly at such a grand scale. Moreover, he had a client inspired by the same ideas and culturally predisposed to help. If the project developed as intended, it would be his masterpiece.

The garden design that McHarg would subsequently seek to implement at Pardisan was deeply entwined with science and religion. As previously discussed, this approach closely corresponds to lectures presented in his seminar course Man and Environment. Robert MacArthur, for example, provided commentary on territorial dynamics, adaptation, and evolution, and he positioned these scientific concepts within a complex network of message signaling, food webs, energy dynamics, and territorial partitioning.[23] Paul Sears taught him the value of traditional stewardship practices and the need for a land ethic grounded in ecology.[24] Morse Peckham tutored McHarg in the symbolic importance of gardens in Western humanist thought.[25] Using the Genesis story of Adam and Eve and their fall from divine grace to frame his discussion, Peckham conceived of gardens as allegories of loss and recovery that dialectically resolved good and evil, man and woman, and thinking and feeling, which, he claimed, made their design a problem of relating an idealized vision of the world and human action to the less-than-ideal realities of life. Gardens, he further noted, mediated this tension through the construction of walls. The world was chaotic and rife with evil intentions, but within these protected and protective spaces the reasoned beauty of Eden resided in all of its glory and it was possible to contemplate human existence without shame, fear, or temptation. Equally memorable for McHarg, was Peckham's description of the miraculous transformation of thought in 18th-century England, when the enclosing walls of tradition, abetted by science and great wealth, crumbled and the whole world became a garden. Entranced by the sublimity of this harmonious union of science, religion, and art, McHarg would imagine a design where it was possible to stand at the summit of the garden, surrounded by a lush natural landscape, and look out over the city of Tehran and know the world to be a place of utter beauty, where the one truly divine thing was reality itself.

Arthur Upham Pope, a historian of Persian art and advisor to the Iranian government, provided an equally important lecture in Man and Environment that grounded the metaphysics of gardens in the unique qualities of the landscape and the guardian spirit of the genius loci.[26] When he spoke to the class in 1959, Pope positioned the Middle Eastern garden in relation to the challenges of intense heat and precarious rainfall where survival required the development of special techniques of living that included underground irrigation systems, storage tanks, deep wells, and migrating populations of nomadic herders who moved their flocks in response to the seasonal availability of water. Pope observed that no matter how harsh and cruel the land, the presence of water, as seen for instance in the ephemeral allure of a lush spring meadow, revealed the hidden, life-enhancing potential of the natural world. When this enticing beauty became a garden, he stated, it was possible to arrange all manner of things in order and loveliness to reveal the inner nature of the land – the sacred cosmological tree, flowers, flowing streams, and singing birds. Pope also explained how the consolidation of political power in ancient Persia transformed these garden oases into vast hunting parks that took years and thousands of workers to build. Once complete, these "artificial kingdoms" fulfilled a heavenly mandate to protect the world from the consuming power of the sun, as eloquently expressed in the rapturous verse of the Sufi poem "Ode to the Garden Carpet" that Pope gave to McHarg. At the end of his talk, Pope related the poem's mystical delights to design:

> Hence the gardener and architect who with poetic feeling, with rationality and imagination plans and nurtures these beneficent creations is as it were, an angelic midwife, bringing to birth and full life the inherent potentials of the earth, aiding man in his long and perilous journey to perfection, giving him surcease, composure, delight, peace and great hope. Is there a nobler profession?[27]

Given his deep reverence of nature, messianic proclivities, and driving ambition, it is fair to suggest that McHarg had no trouble picturing the profession of landscape architecture as a servant of God tasked with the heavenly mandate to bring forth the hidden potential of the earth and provide solace to the world weary. McHarg literally stated this belief in a talk delivered to The Royal Society of Art in London, titled "The Garden as Metaphysical Symbol." During this address, which borrowed heavily from Peckham, he provided an anglicized interpretation of the "creation story of Koran" and the injunction by Mohammed "to dress the garden and keep it." Design, as an act of benevolence that flawlessly mirrored this transcendent aspiration, correspondingly became a "literal plan for godly man to create a paradise garden on earth" that was gentle, abundant, ornate, and inherently divine. When these design principles were properly instituted: "The water welled from the ground and parted into four ways – the four rivers of paradise. In this garden grew all beautiful flowers and godly fruit. There were eight pearl pavilions awaiting the true believer in Paradise."[28]

Feasibility study

In the fall of 1972, WMRT negotiated a contract with the Iranian Department of Environmental Conservation for an initial feasibility study, with McHarg as partner-in-charge, William Roberts as managing partner, Colin Franklin as project captain, and Robison Fisher as design associate. Roberts, Franklin, and Fisher traveled to Iran on a fact-finding mission and they recorded their observations in photographs, slides, and sketches.[29]

The site for the proposed study, a rugged 300-hectare parcel of land northwest of Tehran, had magnificent views of Mount Damavand, Iran's highest mountain. The parcel was part of a larger 1,100-hectare tract under development by Bank Omran, the financial arm of the Pahlavi Foundation.[30] A recently completed *Comprehensive Plan for Tehran*, prepared by Victor Gruen Associates and Firouz's uncle, the architect Abdol Aziz Farman-Farmaian, proposed the conservation and reforestation of land in this section of the city as part of a larger urban greenbelt.[31] The site was dry, largely devoid of vegetation, and bisected by a series of rugged north-south ravines, which made it a less than ideal location for a garden paradise containing all good things the earth provides (Figure 12.2).

A "Descriptive Report" written by Firouz established the planning framework for the feasibility study. First, and foremost, the park would promote a "creative and inseparable" link between people and their surroundings, and it would do so by providing a thorough "familiarity with all elements of life, including the soil, water and all living things, from the basest to the most noble of creatures." He also called for the construction of a zoo, botanic garden, natural history museum, planetarium, and aquarium. Displays in which plants, animals, and people were "exhibited in close proximity to each other as they exist in nature," supplemented this scientific "ecosystem."[32]

Figure 12.2 Photograph of the Pardisan site (circa 1977). Courtesy Dennis Paulson. Reproduced by permission.

A key objective was education. Firouz hoped that park visitors would link what they observed to their personal lifestyles and modify their behavior accordingly. Such lessons, he noted, following the guidelines of the Stockholm Conference, would demonstrate that the "progress and expansion of human communities does not inevitably lead to the destruction of nature."[33] Research facilities devoted to environmental protection, pollution mitigation, and the preservation of endangered species supported this mandate.

Another objective was the celebration of ancient Persian culture, which Firouz claimed was consonant with the principles enumerated at Stockholm because it "conceived of the earth as mother, nature as sacred, and pollution of the environment as sin."[34] This line of thinking led him to also state the park's conception of stewardship would begin at the local level and its vantage point would be Iranian. The design would dramatize the close relationship between his country's multiethnic culture, its religious and philosophical traditions, its art and architecture, and its rich and diverse landscape.

Brian Spooner, the University of Pennsylvania cultural anthropologist who compiled the ethnographic information for the project, identified five possible themes for cultural exhibits in the park: Irrigation and Agriculture, Pastoralism, Housing, Daily Life (videos, not reenactments), and Crafts.[35] Firouz appeared most interested in nomadic pastoralism, which he praised for its self-sufficient production of food (milk, yogurt, and meat), wool (carpets), and hides (tents and saddles) provided by herds of sheep and goats. After noting similarities with the material culture of Native Americans – and arguing (somewhat in contradiction to his earlier references to Stockholm) that Western society no longer valued sustainable traditions – he observed that the nomadic pastoralist did not squander the "resources of nature on whose recurring cycle his very life depended."[36] To promote a similar ethos in park visitors, and prompt them toward more sustainable lifestyles, he called for instructional displays that explained how to balance personal need and resource availability. This approach was both progressive and conservative. As later noted by McHarg, Firouz honored tradition but he did not seek to nostalgically recreate the past. Through his work at the Game Council and the Department of Environmental Conservation, he understood that wasteful and antiquated animal husbandry and farming practices fostered a host of environmental problems in Iran that hampered progress. Pardisan would address these issues through science and education, which, in conjunction with the strategic implemental of environmental policies, would position tradition within a broader modernization and conservation mandate.[37]

The politically savvy Firouz was equally astute in the selection of animals displayed in the Pardisan zoo. The lion, long emblematic of the Persian and Iranian monarchies, but recently hunted to extinction in Iran, would take precedence. Following the wish of the Shah, the lion display at Pardisan would support the recent conservation efforts at the Dasht Arjan Nature Preserve to reintroduce the lion into the wild.[38]

Many of the recommendations outlined by Firouz echoed themes that appear in *Design with Nature*. These themes included a search for cosmic order; a measured balance of development and resource use; a deeper connection with nature made possible by a planning approach that reintegrated science and culture; and, as just discussed, a belief that Middle Eastern garden traditions were fundamentally ecological. Pardisan, in short, was the perfect collaboration for a shared conviction that it was possible to create a better world, as symbolized by the iconic photograph of the earth from space.

An ideogram of the park layout prepared by WMRT adroitly included ideas promoted by Firouz and the Iranian government (Figure 12.3). The scientific venues enumerated in the "Descriptive Report," for example, were reconstituted as the Universal Park (planetarium), the Terrestrial Park (zoo and botanic garden), and the Aquatic Park (aquarium). The ideogram distributed these venues and their associated activities along a central spine labeled the Iranian Bazaar.[39] Perspective sketches of the Iranian Bazaar and the Terrestrial Park by Franklin and Roberts – replete with natural history dioramas, restaurants serving traditional food, shops selling traditional crafts, and animals frolicking in natural habitats – mixed cultural tourism, science, and the consumer-oriented social vision of the officially sanctioned *Comprehensive Plan for Tehran* (Figure 12.4). The *Comprehensive Plan*'s commercial aesthetic, as well as its requirement that future development accommodate the automobile, is similarly evident in the four large parking lots that flank the entrance.[40] Once in the park, visitors could travel via monorail, minibus, electric cart, or on foot – a dream of modern transportation that owes more to the American shopping mall, Disneyland,

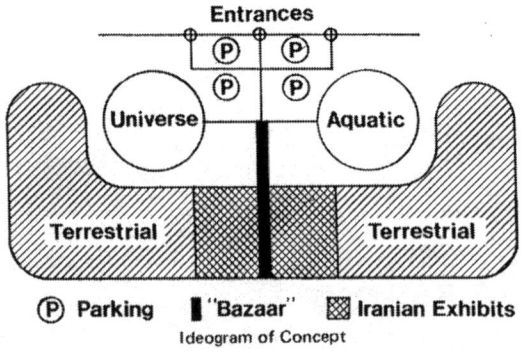

Figure 12.3 Ideogram from the Pardisan feasibility study. Ian and Carol McHarg Collection, The Architectural Archives of the University of Pennsylvania.

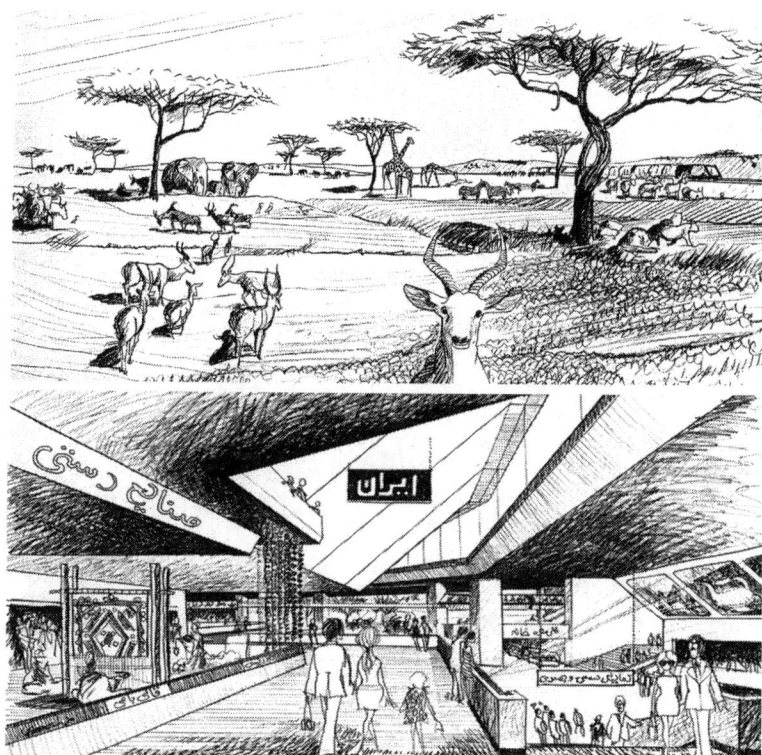

Figure 12.4 The Terrestrial Park by William Roberts and the Iranian Bazaar by Colin Franklin from the Pardisan feasibility study. Ian and Carol McHarg Collection, The Architectural Archives of the University of Pennsylvania.

and the separation of circulation in New York City's Central Park than to indigenous lifestyles.[41]

WMRT also called on the science of ecology to support Firouz's wish that Pardisan represent "the cosmos and the world" from an Iranian perspective.[42] This recommendation is primarily seen in the suggestion to configure the Terrestrial Park to replicate an ecological transect through the middle of Iran. The imaginary line of the transect extended from north to south and contained six bioclimatic (geomorphologic) zones: the Caspian Sea, Broadleaf Forest, Conifer and Scrub, Herbaceous Steppe, Desert, and the Persian Gulf. Comparable biomes from Asia, Africa, and Oceania extended to the right (east), and biomes from Europe and North and South America extended to the left (west).[43] To emphasize the interplay of the land and human settlement patterns, WMRT, in a now routine procedure, created a topographic section of the ecological transect, which they annotated with information on vegetation, physiography, zoology, and culture. The resulting diagram contained five topographic sections and five overlapping narratives of the Iranian landscape (Figure 12.5). The ecologist and former staff scientist at the Museum of Natural History in New York City, Jack McCormick, and the zoologist and director of the National Zoo in Washington, D.C., Theodore Reed, provided annotated plant and animal lists for the proposed biomes based upon Tehran's climate and the site's solar aspect and terrain.[44]

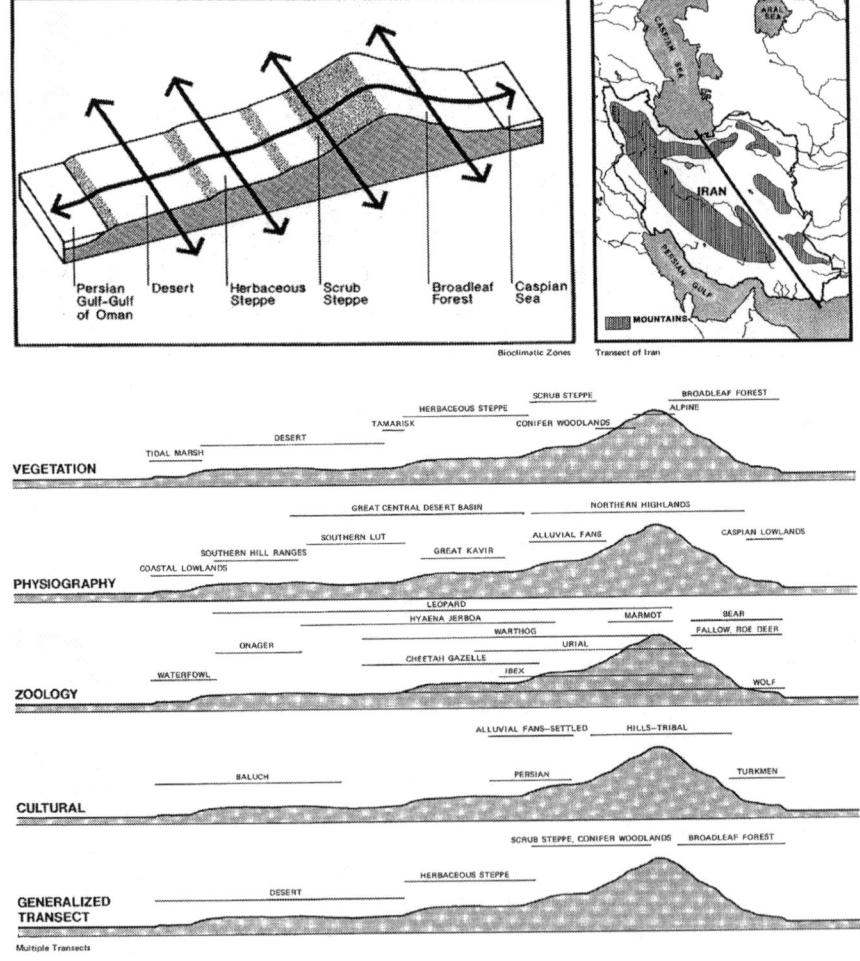

Figure 12.5 Transect diagrams from the Pardisan feasibility study. Ian and Carol McHarg Collection, The Architectural Archives of the University of Pennsylvania.

Needless to say, the attempt to recreate all of the biomes of the world, beginning with Iran, was extraordinarily ambitious. But one of the most striking characteristics of the proposal was the fact that it was as much a flowchart as it was a formal design. This organizational system is notably illustrated in a diagram that conceived the Iranian transect and its global analogues as a series of boxes interconnected by circulation pathways that begin and end with Iran (Figure 12.6). It was possible, it seems, based on the principles of system thinking, to represent Paradise as a matrix connected by iterative circular linkages. The inherent beauty of the diagram involved the way it easily encoded different meanings. At the most basic level, it reflected the organized complexity and territorial dynamics of the natural landscape. On another deeper level, it symbolized an organic cosmology in

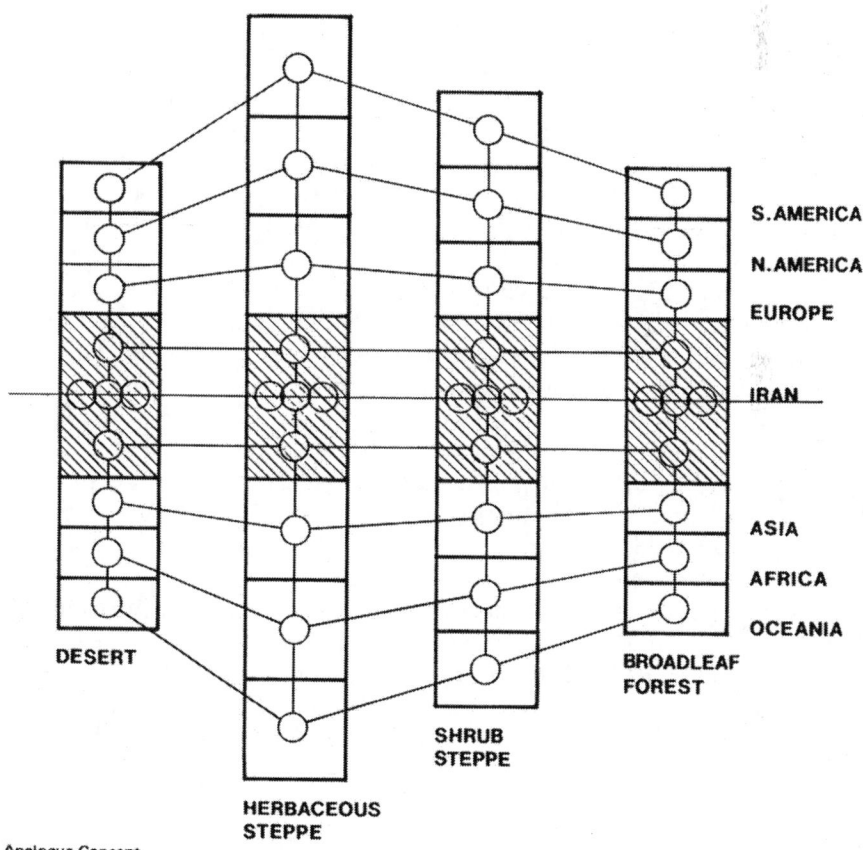

Figure 12.6 Analogue matrix from the Pardisan feasibility study. Ian and Carol McHarg Collection, The Architectural Archives of the University of Pennsylvania.

which an innate order radiates outward and diversifies, but the essence of things always relates back to its origin. To put it another way, the diagram captured the essence of McHarg's evolutionary vision of adaptive change and the essence of Firouz's historical vision of cultural agency, and it turned these thoughts into a network of enticing possibilities.

In March of 1973 – less than a year after the Stockholm Conference, and less than six months after they were hired by Firouz – WMRT completed *A Feasibility Study for an Environmental Park in Tehran, Iran, for the Imperial Government of Iran*. The study proposed a fifteen-year build-out schedule divided into three stages. The $45 million budget included a $2.5 million design fee that reflected 5 percent of the construction cost plus contingency.[45]

Master plan

In May of 1974, shortly after the passage of the Iran Environmental Protection and Enhancement Act that led to the official establishment of the Department of Environment (DOE), Sedaghatfar returned to Iran to manage the Pardisan project and supervise the development of the master plan. A copy of the feasibility study, which he received the previous summer, led to his decision. Equally important, Iran was economically booming – a condition in sharp contrast to the United States following the 1973 OPEC oil embargo.[46]

In addition to WMRT, which was now in partnership with the architect Nader Ardalan and his firm, the Mandala Collaborative, the consultant list for the master plan included Buckminster Fuller and Charles Eames.[47] Their inclusion, as later noted by Sedaghatfar, was a calculated ploy to increase the prestige of the project in the eyes of its royal patrons.[48] The WMRT project team remained essentially the same except for two important changes. Narendra Juneja became the Associate Partner in charge of the project following the departure of Colin Franklin from WMRT to establish his own firm, and McHarg assumed a more active role in thematic design of the project.[49]

The master plan expanded the studies of topography, solar orientation, spatial programming, water requirements, and vehicular and pedestrian circulation. A color-coded map of the world biomes guided these planning efforts. Accompanying charts of the proposed plants, animals, and exhibits situated the biomes within the framework of physical and cultural diversity (Figure 12.7). The configuration of the charts, which began with the "tundra" and ended with the "tropical forest," aligned these studies with the hypothetical global transect that visitors would encounter as they traversed the landscape. An associated spatial diagram illustrated how the biomes would be physically arranged in the park (Figure 12.8). In response to the mandates of Firouz (and McHarg's fascination with evolutionary fitness), this aspect of the work emphasized the adaptive climate strategies of plants, animals, and people, beginning with Iran. Once again, the main

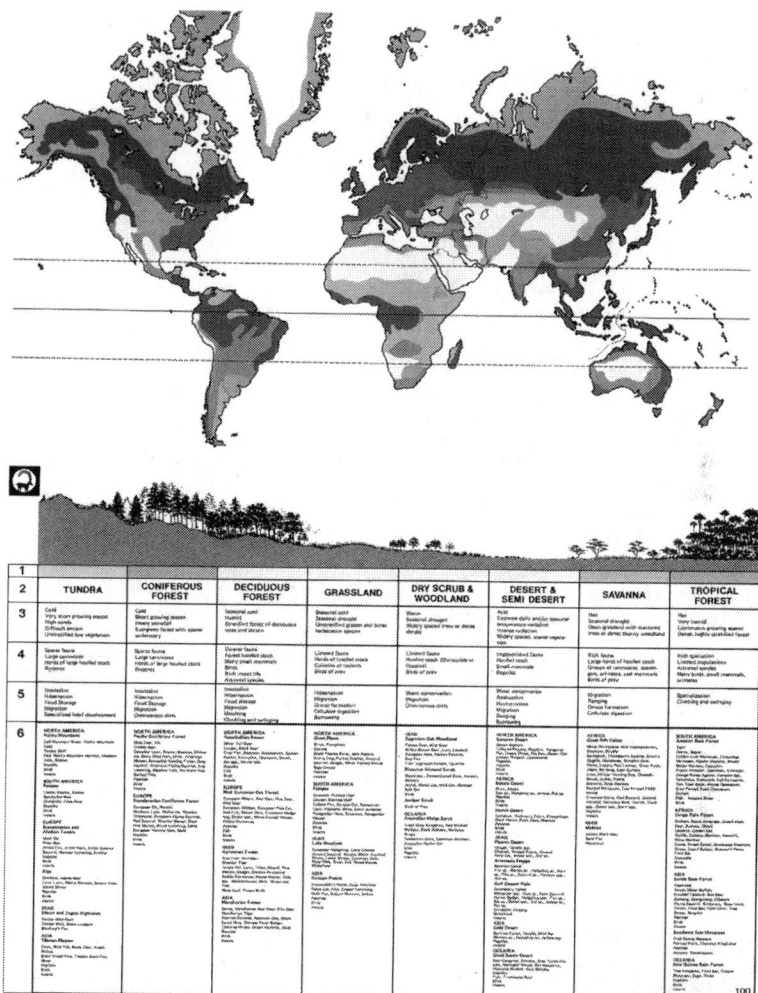

Figure 12.7 World biome map and biome vegetation chart from the Pardisan master plan. Ian L. McHarg Papers, The Architectural Archives of the University of Pennsylvania.

issue was how to optimize the less-than-ideal site conditions and transform the barren landscape into an educational oasis that contained a pleasurable diversity of plants, animals, and activities.

The plants and animals selected for the park biomes reflected the outcome of exhaustive studies that began during the feasibility study with the initial lists made by Theodore Reed and Jack McCormick. The master plan categorized this information into cross-referenced "storylines" and exhibits.[50] The selective agency of the environment and the corresponding adaptive response of organisms structured this organizational framework. One storyline, for

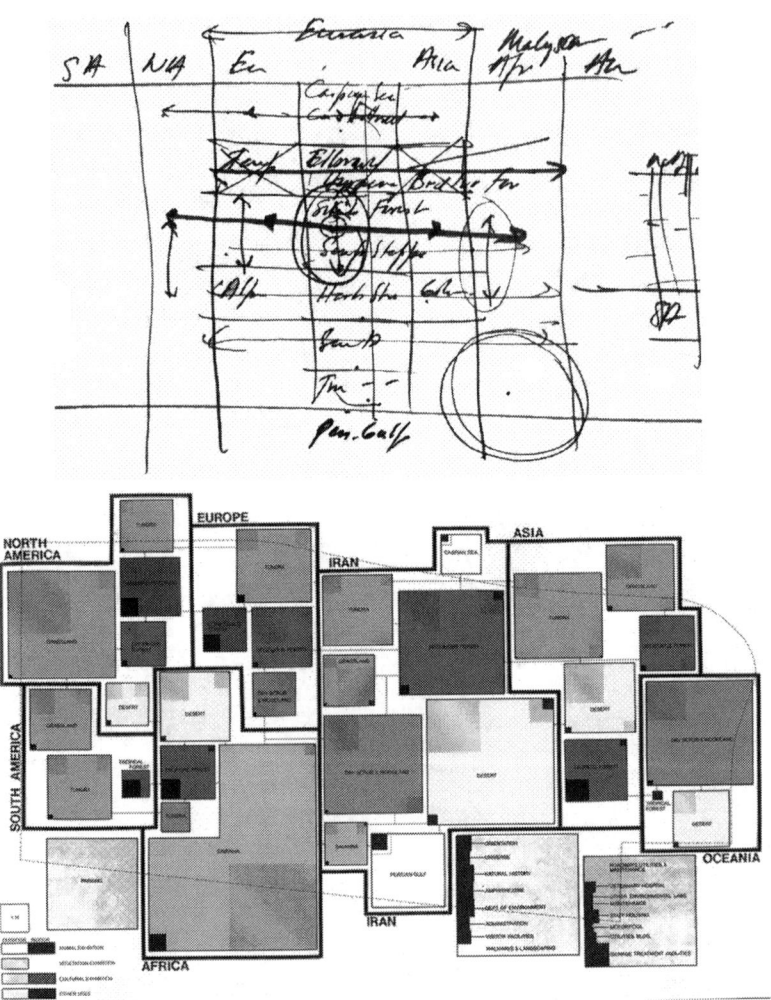

Figure 12.8 Park biome sketch by Ian McHarg (top) and park biome diagram from the Pardisan master plan (bottom). Ian L. McHarg Papers and Ian and Carol McHarg Collection, The Architectural Archives of the University of Pennsylvania.

example, illustrated climate-induced changes in the extremities of animals – in this instance shorter limbs, tails, and ears in northern latitudes to reduce the surface area of heat loss, and longer limbs, tails, and ears in latitudes closer to the equator to increase the surface area of heat loss. Representative species in this prey and predator narrative included the Arctic Hare, Snowshoe Hare, and Jack Rabbit, and the Arctic Fox, Red Fox, and Fennec Fox. A second storyline presented the morphological variations in the beaks of birds as successful adaptive responses to habitat and food sources.

Representative species included the pelican, duck, flamingo, eagle, wood-cock, parrot, cardinal, and hummingbird. A third storyline illustrated plant adaptations to aridity. This storyline included plants with seeds that germinate only when exposed to water, plants with small leaves that have waxy coatings and hairs, plants with deep root systems, and plants with toxins to prevent the near growth of other plants. Representative species included the Saguaro cactus, creosote bush, mesquite, and saltbush of the Sonoran Desert of the United States, and the Tamarix, Russian thistle, and Artemesia species of the plateau desert of Iran. It was generally believed these storylines would allow visitors to the park to become amateur naturalists, and in emulation of Darwin in *The Origin of Species* ponder how a group of animals with common characteristics came into existence, and conversely how a new species diverged from a common ancestor.[51] To relate climate adaptation to human culture, the team devised a storyline that included an igloo, yurt, mud-masonry house, and an open-frame wood building. Correspondingly, the invention of tools became a story of adaptive success and cultural speciation as *Homo sapiens* radiated outward from Africa and the Middle East and colonized the biomes of the world. Drawings that paired human cultural adaptations with comparable adaptations in plants and animals linked the storylines into an overarching narrative (Figure 12.9).

A "world analog matrix" explored potential combinations of storylines and park exhibits.[52] Similar to the analog diagram that structured the exploration of global diversity in the feasibility study, the world analog matrix both illustrated how the environmental adaptations of plants, animals, and human culture operated on multiple levels, and explained how

Figure 12.9 Climate adaptations of plants, animals, and people from the Pardisan master plan. Ian L. McHarg Papers, The Architectural Archives of the University of Pennsylvania.

the design team planned to strategically exploit these interactions to create a rich network of environmental narratives. The design of Pardisan, in
this sense, was less a matter of formal composition, though that remained
important, and more an issue of selecting a diverse set of conditions that
addressed the education mandates of the project program, the site conditions, and the needs of the plants and animals. If we return for a moment
to McHarg's early training, there are strong parallels between the biome
matrix and the CIAM organization grids developed and presented by Le
Corbusier in *The Heart of the City*. Le Corbusier developed these grids
to systematize design thinking and organize the decision-making process.
Analogous to the compartments of an expandable file, each box of the grid
could be rearranged until the organization of the compartments established
the optimum, self-explanatory solution. In defense of the method, Le Corbusier laid down a challenge: "There is no point in just saying this method
doesn't work. Improve it, if you have nothing better to do!"[53] The multiple,
integrated storylines developed for Pardisan took this strategy to a whole
new level of reasoning.

A birds-eye-view of the park's central ravine illustrated how these complex relationships would appear on the site. This illustrative "Valley Section" through the middle of the park contained a temperate forest biome
with a Persian squirrel and a Eurasian badger at the high point, a herd
of Persian donkeys in a savannah biome at mid-slope, and a desert oasis
on the valley floor that contained a lake surrounded by lush gardens and
pavilions. Money was not a consideration. As noted by Robinson Fisher,
there was talk of sending a team on a C-147 transport to Kenya or South
Africa to "roll up the savannah" and transport it back to Tehran for
installation.[54]

The theme of adaptive response is similarly evident in the proposal to use
time-lapse aerial photography from a United States Earth Resource Technology (ERT) satellite to document the transformation of the barren Pardisan
site into a verdant oasis. At the suggestion of Charles Eames, a television
monitor theatrically positioned on the entrance wall of the Natural History
Museum would broadcast the satellite imagery. Conceived as a technologically innovative supplement to the zoological collections and the nativist
dioramas in the Terrestrial Park with their life-like replications of the world
biomes, this high-tech, state-of-the-art "Iranian window of the world"
would highlight the importance of scientific monitoring in the country's conservation efforts.[55] The resultant design narrative, essentially a scientifically
secularized genesis myth with strong moral overtones, linked environmental
awareness to the ancient roots, historical rise, and contemporary salvation of
civilization. As a vision of what ecology could do for a country, this storyline
transformed Pardisan into a neo-colonial terrain of high-minded generosity and politically charged hope that pondered the significance of a world
in which the relations between countries, people, plants, and animals were
closer, more intimate, and far different than what actually existed.

The expanded examination of the biome concept also provided a pretext for McHarg to discuss one of his favorite topics – the biochemical transformation of sunlight into usable form by plants. As he repeatedly argued throughout his career, it was imperative to honor the "negentropic" energetics of the photosynthetic process.[56] For McHarg this meant that human action should not, as he reiterated in the master plan report, "interrupt the self-renewing cycle of natural systems." Pardisan, as an exemplary manifestation of this fundamental ecological principle, would illustrate how natural energy systems operated at the global scale through its pattern of biomes, and at the regional scale through dioramas devoted to the adaptive strategies of indigenous plants, animals and people.[57] In an attempt to make McHarg's ecological digression into the structure of trophic levels more accessible to the project patrons, the report, in a thinly veiled allusion to Iran's hierarchical social structure, contained a diagram in which sunlight falling on the park created an energetic pathway that flowed through the tropic levels of society as it moved upward from plants and herbivores to the noble male lion. The flow of energy culminated with a human figure, in an expression of ecological piety, protecting and preserving a plant and animal[58] (Figure 12.10).

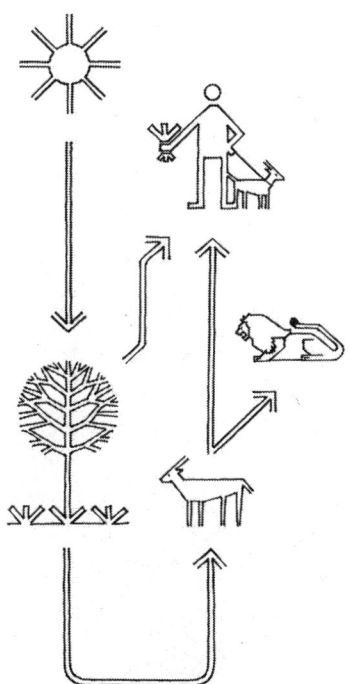

Figure 12.10 Sunlight diagram from the Pardisan master plan. Ian L. McHarg Papers, The Architectural Archives of the University of Pennsylvania.

Firouz and his royal clients could hardly find this parallel narrative of ecological stewardship objectionable since both the sun and the lion were symbols of the Persian and Iranian monarchies.[59]

Another design objective was the experience of the park as visitors left the hot, dry landscape of Tehran and entered "cool shadowed spaces and the sound of water flowing through small channels along terrace banks."[60] While clearly related to the physical attributes of the Persian garden, this aspiration reflected McHarg's belief that health and welfare in urban environments necessitated spaces of retreat, rest, and relaxation (Figure 12.11). At Pardisan this vision, which was materialized as a lush stream valley, came with an ecological cost. The project required over three million cubic feet of water per day, even with a stringent management protocol that emphasized conservation and recycling.[61]

In the summer and fall of 1974, McHarg organized a series of meetings with the project sub-consultants – David Goddard, Yehudi Cohen, Solomon Katz, Brian Spooner, Seyyed Hossein Nasr, Buckminster Fuller, and Charles Eames – to discuss the concept of environmental adaptation

Figure 12.11 Sketch of the interior of the park by Colin Franklin from the Pardisan master plan. Ian L. McHarg Papers, The Architectural Archives of the University of Pennsylvania.

through different disciplinary methods.[62] A project technical report summarized these discussions.[63]

The botanist David Goddard provided practical advice that related the adaptive strategies of plants to habitat ecology. His remarks referenced the existing site conditions, and they emphasized the role of temperature, light, water, humidity, soil, and seed production and dispersal in arid climates.[64]

The cultural anthropologists Yehudi Cohen, Solomon Katz, and Brian Spooner provided complimentary views of human ecology.[65] Boundaries and borders, both imaginary and real, formed the substance of the recommendations made by Cohen, and he stressed the point that these territorial demarcations reflected a multitude of natural and cultural relationships in Iran that could organize the design of the park. The informal agreements, local economies, and porous boundaries of the pastoral nomad, for example, correlated to non-hierarchical structures, while the globalized industrial society of the nation-state of Iran correlated to a stratified society and firm boundary demarcations. Katz, in contrast, stated that Pardisan "must be for people and not about people," and it must advance health. He suggested the team study the following questions: How does a person who lives in Tehran look at the environment of the city? What is lacking in their lives, and how will the park address this omission? For Katz, this meant that Pardisan should be a relaxing place where people could sit in the shade near the sound of running water with picnic lunches. He also stated that people raised outside Iranian society could not fully answer the questions he posed, and he concluded by stating that Iranians should manage the park. Spooner related cultural adaptation to physiography and explained how the flow of water as it traveled from the snow-capped peaks of the Elburz Mountains north to the Caspian Sea and south to Tehran, and the snow-capped peaks of the Zagros Mountains east to the Persian Gulf and west to the central desert determined the location of the major urban centers. To emphasize the critical role of water in Iranian culture, Spooner argued for the use of traditional methods of irrigation. Sketches in the technical report illustrate how the team translated Spooner's remarks on physiography and settlement patterns into a design concept – a ring of high ground encircling a central depression and a series of pavilions positioned to emulate the geographic location of Iran's major cities.

Seyyed Hossein Nasr, the dean of the Faculty of Literature at Tehran University and director of the Imperial Iranian Academy of Philosophy, bluntly stated at the beginning of his remarks that the reduction of the planet to observable laws, as routinely done in Western science, "cuts the hands of God away from the world."[66] Islam, he continued, made no distinction between the natural and supernatural, a "thing is both its scientific description and a symbol of the world," and this principle should govern the design of Pardisan. To both defend his beliefs and countermand

McHarg's insistence that the park must express adaptation through the lens of Darwinian evolutionary theory, he stated that natural order in Islamic thought was not a scientific hypothesis, but instead a living, dynamic combination of intuition, reason, and spiritual beauty that harmoniously united the worldly and otherworldly, mathematics and poetry, and thinking and feeling. Moreover, this belief was often expressed through allegories (particularly in the Sufi traditions of his faith) that related the objects and events of the world to divine revelation and the awe and wonder of God. He also observed that the most profound, life-changing discoveries came from the reflective consideration of commonplace things. "There is enough wisdom in a blade of wheat to contemplate for all of one's life," he stated. To emphasize this point, he noted how simple acts of charity, such as the planting of a tree, cultivate piety and tie human action to divine grace: "The prophet of Islam said, 'if, one day before the day of judgment when the world was to be destroyed, a man should plant a tree, he would be saved and go to paradise for this single act.'"[67] Indeed, for Nasr, the unity at the heart of his conception of environmental order negated the dialectic logic of modern materialism, and, in his discriminating judgment, the decidedly crass consumerist dictates of the West.

Because of these beliefs, Nasr avoided certain subjects, such as evolution, but this did not mean that he rejected science. In fact, he studied mathematics, physics, and geology and wrote a history of science that discussed the important discoveries of Islamic scholars. But he refused to countenance scientific practices that "reduce all things to laws that can be observed," and hypotheses that denied revelation. After repeated prodding by McHarg, he did, however, state that adaptation, as the fulfillment of divine harmony, regularity, and order could be expressed through the beauty of architecture and the tending of the garden. "Perhaps," he stated in an attempt to divert the conversation away from the topic of evolution and toward spiritual enlightenment, "it is better to think of Pardisan as a kind of zoological garden, rather than a scientific venue, that has more to do with poetry than research."[68]

McHarg concurred with Nasr's belief in the unity of existence and the spiritual power of the natural world, but, as might be expected, he felt that the rejection of evolution and the profound beauty of organic adaptation was a scientific sacrilege that denied reality. "Pardisan will show the environments of Iran with the greatest possible illumination and explain how people can live with more deference, more understanding, more productivity," he replied. But to be truly revelatory and humane, illumination had to be paired with knowledge. The critical issue, he observed, is the perception and response of people to the exhibits once Pardisan is filled with plants and animals from around the world. "Zoological gardens," he continued, "are really menageries where animals are put in cages and displayed, and made to sacrifice themselves so that man can learn about nature." The best solution to this dilemma – one that was less selfish and instrument as far as he was concerned – was a park ecosystem that presented geology, physiography, hydrology, soils, plants, and animals as one interacting process. In his judgment, this action would transform Pardisan from a menagerie into a place that illustrated how to live commensurately in the world through symbiotic and complementary relationships. McHarg also reiterated his belief that every creature had the right to exist freely, find its place in the scheme of things, and simply be, as well as his doubts that religion would, or could, promote this objective. To defend this approach, and its incorporation of evolution, he then noted that Nasr was actually promoting an order that had been made by "creatures interacting with each other over long periods of time," and "with natural processes continuously," as they, too, constantly changed. "The question is, for what reason?"[69] Adopting a conciliatory tone, he made a sort of confession:

> Adaptation is a beautiful device for us because it simply says that in order to exist in the environment, you have to understand the way it works, and greater understanding induces a sense of awe, which makes you feel the acts of man and nature are sacramental.[70]

Nasr, in an acknowledgement that at some basic level they sought the same thing, agreed the harmony and beauty of nature's order was best perceived humbly and as a totality. But he cautioned that perception could never be legislated. In an attempt to once again find common ground, he suggested: "Perhaps, it might be better to encourage the patrons of the park to understand a micrograph by writing a poem." McHarg, however, remained skeptical. "Everything I learn about the phenomenal world makes it more transcendent not less," he stated. "Awe is fine, but awe can be ignorance, or enlightenment."[71]

Buckminster Fuller presented a disparate and decidedly technocratic vision of adaptive design, and called for a system of elements – what he referred to as "non-simultaneous, and only partially overlapping events"– that could be encoded with ideas to produce coherent patterns.[72] This "synergy," he stated in his distinctive jargon, required the implementation of a "scenario universe" where scientists with "access to the great eternal metaphysical laws," monitor, "through trial and error," the operative principles of survival:

> We need, in Pardisan, to enable humanity to see, in a big way, that we are not on a closed plane going to infinity. We are on a closed system planet, and we are here to solve problems. We play games, introducing disorder, and then seeing if we can bring order out of it again.[73]

Fuller promoted geometry as one way to construct coherent patterns. Take an angle, he said, play with it, produce triangles, squares, rectangles, and polygons, and encode them to describe "any form in nature."[74]

A subsequent summary of the project objectives compiled by WMRT noted that Fuller provided commentary on "epistemological evolution."[75] Exactly what this meant remained unclear, since each member of the project team interpreted Fuller's theory of nature and contribution to the project differently. McHarg, for instance, associated Fuller's discussion of energy, information, structure, complexity, and pattern integrity with the symbolically empowering vision of natural order promoted by the project's religious adviser – in effect juxtaposing Fuller's technological determinism against Nasr's metaphysical study of Sufi philosophy and science. Sedaghatfar, who heard Fuller speak at Kent State, focused instead on his discussion of geodesic domes and sustainable energy use. Nader Ardalan observed that Fuller's patented Dymaxion worldview inspired the map of contemporary air routes and the map of ancient Persian caravan silk routes that appeared in the master plan report, both of which, it should be noted, situated Tehran at the center of the globe. Robinson Fisher, in a sentence rapidly scrawled during Fuller's presentation, stated that he should perhaps study pattern integrity or he would "miss this on the final exam."[76]

While Fuller's "epistemological" contribution to the discussion of adaptation remained indirect, and situated in the realm of ideas, the contribution of Charles Eames, though of short duration and essentially assigned to his colleague Glen Fleck, produced work that critically advanced this theme and the project.[77] Fleck, in consultation with Eames, created a short animated film, which, in conjunction with an illustrative site plan and model prepared by WMRT, would explain the park and convince Her Royal Majesty Queen Farah Diba-Pahlavi – the cultural patron of the project – that Pardisan, which now cost slightly over $106 million, was worthy of funding. The Queen would then present the scheme to the Shah who would finance the work.[78] To that end, Fleck conceived the film as a two-part lesson in decision-making. The first lesson considered how Iran, as a rapidly modernizing society no longer bound by the dictates of tradition, should address "problems of choice."[79] The second lesson illuminated the knowledge required to make appropriate decisions as Iranian society moved forward (Figure 12.12). Fleck's movie communicated this message with a hypothetical journey through Pardisan that honored innovation and tradition.

In prototypical Eames fashion, the film, titled *Pardisan*, shunned the minimal and portrayed a lot of information in a short period of time. Fleck accomplished this feat by slowly moving the camera across paintings of the proposed design, zooming in on particular details, and dissolving one image into another. During this continuously unfolding sequence of grand ideas and promises, viewers glimpsed a walled courtyard garden, a Persian carpet patterned with the four rivers of paradise, zodiacal signs, a spiral galaxy, a comet, a Dymaxion map, computerized display screens, a monorail, a gazelle, a wetland teaming with birds, a tropical rainforest enclosed

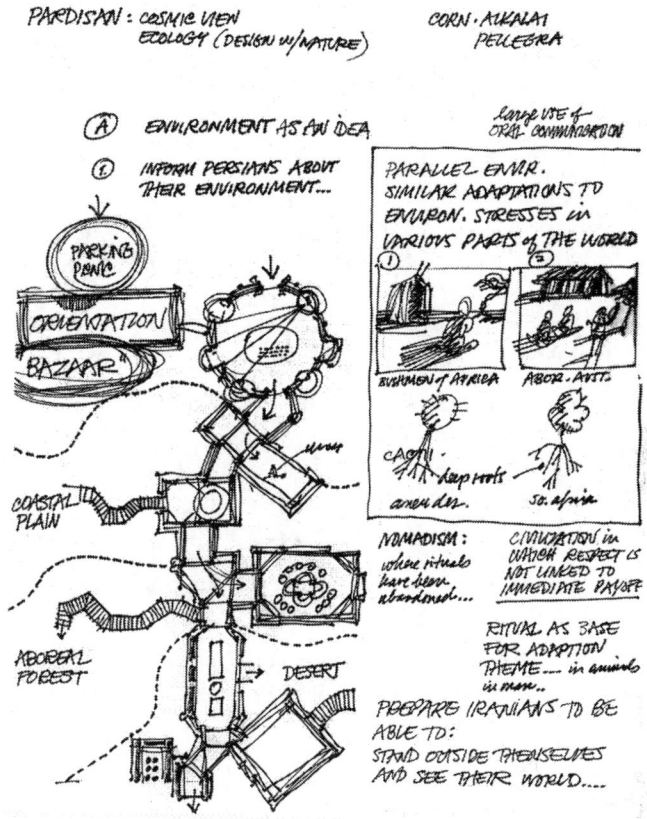

Figure 12.12 Storyboard of the Pardisan movie by Glen Fleck. Obtained by author from the Eames Collection at the Library of Congress. © Eames Office LLC (eamesoffice.com). Reprinted by permission.

in a geodesic dome, a diagram of bioclimatic zones, a map of Iran, and a model of the park. Fleck interspersed this imagery with colorful vignettes of Iranian families enjoying picnic lunches and shopping for local crafts (Figure 12.13).

The grand finale of this mise-en-scène took place in the central hall of the natural history museum, which Fleck presented as an imaginative bricolage containing glass vials and chemistry experiments, an airplane, a sports car, a rocket, a dinosaur skeleton, the handprint of a chimpanzee, hand-controlled industrial robots, digitally displayed photographs of Iran and its people, a satellite, and a view of the earth from space, courtesy of NASA.[80] In this purposefully overwhelming, but carefully packaged and cleverly arranged presentation, each object advanced a different concept

Figure 12.13 A picnic lunch in the park from the Pardisan master plan. Ian L. McHarg Papers, The Architectural Archives of the University of Pennsylvania.

of science, technology, nature, and Iran, while also referencing the cultural themes favored by the Queen, the conservation triumphs of the DOE, and the "One Earth" iconography beloved by Firouz and McHarg.[81] The conveyed message, though slightly sentimental and overtly nationalistic, was clear. Pardisan was a relaxed but stimulating garden that instilled the values and knowledge necessary to make traditionally grounded, divinely ordained, environmentally correct, scientifically informed, and technically sophisticated choices for Iran's future.

Architecture designed by Nader Ardalan provided a neutral backdrop for the multimedia exhibition spectacle proposed by Fleck. This is not to say, however, that the architecture was devoid of meaning. On the contrary, according to Ardalan, its form, color, materials, spatial qualities, and surface patterns symbolically expressed the inherent beauty of nature and Islamic spirituality.

Ardalan previously expressed this vision of modern Iranian architecture in a speech delivered to an international architecture conference in Isfahan, Iran, titled "The Interaction of Tradition and Technology," that he and Laleh Bakhtiar organized to celebrate the 2,500-year anniversary of the Persian Empire. Ardalan's politically astute opening remarks called for the recovery of spiritually ennobling form, and they provided the framework for subsequent discussions on the use of ancient architecture and traditional village plans as models for contemporary practice. Following the scheduled addresses, the invited participants to this gathering, which included the architects Louis Kahn, Paul Rudolf, George Candilis, and Buckminster Fuller, toured the archaeological remains of the city of Persepolis and the Tomb of Cyrus the Great at Pasargadae. The conference has been rightly criticized for the way it purposefully promoted a populous, government-sponsored agenda specifically designed to co-opt those Iranians left out of the modernization process.[82] Yet for Ardalan, who was raised and trained in the United States, this exploration of tradition should also be considered a personal attempt to understand himself as a member of a global architecture community and as an Iranian working within a regional context.[83]

Ardalan and Bakhtiar further developed the ideas discussed at the Isfahan conference in *The Sense of Unity: The Sufi Tradition in Persian Architecture*. In this text, they reiterated their belief that modern Iranian architecture had to reference indigenous craft, incorporate local materials and colors, and account for the harsh climate, but they also claimed that truly meaningful designs integrated these spatial explorations with science, philosophy and art.[84] To organize this all-embracing cosmology into a manageable design system, Ardalan and Bakhtiar developed eight motifs derived from their interpretation of traditional Iranian architecture and Islamic religious symbolism: garden (recapitulation of paradise), socle (sacred mountain), porch (transition – the way), gateway (hierarchic demarcation), room (multiplicity), sphere or dome (unity), *chahār tāq* (reintegration), and column or minaret (ontological axis). Drawings and photographs tied this system of architectural elements to the Iranian landscape, the archaeological discoveries at Persepolis and Pasargadae, and the urban plan of Isfahan. When Ardalan applied these ideas to Pardisan, he created a matrix that aligned his ideas with the structural logic of Buckminster Fuller (Figure 12.14). The matrix assigned geometric shapes to seven of the above elements (the garden was omitted, the gateway was replaced by the courtyard, and macrons were removed from the spelling). The resultant buildings consisted of different combinations of these encoded elements. Louis Kahn – an architect admired by the Queen and the lead consultant of Shahestan Pahlavi, a proposed monumental government center for Tehran then under design, deployed a similar structural approach to great effect.[85]

There is, however, another important logic embedded within the Ardalan-Bakhtiar argument – namely, the elements that comprised their system of modern, yet indigenous, Iranian architecture would, in creative combination, reveal the hidden meaning of the world and thus be inherently beautiful and transcendent. Following the teachings of Nasr, this aspect of their argument positioned the interplay of intuition, reason, and instruction

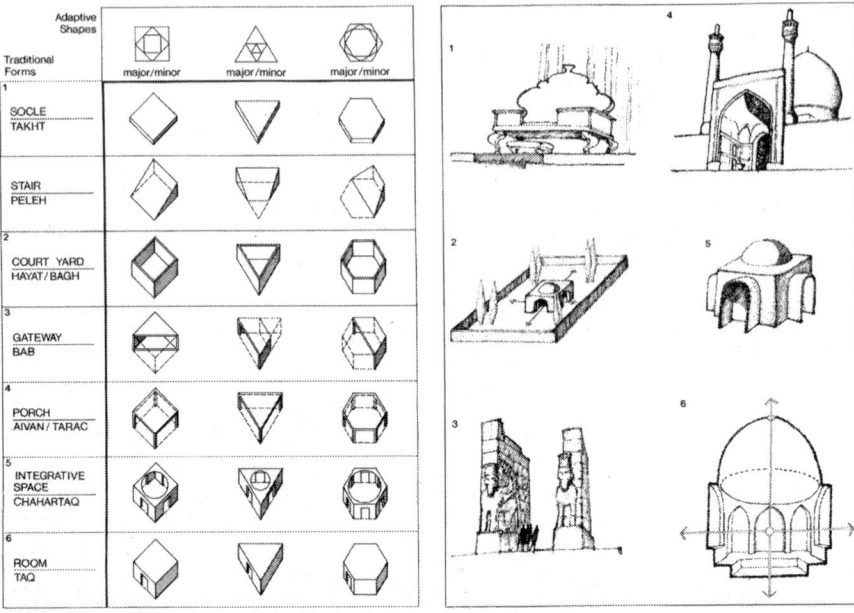

Figure 12.14 Architecture typology matrix by Nader Ardalan from the Pardisan master plan. Ian L. McHarg Papers, The Architectural Archives of the University of Pennsylvania.

as a fruitful subject for imaginative exploration, and it presented architecture as the consummate means to symbolically express God's handiwork in the natural order of things. Yet as ultimately expressed in the Pardisan design report, it was the playful pattern of objects in the surrounding landscape and within the buildings that provoked imaginative associations. In this regard, Fleck's cinematic interpretation of the architecture is telling. The Pardisan film emphasized the structural, or generative, logic of Ardalan's architectural elements and their combinatory possibilities. Attendant spiritual meaning became evident only when these geometric forms were juxtaposed against images of Persian gardens, zodiacal signs, and scientific equipment.[86] The architecture (at least for the consultant firms from the United States) provided a neutral framework that allowed objects and ideas to interact and attendant meaning to emerge – just as the Cartesian grid of biomes proposed by McHarg made the ecological concept of adaptation visible and understandable.

Critical here is the fact that Ardalan and McHarg framed their approaches differently, but they both deployed archetypal motifs to create functional and enduring designs. As McHarg repeatedly stated in discussions of his work, this approach countered normative and less rational patterns of organization by formulating a set of rules – natural and social – that he could combine and layer to create more complex landscapes with deeper attendant meaning. The intent, similar to the diplomatic language formalized at the Stockholm Conference, was the creation of a locally resonate spatial language that would also extend outward to form larger patterns of global behavior.

An image of the earth from space, as mentioned previously, served as the central archetype of Pardisan. The proposed design reconstituted this starkly beautiful object as a paradise garden that contained all manner of things arranged in order and loveliness to reveal the inner nature of the land. The intricate pattern of biomes and buildings anchored the tectonics of the scheme and furnished a structural model for global expansion. Regional specifics emerged in the plants and animals selected to inhabit the park, in the topographic manipulations that recreated Iran's physiography, and in the entry sequence of buildings and gardens that emulated the urban plan of Isfahan (Figure 12.15). The park's natural history exhibits, which presented the country's progress under the Pahlavi regime as evolutionary, adaptive, and environmentally beneficial, reaffirmed the physical plan's nationalist hagiography, imperialist visions, and colonizing rubric.

Figure 12.15 Illustrative site plan from the Pardisan master plan. Ian L. McHarg Papers, The Architectural Archives of the University of Pennsylvania.

In March of 1975, WMRT completed the master plan report. To celebrate this achievement, Firouz, McHarg, Ardalan, Sedaghatfar, and Narendra Juneja in an act of piety, ceremoniously planted the first tree in the park.[87] That same year Firouz became the director of the DOE.

The DOE subsequently presented a report titled *Pardisan: Plan for an Environmental Park in Tehran*, along with an illustrative site plan, model, and Fleck's movie to the Queen.[88] The report grandly stated that Pardisan was an international symbol of enlightened stewardship that would outshine all other environmental parks. At the national level, it would integrate ecology into Iran's modernization schemes and provide a platform to showcase the country's rich natural and cultural history.[89] At the international level, it would position Iran as a global intellectual leader and highlight the progressive vision of the Imperial Government. The report's conservation objectives, education mandate, and desire to provide an oasis-like recreation setting for the Iranian middle class reflected the vision of modernization promoted by the Queen; and since the project required the approval and financial backing of the Shah, the report referenced Persian architectural grandeur, Islamic spiritualism, and traditional customs.

Schematic design

Following the official presentation of the master plan, negotiations began for the control and ownership of the Pardisan site. Two years later, the Iranian government purchased the property for the DOE and detailed design work began. Firouz, however, fell out of favor with the increasingly autocratic Pahlavi regime and resigned from his post. In the absence of Firouz, Sedaghatfar, who was now a deputy director at the DOE, continued to manage the project.[90] The support of the Queen assured the continuation of the work. Meanwhile, and despite his recent completion of a lengthy management plan for Pardisan, Ardalan returned to the United States and founded Mandala International.[91]

On November 12, 1977, WMRT signed a contract with the DOE for schematic design services. The cost of the park was now estimated at $500 million, with design fees of $1.6 million paid on an interim basis according to the percentage of work completed. In response to the departure of Firouz and the increasing social unrest in Tehran, the partners at WMRT (at the urging of David Wallace) asked McHarg to sign an agreement that held him solely responsible for any losses stemming from the project.[92]

During this phase of the design, the DOE required WMRT to establish an office in Tehran. Narendra Juneja moved to the city to act as the on-site liaison between Tom Atkins, the WMRT project manager in Philadelphia, and Sedaghatfar at the DOE. A rotating staff of designers that included Neal Belanger, Hank Bishop, Gary Lee, and Victoria Steiger traveled to Iran to work on the project.[93] One of the first tasks undertaken by the Tehran office involved the design of a wall that would enclose the park and prevent the intrusion of urban development. The park was going to take years to materialize, and during that time it was necessary to demarcate the desert terrain as parkland. This set of drawings, produced by Belanger, presented options that ranged from a chain-link fence to a stone wall.[94]

To address WMRT's lack of knowledge in zoo design, McHarg, at the suggestion of David Hancocks, a proponent of immersive and naturalistic zoo designs, hired the firm Jones and Jones as a project consultant. The firm, under the direction of Grant Jones, had recently completed an innovative design for the Woodland Park Zoo in Seattle, Washington that ingeniously disguised the walls of the enclosures so that the animals no longer looked like caged beasts in a menagerie, but instead appeared to live in natural surroundings. As later noted by Jones, WMRT retained the final responsibility for design implementation and construction. His office was to supply detailed layouts of the topography, water forms, and plantings for the animal habitats, as well as designs for the animal barriers, animal holding areas, and public viewing areas and overlooks. Jones hired the ecologist Dennis Paulson, who had worked on the design of the Woodland Zoo, to assist his efforts.[95]

The scientific monograph *Life Zone Ecology* by L. R. Holdridge provided the conceptual foundation of the Jones and Jones work. This text outlined a plant classification system that consisted of thirty ecologically distinct formations, or world life zones. A triangular five-axis diagram that resembled a soil texture classification chart positioned these formations in relation to latitude, longitude, evapotranspiration, humidity, and precipitation.[96] Jones and Jones had used this classification system at Woodland Park Zoo to create a range of habitats for animals from around the world, and they were asked to do the same for Pardisan.[97]

As later noted by Dennis Paulson, the geographically deterministic classification system detailed in *Life Zone Ecology* made it possible to select appropriate species for each of the Pardisan world biomes – forest, woodland, and savannah for example – using morphologically similar plants that could grow in the climate of Tehran.[98] The design team grouped these climate-appropriate plant formations into relatively large areas and interspersed them with animal enclosures. Manipulations of topography and irrigation established the physical parameters of these artificial biomes and their placement within the park. The resulting layout, as already mentioned, mimicked Iran's natural physiography, and it included paths and pavilions positioned to replicate the country's major roads and cities. In keeping with the client's wish to create an experience that began with Iran, the design also contained a lake that represented the Persian Gulf, an Iranian fishing village, a Persian Lion enclosure, a walled garden named the "Isfahan Garden" as well as an associated European analog named "The Court of the Lions," and exhibits of the common domesticated animals of Iran. To reach these delights from the parking lot park visitors had to transverse a bridge, similar to the path followed by ancient travelers when they entered Isfahan. This particular design element established a metaphysical connection between the material concerns of the outside world and the contemplative pleasures of the interior garden.

While Jones and Jones refined the species composition and layout of the zoological exhibits, WMRT developed the design of the Pardisan parking lot.

This endeavor transformed the terraced rows of cars depicted in an earlier concept sketch by Colin Franklin into a grand entry sequence and lush garden.[99] The central axis of this scheme began near the highway in a small park with a circular basin of water. In front of the water basin, near the road, a sculpture, which from its shadow displays the features of a winged Faravahar, announced the park and called attention to the benevolent presence of its royal patronage.[100] As this entry sequence moved up through the parking lot toward the main entrance it ascended through a plaza that contained a fountain, a plaza with an octagon-shaped pool, and a plaza with a square pavilion topped by a circular dome. The last plaza in this sequence, the one immediately in front of the Visitor's Center, contained a large metal sculpture of the globe that resembled the Unisphere from the 1964 World's Fair. In combination, the entry sculptures and plazas signaled the journey

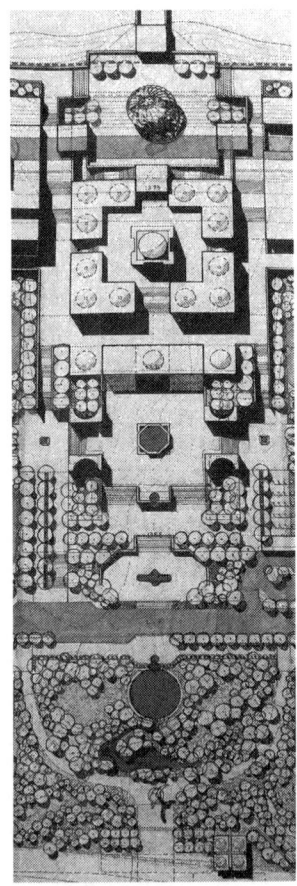

Figure 12.16 Parking lot "Central Axis" from the Pardisan schematic design documents. Ian L. McHarg Papers, The Architectural Archives of the University of Pennsylvania.

through Pardisan would begin with Iran, wisdom, and good thoughts, and end with divine transcendence and world peace (Figure 12.16).

The same care was bestowed on the two large parking lots that flanked the central axis. The lots were subdivided into small courts lined with trees, and the courts were organized around linear parks, or "picnic valleys" that contained runnels of water in the shape of a stream. A series of formal gestures allowed the water to meander through the rows of park cars and sidewalks. Trees, plazas, and pavilions lined the flowing water and provided a place to relax and enjoy a picnic lunch prior to entering the park. Run-off from the parking lot irrigation system provided the water supply. The overall visual presence of this intelligent merger of aesthetics and function was harmonious and unexpected[101] (Figure 12.17).

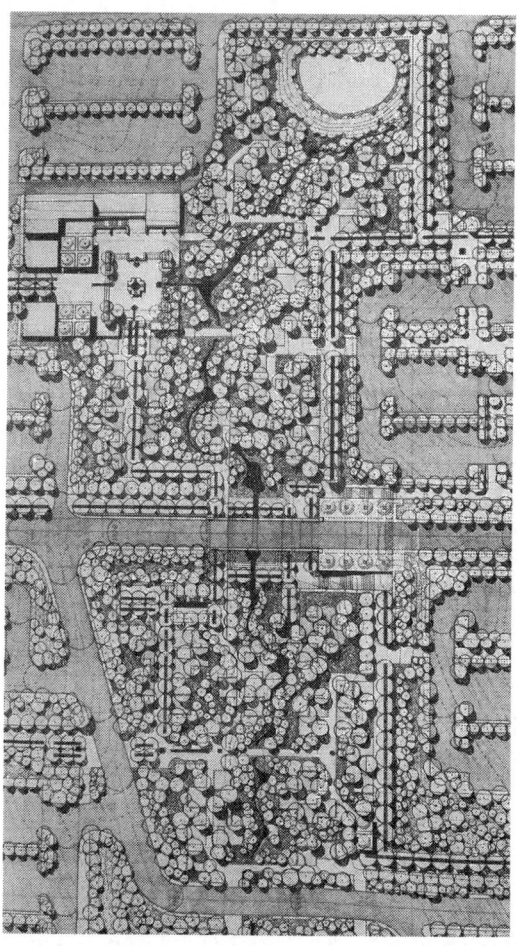

Figure 12.17 Parking lot "Picnic Valley" from the Pardisan schematic design documents. The Ian L. McHarg Papers, The Architectural Archives of the University of Pennsylvania.

In November, several weeks after the completion of the schematic design drawings, Sedaghatfar directed WMRT to suspend work on Pardisan due to the uncertain political climate. In December, shortly before the mass demonstrations that toppled the Pahlavi monarchy, he approved payment for all outstanding WMRT invoices and went on indefinite vacation leave. Thirty days later, after receiving no clear word regarding payment, WMRT terminated their contract with the DOE. Now facing a huge monetary loss, long-simmering disputes between McHarg and his partners, which revolved around time spent in the office, fee structure, and design recognition, boiled to the surface. McHarg was asked to leave the firm. A lengthy legal battle eventually led to the recovery of all monies as mandated by a ruling of the Iran-US Tribunal at The Hague. The award totaled $654,644.80 for unpaid fees, plus interest and arbitration costs.[102]

The Islamic Republic of Iran did not implement the Pardisan design conceived by Firouz and developed by McHarg and his collaborators. Instead, they reforested the site following the earlier (and significantly less expensive) reforestation recommendation of the *Comprehensive Plan*. The park's official name is Pardisan Forest Park, though it is commonly referred to as Shahrak-e-Garb's Forest Park, reflecting its location south of the upscale community of that name.[103] A small natural history museum, wildlife park, and climate research facility are the only indicators of the original proposal.

Coda

Reflecting on these events, perhaps the best way to summarize Pardisan is to begin by noting that the members of the design team shared the belief that modernization could honor tradition and be environmentally friendly, and they expressed this faith through a visually seductive design that was purposefully configured to educate and morally uplift. Following the traditions of Middle Eastern garden design, a walled enclosure graced with streams, fountains, ornate pavilions, beautiful plants, and wild animals provided the framework of these efforts. Ecology, overlaid with the precepts of Darwinian natural selection, adaptation, and evolution updated this organizational structure to align with 20th-century science. The adroit use of systems diagrams and matrices determined how to implement the project in a barren desert, and a state-of-the-art transportation infrastructure opened the design to the city, country, and world. Entertainment and shopping enticed visitors to the park and allowed Iranians to enjoy the fruits of Western industrial modernization. The overall effect was ennobling, forward thinking, stereotypical, and brash – although, and not inconceivably because this ideological bricolage embraced 19th-century natural history dioramas and 20th-century ecosystem infrastructures, archetypal symbolism and architectural structuralism, Islamic spiritualism and scientific secularism, and Persian cultural tradition and international globalization.

There was, of course, an undeniable irony underlying the project's ambition. The design stemmed from power and money – the two commodities the Shah had in excess at the start of the project. This power and money actively supported a nationalist agenda that sought legitimacy by embracing and exploiting Iran's rich environmental and cultural heritage.

A particularly intriguing aspect of the proposal is the fact that it was as much a symbolic gesture as it was a physical design, both engaged with and disconnected from reality. McHarg, in particular, remained loyal to his romanticized vision of Pardisan – even when confronted with potential monetary ruin, mounting social unrest, and dwindling political support – to the point where his unwavering faith in the project appears as hubris. For McHarg, this act was deeply political – part activist statement of intent as well as an ideal way of being in the world, and, therefore, an act of unmitigated love. But here again his actions were rife with mixed messages and internal contradictions. By positioning his love affair with nature in a pleasurable sanctuary separate from the world, he could advocate for the survival of species, the ecosystem, and the planet, and he could avoid issues of power, money, and control.

In terms of the Pardisan's impact on the field of landscape architecture and its current vision of environmental design, the dissolution of the project brought an end to the most productive and innovative period of McHarg's career. But this ignominious ending also signaled the beginning of a new series of explorations by his colleagues and students. These intellectual journeys of discovery have subsequently refined, revised, and filled the voids in McHarg's thinking and this, in turn, has led to alternative paradigms of ecological design and environmental stewardship that make it impossible to return to what existed before. In contrast to the worldview promoted by McHarg, these proposed alternatives are, for the most part, less generalizing and more specific, and they actively embrace the city, social diversity, and the agency of non-human systems. In keeping with the worldview of McHarg, ecology, process, reciprocity, adaption, practice-based research, surveys, systems, redundancy, boundaries, limits, and the modernist faith in science remain foundational principles. But the perplexing problem – the one that eluded McHarg and continues to frustrate contemporary attempts to live commensurately in the world and design with nature – relates to the question of purpose: Who and what are we designing for? Why do we seek order? What actions achieve the ends we seek? And can we admit, as Jane Bennett has eloquently argued in *Vibrant Matter*, that the world we wish to organize contains geological, biological, and climatic forces that, by their very nature and complexity, contain emergent properties that defy predictability and resist control?[104] McHarg, at his most honest, acknowledged the world would never be completely known and design would never address all of the environmental issues that he sought to resolve. And yet he persisted and never gave up trying to foment change until he ultimately succumbed to failure. There is a moral here, but in sympathy with the landscapes that McHarg diligently surveyed, categorized, and assembled into a patchwork mosaic of ideas, actions, events, and things, I resist the urge to draw a singular conclusion. Instead, I suggest that design, similar to McHarg's declared love affair with nature, is never definitive and it defies reason. Order, in this sense, is an act of faith, a statement of resolve, and foolishly impractical, especially when ideals are involved; however, the defining, and perhaps most memorable, and diconcerting, trait of this hopeful and risky pursuit is the expectant way it looks to the past for answers that only the future can give.

Notes

1 McHarg received this poem from Arthur Upham Pope in conjunction with a 1959 lecture delivered by this Persian art and architectural scholar in the seminar course "Man and Environment." See The Architectural Archives of the University of Pennsylvania, Ian L, McHarg Collection, "Arthur Upham Pope Islam and Environment," 109.II.E.2.57.

2 The United Nations, "Declaration of the United Nations Conference on the Human Environment" in *Report of the United Nations Conference on the Human Environment, Stockholm, 5–16 June 1972* (New York: The United Nations, 1973), 3.

3 Felix Dodds, Michael Strauss and Maurice Strong, *Only One Earth: The Long Road Via Rio to Sustainable Development* (London: Routledge, 2012), 11–12. See also, Barbara Ward and René Dubois, *Only One Earth: The Care and Maintenance of a Small Planet* (New York: W.W. Norton & Company, 1972).

4 Ibid., 6–8.

5 G.V.T. Matthews, *The Ramsar Convention on Wetlands: Its History and Development* (Gland: Ramsar Convention Bureau, 1993), www.ramsar.org/sites/default/files/documents/pdf/lib/Matthews-history.pdf.

6 Firouz, the grandson of the prominent politician and Qajar Prince Abdol Hossein Mirza Farmanfarma, was born His Highness Shahzada Iskandar Iskandar Mirza in Shiraz, Iran, in 1926. He was educated in Germany and then in the United States – first at the Lawrenceville Preparatory School and then at Yale University, where he studied civil engineering. Before he devoted his energies to wildlife conservation and environmental protection, he served as a consulting engineer for the 1958–1961 Karaj Dam Project. The United States funded the Karaj Dam Project, which brought water and hydroelectric power to Tehran and helped fuel Iran's industrial modernization. See Cyrus Schayegh, "Iran's Karaj Dam Affair: Mass Consumerism, the Politics of Promise, and the Cold War in the Third World" *Comparative Studies in Society and History* 54 no. 3 (2012): 612–643. See Persia: The Qajar Dynasty (Firouz, Farmanfarmaian, Farman-Farmaian and Mossadeq) Genealogy www.royalark.net/Persia/qajar9.htm; Eskandar Firouz, *The Complete Fauna of Iran* (New York: I.B. Tauris, 2005), 6–10; and Sattareh Farman Farmaian and Dona Munker, *Daughter of Persia: A Woman's Journey from Her Father's Harem through the Islamic Revolution* (New York: Anchor Books, 1992).

7 See Eskandar Firouz, *Environment Iran* (Tehran: Offset Press, 1974); and Firouz, *The Complete Fauna of Iran*, 10–11.

8 International Union for Conservation of Nature (IUCN), *Proceedings of a Technical Meeting on Wetland Conservation*, October 9–16, 1967 (Morges: UNESCO and the Commission on Ecology of the IUCN, 1968), 211–213, 239–241. For further discussion of the IUCN see Alexis Schwarzenbach, *Saving the World's Wildlife: WWF – The First 50 Years* (London: Profile Books,

2011), 73–85; and Nancy Lee Nash, "Conservation: Bringing Back Wildlife," *Nature* 472 (21 April 2011): 290–291.

9 Firouz, *The Complete Fauna of Iran*, vi. The establishment of the World Wildlife Fund was an outcome of the meeting. See: World Wildlife Fund History: www.worldwildlife.org/about/history. For further discussion of mid-20th-century conservation, see F. Fraser Darling, "Introduction," in *Future Environments of North America*, eds. F. Frazer Darling and John P. Milton (Garden City, NY: Natural History Press, 1966), 3. Darling promoted land management practices that followed the aristocratic ideal "of being the servant of those less able to care for themselves."

10 Matthews, *The Ramsar Convention on Wetlands: Its History and Development*, 25.

11 Firouz, *Environment Iran*, 18.

12 Ibid., 38–51; and Firouz, *The Complete Fauna of Iran*, 3–38.

13 Wallace, McHarg, Roberts and Todd, *A Feasibility Study for an Environmental Park in Tehran, Iran, for the Imperial Government of Iran* (Philadelphia, PA: WMRT, 1973), 5. The Architectural Archives of the University of Pennsylvania, Ian L. McHarg Collection, 109.IV.C.4.

14 Translation graciously provided by Aleksandar Sopov, Tyler Fellow, Dumbarton Oaks, Washington, D.C.

15 Wallace, McHarg, Roberts and Todd, *A Feasibility Study for an Environmental Park in Tehran*, 5.

16 Born in Tehran, Sedaghatfar left Iran in 1963 to study architecture at Queens College in New York. He transferred to Kent State University in Ohio, where he received an undergraduate degree in architecture in 1970 and a graduate degree in urban design in 1971. Following graduation, he became a faculty member of Hampton College, where he taught architecture and urban design. Personal communication, Jahangir Sedaghatfar, November 11, 2013, and April 28, 2014; and Personal communication, John Spencer, Chair of the Architecture Department, Hampton College in the 1970s, April 28, 2014.

17 Jahangir Sedaghatfar, *Entopia* (Kent, OH: Kent State School of Architecture, 1971). See also Constantinos Doxiadis, *Between Dystopia and Utopia* (Hartford, CT: The Trinity College Press, 1966).

18 For further discussion of the anti-modern sentiment permeating intellectual discourse in Iran at that time and its expression in architecture and urban design see Talinn Grigor, *Building Iran: Modernism, Architecture, and National Heritage Under the Pahlavi Monarchs* (New York: Periscope Publishing, 2009), 144–173; Talinn Grigor, "Recultivating 'Good Taste': The Early Pahlavi Modernists and their Society for National Heritage" *Iranian Studies* 37 no. 1 (March 2004): 17–45; and Sandy Isenstadt and Kishwar Rizvi, *Modernism and the Middle East: Architecture and Politics in the Twentieth Century* (Seattle: University of Washington Press, 2008), 3.

19 Personal communication, Jahangir Sedaghatfar, November 11, 2013 and April 28, 2014.

20 Ibid. The interview with McHarg occurred over the phone.

21 McHarg, *Quest for Life*, 290–296. See also David A. Wallace, *Urban Planning My Way* (Chicago, IL: Planners Press, 2004), 191–196.

22 Personal communication, Jahangir Sedaghatfar, November 11, 2013.

23 See "Dr. Robert MacArthur, Ecology."

24 See "Paul Sears, An Ethic for Man and Environment."

25 See "Morse Peckham 18th and 19th Century Philosophies," and "Morse Peckham, The Garden as a Metaphysical Symbol."

26 See "Arthur Upham Pope, Islam and the Environment." For further information on Arthur Upham Pope and his relationship to Persian art, architecture, and garden design see: Noel Silver, *Encyclopaedia Iranica*, "Pope, Arthur

Upham," July 20, 2005, www.iranicaonline.org/articles/pope-arthur-upham; and Kishwar Rizvi, "Art History and the Nation: Arthur Upham Pope and the Discourse on 'Persian Art' in the Early Twentieth Century" *Muqarnas an Annual of the Visual Culture of the /Islamic World* 24, History and Ideology: Architectural Heritage of the "Lands of Rum" (2007): 45–67.

27 "Arthur Upham Pope, Islam and the Environment," 12.

28 McHarg, "The Garden as Metaphysical Symbol (1979)," 10.

29 Wallace, McHarg, Roberts and Todd, *Pardisan, A Feasibility Study for an Environmental Park in Tehran*, 47. Ian and Carol McHarg Collection, The Architectural Archives of the University of Pennsylvania, 365.II.89. Email communication, Carol Franklin, December 2, 2013. Personal conversation, Robinson Fisher, June 21, 2016.

30 Wallace, McHarg, Roberts and Todd, *Pardisan, A Feasibility Study for an Environmental Park in Tehran*, 1–11. The Shah owned The Bank of Omran, and it served as the financial arm of the Pahlavi Foundation. For further discussion of the Bank of Omran, see Abbas Milani, *The Shah* (New York: Palgrave MacMillan, 2011), 239, 387.

31 "The Comprehensive Plan for Tehran First Stage – Concept Development Study and Evaluation – Volume III," Manuscript Division, Library of Congress, Victor Gruen Papers, Box 46, Folders 1 and 14, III-D-4-13.

32 Eskandar Firouz, "Descriptive Report" in *Pardisan, A Feasibility Study for an Environmental Park in Tehran, Iran, for the Imperial Government of Iran, Appendix*, eds. Wallace, McHarg, Roberts and Todd (Philadelphia, PA: WMRT, 1973), 1–11. The Architectural Archives of the University of Pennsylvania, Ian and Carol McHarg Collection, 365.II.91.

33 Ibid.

34 Ibid.

35 "Dr. Brian Spooner Cultures of Iran," in *A Feasibility Study for an Environmental Park in Tehran, Iran, Appendix*, 123–130. In a later discussion of Pardisan, Spooner stressed that the cultural exhibits were to be presented through videos and displays rather than cultural reenactments. McHarg stated the same in a discussion of the park with Charles Eames. Personal communication, Brian Spooner, October 25, 2013; See Mandala Collaborative/Wallace, McHarg, Roberts and Todd, *Plan for an Environmental Park in Tehran, Iran for the Imperial Government of Iran, Technical Report, 1975*," 39–42. Ian and Carol McHarg Collection, The Architectural Archives of the University of Pennsylvania, 365.II.92. For further information on the work of Brian Spooner, see "Cultural Anthropology in Iran: Beginnings and Prospects" *Expedition* 13 no. 3 (Spring–Summer 1971): 66–71; Brian Spooner, "The Cultural Ecology of Pastoral Nomads" *Addison Wesley Module in Anthropology* 45 (1973): 1–53.

36 Wallace, McHarg, Roberts and Todd, *Pardisan, A Feasibility Study for an Environmental Park in Tehran, Appendix*, 1–11.

37 The Architectural Archives of the University of Pennsylvania, Ian and Carol McHarg Collection, "Ian's Notes on Pardisan Concept," 109.II.C.19.

38 Ibid.

39 Wallace, McHarg, Roberts and Todd, *Pardisan, A Feasibility Study for an Environmental Park in Tehran*, 6.

40 The Pardisan project files contain land use maps from *The Comprehensive Plan for Tehran*. WMRT would have known of this work through Nader Ardalan, the architectural consultant for the feasibility study. Ardalan had worked at Abdol-Aziz Farman-Farmaian Associates, the architecture firm owned by an uncle of Eskandar Firouz, during the period when the firm was involved

in the development of *The Comprehensive Plan*. Recommendations enumerated in *The Comprehensive Plan* include actions to control urban growth, relieve congestion, modernize inefficient infrastructure, ameliorate rural-urban migration, and mitigate air and water pollution. See R. Stephen Sennott, *Encyclopedia of Twentieth Century Architecture: Volume 1 A–F* (London: Fitzroy Dearborn, 2004), 62–65; and "Major Planning and Development Issues Affecting Tehran's Future," Victor Gruen Papers, Box 48, Folder 12, (Plan and Budget Organization Urban Development and Housing Department Tehran Development Council Secretariat, January 1976), Manuscript Division, Library of Congress; "The Comprehensive Plan for Tehran First Stage – Concept Development Study and Evaluation – Volume II," Box 45, Folder 13, Manuscript Division, Library of Congress; "The Comprehensive Plan for Tehran First Stage – Concept Development Study and Evaluation – Volume III," Box 46, Folders 1 and 14, Manuscript Division, Library of Congress; and Abdol-Aziz Farmanfarmaian, www.royalark.net/Persia/qajar9.htm.

41 Wallace, McHarg, Roberts and Todd, *A Feasibility Study for an Environmental Park in Tehran*, 6 and 34.

42 Ibid., 8.

43 Ibid.

44 Ibid., 12–82.

45 Ibid., 155.

46 Personal communication, Jahangir Sedaghatfar, April 28, 2014. The Department of Conservation became the Department of Environment following the 1974 passage of the Environmental Protection and Enhancement Act by the Iranian government. See Firouz, *The Complete Fauna of Iran*, 12–13. See also, Ronald Ferrier, "18 – The Iranian Oil Industry" *The Cambridge History of Iran from Nadir Shah to the Islamic Republic* 7 (1991): 639–702, doi:10.1017/CHOL9780521200950.

47 The Mandala Collaborative/Wallace, McHarg, Roberts and Todd, *Pardisan, Plan for an Environmental Park in Tehran, for the Imperial Government of Iran, Department of Environment*, (Philadelphia, PA: WMRT, 1975), 98. Ian and Carol McHarg Collection, 365.II.94.

48 Personal communication, Jahangir Sedaghatfar, April 28, 2014.

49 Franklin departed part way through this phase of the project, and as noted by Robinson Fisher, much of the master plan analysis work had been completed during the feasibility study. Personal communication Robinson Fisher, June 21, 2016.

50 The Architectural Archives of the University of Pennsylvania, Ian and Carol McHarg Collection, "Pardisan Storylines," 365.II.84.

51 A reference to Charles Darwin and the Galapagos finches appears in the storyline material. See The Architectural Archives of the University of Pennsylvania, Ian and Carol McHarg Collection, "World Wide Adaptive Strategies," 365.II.85, 117.

52 Robinson Fisher provided a copy of the world analog matrices that appear in the "World Wide Adaptive Strategies" and an explanation of how they worked. Personal communication, Robinson Fisher, July 21, 2016.

53 See Le Corbusier, "Description of the CIAM Grid, Bergamo 1949" in *The Heart of the City: Towards the Humanization of Urban Life*, eds. J. Tyrwhitt, J. L. Sert, E. N. Rogers (London: Lund Humphries & Co. Ltd., 1952), 175.

54 Personal communication, Robinson Fisher, July 21, 2016.

55 Mandala Collaborative/Wallace, McHarg, Roberts and Todd, *Pardisan, Plan for an Environmental Park in Tehran*, 10 and 29. The first page of the master plan report contains an image of Iran taken from an ERT satellite.

56 McHarg, *Design with Nature*, 43–53.

57 Wallace, McHarg, Roberts and Todd, "The Concept of Pardisan," The Papers of Charles & Ray Eames: Pardisan Environmental Park, Printed Matter, Box 83, Folder 13 (n.d.), Manuscript Division, Library of Congress.

58 The Mandala Collaborative/Wallace, McHarg, Roberts and Todd, *Pardisan, Plan for an Environmental Park in Tehran*, 45. McHarg's concept of sunlight, energy, and organization relates to trophic levels, an ecological concept that represents ecosystem productivity as a food chain that extends from producer organisms to consumer organisms. For further discussion of the food chain from a textbook of the era, see Paul Colinvaux, *Introduction to Ecology*, 129–134.

59 See Kathryn Babayan, *Mystics, Monarchs and Messiahs: Cultural Landscapes of Early Modern Iran* (Cambridge, MA: Harvard University Press, 2002), 491–492.

60 Mandala Collaborative/Wallace, McHarg, Roberts and Todd, *Pardisan, Plan for an Environmental Park in Tehran* 13.

61 Ibid., 59.

62 Personal communication, Robinson Fisher, July 21, 2016. As noted in a telephone memo from Glen Fleck to Charles Eames, the initial meeting was rescheduled due to the death of McHarg's wife Pauline.

63 The Mandala Collaborative/Wallace, McHarg, Roberts and Todd, *Technical Report*, 1975.

64 "Abstract of Meeting with Dr. David R. Goddard," *Technical Report*, 9–19.

65 Ibid., "Abstract of Meeting with Doctor Yehudi A. Cohen," 20–24, "Abstract of Meeting with Dr. Solomon H. Katz," 25–29, "Abstract of Meeting with Dr. Brian John Spooner," 30–36.

66 Ibid., "Abstract of Meeting with Dr. Seyyed Hossein Nasr Chancellor, Agra Mehr University, Tehran," 42–53. Nasr was born in Tehran. His father served as the physician to the Iranian royal family. He studied in the United States at the Peddie School in Hightstown, New Jersey, and later at M.I.T. He also received a MS in geology and geophysics, and a Ph.D. in the history of science from Harvard University. When he spoke to McHarg and the WMRT office regarding Pardisan, the Queen had recently appointed him the Director of the Iranian Imperial Academy of Philosophy. See The Seyyed Hossein Nasr Foundation, "About Seyyed Hossein Nasr," www.nasrfoundation.org/biography.html See also Seyyed Hossein Nasr, *The Encounter of Man and Nature: The Spiritual Crisis of Modern Man* (London: George Allen and Unwin Ltd., 1968); Seyyed Hossein Nasr, *Science and Civilization in Islam* (Cambridge, MA: Harvard University Press, 1968), and Tarik M. Quadir, *Traditional Islamic Environmentalism* (New York: University Press of America, 2013).

67 Ibid., 42–45.

68 Ibid.

69 Ibid., 47–50.

70 Ibid., 50.

71 Ibid., 52.

72 Ibid., "Abstract of Meeting with R. Buckminster Fuller," 1–8.

73 Ibid., 2–3. The wording of the passage, in keeping with Fuller's distinct jargon and advocacy of multiple pathways of perception, uses the term "seeing" instead of see.

74 Ibid., 4–5, 8.

75 "The Concept of Pardisan."

76 Personal communication, Jahangir Sedaghatfar, November 11, 2013. Personal communication, Nader Ardalan, September 17, 2013. Personal communication, Robinson Fisher, June 21, 2016.

77 The Eames Office hired Fleck in 1950. For further information see, Paul Schrader, "Poetry of Ideas: The Films of Charles Eames" *Film Quarterly* 23, no. 3 (Spring 1970): 2–19.

78 Pardisan was commonly known among the consultants as the Queen's project. Personal communication, Jahangir Sedaghatfar, April 28, 2014.

79 Storyboard sketches by Glenn Fleck. Box 83, Folder 13, Charles and Ray Eames Papers, Manuscript Division, Library of Congress, Washington.

80 The Architectural Archives of the University of Pennsylvania, Ian L. McHarg Collection, "Mandala Collaborative/Wallace, McHarg, Roberts Todd, and Glen Fleck, Project Film *Pardisan*, 1975," 109.V.B.2.2.

81 For Farah Diba Pahlavi's role in the promotion and popularization of art and culture in Iran, see Talinn Grigor, *Building Iran*; Bob Colacello, "Farah Pahlavi," *New York Magazine*, September (2013): 153–161, 188–190; and Robert Gluck, "The Shiraz Arts Festival: Western Avant-Garde Arts in 1970s Iran" *Leonardo* 40, no. 1 (2007): 20–28.

82 For further discussion of Persian heritage in Iranian architectural discourse, see Talinn Grigor, *Building Iran*; and Talinn Grigor, "Recultivating 'Good Taste: The Early Pahlavi Modernists and their Society for National Heritage" *Iranian Studies* 37 no. 1 (March 2004): 17–45.

83 Ardalan was born in 1939. As a child he moved from Tehran to Washington, D.C., following his father's diplomatic assignment to the Iranian Embassy in the United States. He studied architecture at the Carnegie Institute of Technology (Carnegie Mellon University) in Pittsburgh, Pennsylvania and received a BA degree from that institution in 1961. At this time, he also met and married Laleh Bakhtiar, an Iranian-American attending Chatham College. In 1962, he received a Master of Architecture degree from the GSD. After a short stint at the firm Skidmore, Owings and Merrill, he returned to Iran to work for the National Iranian Oil Company. In 1966, he entered into partnership with Abdol Aziz Farman – Farmaian. In 1972, the year that Firouz conceived of Pardisan and Sedaghatfar began his search for the lead consultant, Ardalan founded the Mandala Collaborative and became the architectural adviser for the project. During the preparation of the feasibility report, his contribution was minimal. Following the 1973 publication of *The Sense of Unity: The Sufi Tradition in Persian Architecture*, which he co-wrote with Bakhtiar, he assumed a more prominent role in the project. See *Encyclopedia of Twentieth Century Architecture A-F*, ed. R. Stephen Sennott (London: Fitzroy Dearborn, 2004), 62–65. See also, Davar Ardalan, *My Name is Iran* (New York: Henry Holt and Co., 2007); and Ardalan Associates, LLC, Consultants in Architecture, http://ardalanassociates.com/about/about-nader-ardalan/.

84 Nader Ardalan and Laleh Bakhtiar, *The Sense of Unity: The Sufi Tradition in Persian Architecture* (Chicago, IL: University of Chicago Press, 1973). For Nasr's relationship to Ardalan and Bakhtiar, see Grigor, *Building Iran*, 164–165. Though not officially recognized as a consultant in any of the Pardisan reports, Laleh Bakhtiar, in conjunction with Seyyed Hossein Nasr, developed a document titled "Persian Cosmologies." McHarg referenced this document in a letter dated April 10, 1979, addressed to Shahriar Rouhani, the *chargé d'affaires* of the Iranian Embassy in Washington. In the letter, he states that the Bakhtiar and Nasr document ensured Pardisan was "consonant with the Islamic concept of God-Man-Nature." See Letter from Ian McHarg to Shahriar Rouhani, The Architectural Archives of the University of Pennsylvania, Ian L. McHarg Collection, 109.III.C.15; and The Architectural Archives of the University of Pennsylvania, Ian and Carol McHarg Collection, Laleh Bakhtiar "Persian Epistemologies." 365.II.96. See also Laleh Bakhtiar, *Sufi Expressions of the Mystic Quest* (London: Thames and Hudson, 1976).

85 For further discussion of the dialogue between Ardalan and Kahn, and Kahn's subsequent commission for the new government and cultural center at Abbasabad (Shahestan Pahlavi), see Farshid Emami, "Civic Visions, National Politics, and International Designs: Three Proposals for a National Center in Tehran (1966–1976)," Master of science in architecture, Massachusetts Institute of Technology, 2011; and Farshid Emami, "Urbanism of Grandiosity: Planning a New Urban Centre for Tehran (1973–1976)" *International Journal of Islamic Architecture* 3 no. 1 (March 2014): 47–54. For further discussion of mid-20th-century structuralism, see Claus Dreyer, "Structural Approaches in Architectural Theory of the 1960s and 1970s" in *Structuralism Reloaded: Rule-Based Design in Architecture and Urbanism*, eds. Tomáš Valena, Tom Avermaete, and Georg Vrachliotis (London: Axel Menges, 2011), 40–45.

86 "Mandala Collaborative/Wallace, McHarg, Roberts Todd, and Glen Fleck, Project Film *Pardisan*, 1975."

87 Personal communication, Jahangir Sedaghatfar, April 28, 2014.

88 Personal communication, Jahangir Sedaghatfar, April 28, 2014.

89 Mandala Collaborative/Wallace, McHarg, Roberts and Todd, *Pardisan, Plan for an Environmental Park in Tehran*, 1975.

90 Personal communication, Jahangir Sedaghatfar, April 28, 2014. See also Persia: The Qajar Dynasty (Firouz, Farmanfarmaian, Farman-Farmaian, and Mossadeq).

91 See Nader Ardalan and Eskandar Firouz, *Management Plan Report for Pardisan. 1* (Tehran: The Mandala Collaborative, 1976). See also Sennott, *Encyclopedia of Twentieth Century Architecture*, 62–65.

92 The Architectural Archives of the University of Pennsylvania, Ian L. McHarg Collection, "Pardisan Project," 109.III.C.17, and 109.III.C.19. Ian L. McHarg Collection, The Architectural Archives of the University of Pennsylvania, 109.III.C.19; and Wallace, *Urban Planning My Way*, 193–196.

93 Personal communication Victoria Steiger, October 29, 2013. Personal communication Grant Jones, April 6, 2016. Jones noted that many of the staff members working on the schematic design were on loan from Collins, DuTot & Associates, a local Philadelphia firm that would later be known as The Delta Group.

94 Personal communication, Neal Belanger, June 21, 2016. See also, Pardisan Drawing Folder, Ian L. McHarg Collection, The Architectural Archives of the University of Pennsylvania, 109.V.C.16.

95 Personal communication Grant Jones, April 6, 2016. See also, The Architectural Archives of the University of Pennsylvania, Ian L. McHarg Collection, "Pardisan Drawing Folder," 109.V.C.16. See also David Hancocks, *A Different Nature: The Paradoxical World of Zoos and their Uncertain Future* (Berkeley: University of California Press, 2001), xv; and Jeffrey Hyson, "Jungles of Eden: The Design of American Zoos," in *Environmentalism in Landscape Architecture*, ed. Michel Conan (Washington, DC: Dumbarton Oaks Research Library and Collection, 2000).

96 See L. H. Holdridge, *Life Zone Ecology* (San Jose: Tropical Science Center, 1967), 1–51.

97 Personal communication, Grant Jones, March 7, 2014. Personal communication, Denis Paulson, March 7, 2014.

98 Personal communication, Denis Paulson, March 7, 2014.

99 Manuscript Division, Library of Congress, Washington, Charles and Ray Eames Papers, "October, 1974 Sketch by Colin Franklin of Parking Lot Terraces," Box 83, Folder 13.

100 The Faravahar is an ancient Persian symbol that appears prominently in the Achaemenid ruins of Persepolis. The Faravahar represents majesty, charismatic command, and patriarchal sovereignty. Today it is a fervent nationalist symbol. For further information on the Faravahar and its symbolic meaning and political significance see: Andrew Davidson. *The Word is my Home: A Hamid Dabashi Reader* (New York: Routledge, 1970).

101 See "Pardisan Drawing Folder."

102 See "Pardisan Project." See also, M. E. Macglashan and E. Lauterpacht, *Iran-United States Claims Tribunal Reports* 13 (Cambridge: Grotius Publications Ltd., 1988), 286–323; and Maryam Moradi, "Cooperation and Conflict Between Iran and America at the Iran-United States Claims Tribunal," Ph.D. diss., University of Exeter, 2010. The Algiers Agreement, which led to the release of the American embassy hostages, called for the creation of a tribunal to arbitrate disputes between the United States and Iran, and to ensure conflict resolution by peaceful means. The tribunal will cease to exist when Iran and the United States resume official diplomatic relations.

103 Wikimapia, "Pardisan Forest Park (Tehran)," http://wikimapia.org/1538316/Pardisan-Forest-Park.

104 See Jane Bennett, *Vibrant Matter: A Political Ecology of Things* (Durham, NC: Duke University Press, 2010).

Bibliography

Archives and Special Collections

Ford Foundation, Information Management Services.
Glen L. Martin Maryland Aviation Museum.
Rockefeller Archive Center.
Special Collections Department, Francis Loeb Library, Harvard Design School.
The Architectural Archives of the University of Pennsylvania, G. Holmes Perkins Papers.
The Architectural Archives of the University of Pennsylvania, Ian L. McHarg Collection.
The Architectural Archives of the University of Pennsylvania, Ian and Carol McHarg Collection.
The Architectural Archives of the University of Pennsylvania, Louis I. Kahn Collection.
The Graduate School of Design, Courses in Architecture, Landscape Architecture, City and Regional Planning, Archival Collections.
The Library of Congress Papers, Manuscript Division, Charles and Ray Eames Papers.
The Library of Congress, Moving Image Research Center, *The House We Live In*.
The University of Pennsylvania, School of Design Department of Landscape Architecture.

Books and Articles

Albert, Richard C. *Damming the Delaware: The Rise and Fall of Tock's Island*. University Park: The Pennsylvania State University Press, 1987.
Alofsin, Anthony. *The Struggle for Modernism: Architecture, Landscape Architecture, and City Planning at Harvard*. New York: W. W. Norton & Company, 2002.
Alsop, Stewart. "Dr. Calhoun's Horrible Mousery," *Newsweek* (August 17, 1970): 96.
Altschuler, Glenn C., and Stuart M. Blumin. *The GI Bill: The New Deal for Veterans*. New York: Oxford University Press, 2009.
Archives at Yale. "Paul Bigelow Sears, 1891–1990," Guide to the Paul Bigelow Sears Papers MS 663, https://archives.yale.edu/repositories/12/resources/4448, accessed March 4, 2019.
Ardalan Associates, LLC, Consultants in Architecture. http://ardalanassociates.com/about/about-nader-ardalan/, accessed March 4, 2019.

Ardalan, Davar. *My Name Is Iran*. New York: Henry Holt and Co., 2007.

Ardalan, Nader, and Eskandar Firouz. *Management Plan Report for Pardisan*. Tehran: The Mandala Collaborative, 1976.

Ardalan, Nader, and Laleh Bakhtiar. *The Sense of Unity: The Sufi Tradition in Persian Architecture*. Chicago, IL: University of Chicago Press, 1973.

Ashworth, William. *The Genesis of Modern British Planning*. London: Routledge & Kegan Paul Ltd., 1954.

Association for Planning and Regional Reconstruction. *Maps for the National Plan: A Background to The Barlow Report, The Scott Report, The Beveridge Report*. London: Lund Humphries & Co Ltd, 1945.

Babayan, Kathryn. *Mystics, Monarchs and Messiahs: Cultural Landscapes of Early Modern Iran*. Cambridge, MA: Harvard University Press, 2002.

Baigent, Elizabeth. "Patrick Geddes, Lewis Mumford and Jean Gottmann: Divisions Over Megalopolis," *Progress in Human Geography* 28 no. 6 (2004): 687–700.

Bakhtiar, Laleh. *Sufi Expressions of the Mystic Quest*. London: Thames and Hudson, 1976.

Bartlett, Jason. "Model Cities," *The Encyclopedia of Greater Philadelphia*, https:// philadelphiaencyclopedia.org/archive/model-cities/.

Bayer, Herbert, Walter Gropius, and Ise Gropius. *Bauhaus 1919–1928*. New York: The Museum of Modern Art, 1938.

Bellamy, Edward. *Looking Backward*. New York: Oxford University Press, 1888.

Beneš, Mirka. "Inventing a Modern Sculpture Garden in 1939 at the Museum of Modern Art, New York," *Landscape Journal* 13 no. 1 (Spring 1994): 1–20.

Bennett, Jane. *Vibrant Matter: A Political Ecology of Things*. Durham, NC: Duke University Press, 2010.

Bertalanffy, Ludwig von. *General System Theory: Foundations, Development, Applications*. New York: George Braziller, 1968.

Bibby, Cyril. "Huxley and the Reception of the 'Origin'," *Victorian Studies* 3 no. 1 Darwin Anniversary Issue (September, 1959): 76–86.

Blake, Peter. *God's Own Junkyard: The Planned Deterioration of America's Landscape*. New York: Holt, Rinehart and Winston, 1964.

Bourdeau, Michel. "Auguste Comte," in *The Stanford Encyclopedia of Philosophy*, ed. Edward N. Zalta (Summer 2018 Edition), https://plato.stanford.edu/ archives/sum2018/entries/comte/.

Brand, Stewart. *The Whole Earth Catalogue*. Menlo Park, CA: Nowels Publications, Fall 1968.

British War Office. "Correspondence Courses for Members of H. M. Forces. Part 1: Background to Planning, Part II: Planning Factors, and Part III: Planning Practice." British War Office, 1944 and 1945.

Bunster-Ossa, Ignacio F. *Reconsidering Ian McHarg: The Future of Urban Ecology*. Chicago, IL: American Planning Association Planners Press, 2014.

Burke, Edmund. *A Philosophical Inquiry into the Origin of Our Ideas of the Sublime and Beautiful*. New York: Oxford University Press, 1757, 2015.

Butler, Tom, Daniel Lerch, and George Wuerthner. *The Energy Reader: Over Development and the Delusion of Endless Growth*. Sausalito, CA: The Center for Deep Ecology, 2012.

Calhoun, John B. "Design for Mammalian Living," *The Architectural Association Quarterly* 1 no. 3 (1969): 24–35.

———. "Population Density and Social Pathology," *Scientific American* 206 no. 2 (February 1962): 139–146, 148.

———. "Population Density and Social Pathology," in *The Urban Condition: People and Policy in the Metropolis*, ed. Leonard Duhl. New York: Basic Books, Inc., 1963, 33–43.

———. "Space and the Strategy of the City," *Ekistics* 29 no. 175 City of the Future (July 1970): 425–437.

Carson, Rachel. *Silent Spring*. Boston, MA: Houghton Mifflin Company, 1962.

Cobb, Edith. "The Ecology of Imagination in Childhood," in *Daedalus* 8, no. 3 (Summer 1959): 537–548.

———. "The Ecology of Imagination in Childhood," in *The Subversive Science: Essays Toward an Ecology of Man*, eds. Paul Shepard and Daniel McKinley. Boston, MA: Houghton Mifflin Company, 1969, 122–132.

Colacello, Bob. "Farah Pahlavi," *New York Magazine* (September 2013): 153–161, 188–190.

Colinvaux, Paul A. *Introduction to Ecology*. New York: John Wiley & Sons, Inc.

Conklin, William, Robert Geddes, Ian L. McHarg, and Martin Sevely, *Collaborative Thesis Program*, 1949–1950. Special Collections Department, Francis Loeb Library, Harvard Design School.

Darling, F. Frazer. "Introduction," in *Future Environments of North America*, eds.F. Fraser Darling and John P. Milton, Garden City, NY: Natural History Press, 1966, 1–7.

Darwin, Charles. *The Origin of Species: A Variorum Text*, ed. Morse Peckham. Philadelphia: University of Pennsylvania Press, 1959.

———. *The Origin of Species by Means of Natural Selection or the Preservation of Favored Races in the Struggle for Life*. New York: Collier Books, 1974.

Davidson, Andrew. *The Word Is My Home: A Hamid Dabashi Reader*. New York: Routledge, 1970.

Davis, Timothy. "Rock Creek and Potomac Parkway, Washington, D.C.: The Evolution of a Contested Urban Landscape," *Studies in the History of Garden & Designed Landscapes* 19 no. 2 (1999): 123–237.

Delaware Online. "Nicholas Muhlenberg," www.legacy.com/obituaries/delaware-online/obituary.aspx?n=nicholas-muhlenberg&pid=145521555, accessed March 4, 2019.

DeVoto, Bernard. "Sacred Cows and Public Lands," *Harper's Magazine* 197 no. 1178 (July 1948): 44–55.

Dewey, John. *Art as Experience*. New York: The Berkeley Publishing Group, 1934, 1980.

Dewhurst, J. Frederick, ed. *America's Needs and Resources*. New York: The Twentieth Century Fund, 1947.

Division of Christian Education of the National Council of Churches in the United States of America. *The Holy Bible*, translated from the original tongues being the version set forth in A. D. 1161, and revised A. D. 1952. New York: Thomas Nelson & Sons, 1952.

Dodds, Felix, Michael Strauss, and Maurice Strong. *Only One Earth: The Long Road Via Rio to Sustainable Development*. London: Routledge, 2012.

Doxiadis, Constantinos. *Between Dystopia and Utopia*. Hartford, CT: The Trinity College Press, 1966.

Dreyer, Claus. "Structural Approaches in Architectural Theory of the 1960s and 1970s," in *Structuralism Reloaded: Rule-Based Design in Architecture and*

Urbanism, eds. Tomáš Valena, Tom Avermaete, and Georg Vrachliotis. London: Axel Menges, 2011, 44–45.

Duhl, Leonard. *The Urban Condition: People and Policy in the Metropolis.* New York: Basic Books, Inc., 1963.

Eames, Charles. "What Is a House?" *California Arts & Architecture Magazine* 61 (July 1944): 22–17.

Edinburgh University Library Special Collections. The Papers of Thomas Henry Huxley (1825–1895), https://archiveshub.jisc.ac.uk/data/gb237-coll-376, accessed March 4, 2019.

Edwards, Jonathan. "On Being," in *Jonathan Edwards's Sinners in the Hands of an Angry God: A Case Book*, eds. Wilson H. Kimnach, Caleb J. D. Maskell and Kenneth P. Minkema. New Haven, CT: Yale University Press, 2010, 52–53.

Eiseley, Loren. *Darwin's Century: Evolution and the Men Who Discovered It.* New York: Barnes & Noble, 1958, 2009.

Elliot, Charles William. *Charles Eliot Landscape Architect: A Lover of Nature and of His Kind Who Trained for a New Profession.* New York: Houghton Mifflin, 1902.

Elton, Charles. *Animal Ecology.* New York: Macmillan Co., 1927.

Emami, Farshid. "Civic Visions, National Politics, and International Designs: Three Proposals for a National Center in Tehran (1966–1976)," Master of Science in architecture thesis, Massachusetts Institute of Technology, 2011.

———. "Urbanism of Grandiosity: Planning a New Urban Centre for Tehran (1973–1976)," *International Journal of Islamic Architecture* 3 no. 1 (March 2014): 47–54.

Encyclopaedia Britannica. "Charles Lyell and Uniformitarianism," www.britannica.com/science/Earth-sciences/William-Smith-and-faunal-succession#ref292518, accessed March 4, 2019.

Fagg, C. C., and G. E. Hutchings. *An Introduction to Regional Surveying.* Cambridge: Cambridge University Press, 1930.

Farman Farmaian, Sattareh, and Dona Munker. *Daughter of Persia: A Woman's Journey from Her Father's Harem through the Islamic Revolution.* New York: Anchor Books, 1992.

Farney, Dennis. "Father Nature: How an Exuberant Scot, A Landscape Architect, Hopes to Shape the World," *The Wall Street Journal* (Monday, August 30 1971).

Ferrier, Ronald. "18 – The Iranian Oil Industry," *The Cambridge History of Iran from Nadir Shah to the Islamic Republic* 7 (1991): 639–702, doi:10.1017/CHOL9780521200950, accessed March 4, 2019.

Festinger, Leon, Stanley Schachter, and Kurt Black. *Social Pressures in Informal Groups: A Study of Human Factors in Housing.* Stanford, CA: Stanford University Press, 1950.

Firouz, Eskandar. *The Complete Fauna of Iran.* New York: I.B. Tauris, 2005.

———. *Environment Iran.* Tehran: Offset Press, 1974.

Fisher, R. A. *The Genetical Theory of Natural Selection.* Oxford: Oxford University Press, 1930.

Fitch, James Marston. *American Building: The Forces that Shape It.* Boston, MA: Houghton Mifflin Co., 1948.

Flower, Joe. *Healthcare Forum* "Building Healthy Cities: Excerpts from a conversation with Leonard J. Duhl, M.D," https://people.well.com/user/bbear/duhl.html, accessed March 4, 2019.

Frampton, Kenneth. *Modern Architecture: A Critical History.* London: Thames and Hudson, 1992.

Frank, Lawrence K. "The World as Communication Network," in *Sign, Symbol, Image,* ed. Gyorgy Kepes. New York: George Braziller, 1966, 1–14.

Galbraith, John Kenneth. *The Affluent Society.* Boston, MA: Houghton Mifflin Company, 1958.

———. *The Good Society: The Humane Agenda.* New York: Houghton Mifflin Company, 1996.

Gans, Herbert. *Urban Villagers: Group and Class Life in the Italian-Americans.* New York: Simon & Schuster, 1962.

Gargan, Edward, T. "Toynbee Revisited," *Time Inc.* 77 no. 20 (12 May 1961): 99–100.

Geddes, Patrick. *Cities in Evolution.* London: Williams & Norgate, LTD., 1949.

Giedion, Sigfried. *The Beginnings of Architecture: The A. W. Mellon Lectures in the Fine Arts, 1957, The National Gallery of Art, Washington, D.C.* Princeton, NJ: Princeton University Press, 1964.

———. "CIAM at Sea: The Background of the Fourth (Athens) Charter," in *Architects Yearbook 3,* ed. Trevor Dannatt. London: Paul Elek, 1949, 36–39.

———. *Space, Time and Architecture: The Growth of a New Tradition.* Cambridge, MA: The Harvard University Press, 1944.

Glacken, Clarence J. "Changing Ideas of the Habitable World," in *Man's Role in Changing the Face of the Earth,* ed. William L. Thomas, Jr. Chicago, IL: University of Chicago Press, 1956, 70–92.

Gleick, James. *The Information: A History, A Theory, A Flood.* New York: Pantheon Books, 2011.

Gluck, Robert. "The Shiraz Arts Festival: Western Avant-Garde Arts in 1970s Iran," *Leonardo* 40 no. 1 (2007): 20–28.

Goodman, Paul, and Percival Goodman. *Communitas: Ways of Livelihood and Means of Life.* New York: Vintage Books, 1947.

Gottmann, Jean. *Megalopolis: The Urbanized Northeastern Seaboard of the United States.* New York: The Twentieth Century Fund, 1961.

Gould, Stephen Jay. *Time's Arrow Time's Cycle: Myth and Metaphor in the Discovery of Geologic Time.* Cambridge, MA: Harvard University Press, 1987.

Grigor, Talinn. *Building Iran: Modernism, Architecture, and National Heritage Under the Pahlavi Monarchs.* New York: Periscope Publishing, 2009.

———. "Recultivating 'Good Taste': The Early Pahlavi Modernists and their Society for National Heritage," *Iranian Studies* 37 no. 1 (March 2004): 17–45.

Gropius, Walter. *Apollo in the Democracy: The Cultural Obligation of the Architect.* New York: McGraw Hill, 1957.

———. "Education," *Task* 1. Cambridge, MA: Harvard Graduate School of Design, 1941: 34–35.

———. *Rebuilding Our Communities.* Chicago, IL: Paul Theobald, 1945.

Grunwald, Henry Anatole. "Arnold Toynbee: Mapping of a Great Mind," *Life* 37 (29 November 1954): 87.

———. "Toynbee's Best Seller," *Life* 24 (23 February 1948): 118–124, 126, 128, 130, 133.

Haldane, J.B.S. "A Mathematical Theory of Natural and Artificial Selection, Part V: Selection and Mutation," *Mathematical Proceedings of the Cambridge Philosophical Society* 23 no. 7 (1927): 838–844.

Halpern, Orit. *Beautiful Data: A History of Vision and Reason Since 1945*. Durham, NC: Duke University Press, 2014.

Hamalian, Linda. *A Life of Kenneth Rexroth*. New York: W.W. Norton & Company, 1991.

Hancocks, David. *A Different Nature: The Paradoxical World of Zoos and their Uncertain Future*. Berkeley: University of California Press, 2001.

Haraway, Donna J. "A Manifesto for Cyborgs: Science, Technology and Social Feminism in the 1980s," in *Simians, Cyborgs and Women: The Reinvention of Nature*, ed. Donna Haraway. New York: Routledge, 1991, 149–181.

Harwood, William B. *Raise Heaven and Earth: The Story of Martin Marietta People and their Pioneering Achievements*. New York: Random House, 1993.

Henderson, Lawrence J. *The Fitness of the Environment: An Inquiry into the Biological Significance of the Properties of Matter*. Boston, MA: Beacon Press, 1913, 1958.

———. "The Fitness of the Environment, an Inquiry into the Biological Significance of the Properties of Matter," *The American Naturalist* 47 no. 554 (February 1913): 105–115.

———. *The Order of Nature*. Cambridge, MA: Harvard University Press, 1917.

Herber, Lewis. *Our Synthetic Environment*. New York: Alfred A. Knopf, 1962.

Ho, Mae-Wan. "What Is (Schrödinger's) Negentropy," (Modern Trends in Bio-ThermoKinetics 1994, Science in Society Archive, www.i-sis.org.uk/negentr. php, accessed March 4, 2019.

Holdridge, L. H. *Life Zone Ecology*. San Jose: Tropical Science Center, 1967.

Howard, Ebenezer. *Garden Cities of To-Morrow*. Cambridge, MA: The M.I.T. Press, 1898, 1965.

Hughes, Thomas P., and Agatha C. Hughes. *Lewis Mumford: Public Intellectual*. New York: Oxford Press, 1990.

Hutton, James. *The Theory of the Earth, with Proof and Illustrations Vol. III*. London: Geologic Society, Burlington House, 1788.

———. "The Theory of the Earth by James Hutton," www.gutenberg.org/ebooks/12861, accessed March 4, 2019.

Huxley, T. H. *Physiography: An Introduction to the Study of Nature*. London: MacMillan and Co., 1884.

———. "Science and Culture," in *Science and Education*, ed. Thomas Huxley. New York: D. Appelton and Company, 1898, 134–159.

Hyson, Jeffrey. "Jungles of Eden: The Design of American Zoos," in *Environmentalism in Landscape Architecture*, ed. Michel Conan. Washington, DC: Dumbarton Oaks Research Library and Collection, 2000, 23–44.

International Union for Conservation of Nature (IUCN). *Proceedings of a Technical Meeting on Wetland Conservation*, October 9–16, 1967 (Morges: UNESCO and the Commission on Ecology of the IUCN, 1968).

Isenstadt, Sandy, and Kishwar Rizvi. *Modernism and the Middle East: Architecture and Politics in the Twentieth Century*. Seattle: University of Washington Press, 2008.

Jacobs, Jane. *The Death and Life of Great American Cities*. New York: Random House, 1961.

James, William. "Great Men Great Thoughts, and the Environment," *Atlantic Monthly* 66 (1880): 441–459.

Jenkins, Stover, and David Mohney. *The Houses of Philip Johnson*. New York: Abbeville Press Publishers, 2001.

Jensen, Jens. *Shiftings*. Chicago, IL: Ralph Fletcher Seymour, Publisher, 1939.

Johnson, Lyndon Baines. "Special Message to the Congress on Conservation and Restoration of Natural Beauty, February 8, 1965," *Public Papers of the Presidents of the United States: Lyndon B. Johnson, 1965*. Volume I, entry 54 (Washington, DC: Government Printing Office, 1966). The LBJ Presidential Library, www.lbjlibrary.net/collections/selected-speeches/1965/02-08-1965.html, accessed March 4, 2019.

Kahn, Louis. "Form and Design," in *Louis Kahn: Essential Texts*, ed. Robert Twombly. New York: W.W. Norton & Company, 2003, 62–74.

———. "The Nature of Nature," in *Louis Kahn: Essential Texts*, ed. Robert Twombly. New York: W.W. Norton & Company, 2003, 119–122.

Kakuzo, Okakura. *The Book of Tea*. Rutland, VT: Charles E. Tuttle Company, 1956.

Kass, Ray. *Morris Graves: Vision of the Inner Eye*. New York: George Braziller, Inc., 1983.

Kelly, Michel, ed. *Encyclopedia of Aesthetics* 2. New York: Oxford University Press, 1998.

Kepes, Gyorgy, ed. *Module, Proportion, Symmetry, Rhythm*. New York: George Braziller, 1966.

Kingsland, Sharon E. *The Evolution of American Ecology 1890–2000*. Baltimore, MD: The Johns Hopkins University Press, 2005.

Kriesis, Paul. "A New Town Pattern is Born," in *Architects' Yearbook 6*, ed. Trevor Dannatt. London: Elek Books Limited, 1955, 53–74.

Kuhn, Thomas S. *The Structure of Scientific Revolutions*. Chicago, IL: The University of Chicago Press, 1962, Third Edition 1996.

Langer, Susanne K. *Feeling and Form: A Theory of Art*. New York: Charles Scribner's and Sons, 1953.

———. *Philosophy in a New Key*. Cambridge, MA: Harvard University Press, 1957.

Lâo-Tse. *Tao Te Ching*, trans D. C. Lau. New York: Penguin Books, 1963.

Laurence, Peter L. "The Death and Life of Urban Design: Jane Jacobs, The Rockefeller Foundation and the New Research in Urbanism, 1955–1965," *Journal of Urban Design* 11 no. 2 (June 2006): 145–172.

———. *Jane Jacobs, American Architectural Criticism and Urban Design Theory, 1935–1965*, PhD diss., (The University of Pennsylvania, 2009).

Le Corbusier. *The City of To-morrow and Its Planning*. New York: Dover Publications, Inc., 1929, 1987.

Leopold, Aldo. *A Sand County Almanac and Sketches Here and There*. New York: Oxford University Press, 1949.

Lewens, Tim. "Cultural Evolution," in *The Stanford Encyclopedia of Philosophy* (Summer 2018 Edition), ed. Edward N. Zalta, https://plato.stanford.edu/archives/sum2018/entries/evolution-cultural/, accessed March 4, 2019.

Lewis, Philip H. "The Landscape Resources of Wisconsin," in *The Wisconsin Blue Book*. ed. H. Rupert Theobald. Wisconsin Department of Resource Development, 1964. http://digicoll.library.wisc.edu/cgi-bin/WI/WI-idx?type=turn&id=WI.WIBlueBk1964&entity=WI.WIBlueBk1964.p0145&q1=Lewis, accessed March 4, 2019.

Lindeman, Raymond, "The Trophic-Dynamic Aspect of Ecology," *Ecology* 23 (October 1942): 399–418.

Lotka, Alfred J. *Elements of Physical Biology*. Baltimore, MD: Williams & Wilkins Company, 1925.

Lowenthal, David. "George Perkins Marsh and the American Geographical Tradition," *Geographic Review* 43 no. 2 (April 1953): 201–213.

———. "Nature and Morality from George Perkins Marsh to the Millennium," *Journal of Historical Geography* 26 no. 1(2000): 3–27.

Lyell, Charles. *Principles of Geology: Being an Attempt to Explain the Former Changes of the Earth's Surface by Reference to Causes Now in Operation, Vol. I and II.* London: William Clowes, 1830.

Lynch, Kevin. *The Image of the City.* Cambridge, MA: The M.I.T. Press, 1960.

MacArthur, Robert. "Fluctuations of Animal Populations and a Measure of Community Stability," *Ecology* 3 no. 36 (July 1955): 533–536.

MacArthur, Robert M., and E. O. Wilson. *The Theory of Island Biogeography.* Princeton, NJ: Princeton University Press, 1967.

Macglashan, M. E., and E. Lauterpacht. *Iran-United States Claims Tribunal Reports* 13. Cambridge: Grotius Publications Ltd., 1988, 286–323.

MacKaye, Benton. "Regional Planning and Ecology," *Ecological Monographs* 10 no. 3 (July 1940): 349–353.

———. *The New Exploration: A Philosophy of Regional Planning.* Urbana-Champaign: The University of Illinois Press, 1928, 1990.

———. "The New Exploration, Charting the Industrial Wilderness," *The Survey Graphic* LIV no. 1 (May 1 1925): 153–157.

———. "The Townless Highway," *The New Republic* (March 12, 1930): 93–95.

MacKaye, Benton, and Lewis Mumford. "Townless Highways for the Motorist: A Proposal for the Automobile Age," *Harpers Magazine* (August 1, 1931): 347–356.

Margulis, Lynn, Adam MacConnell, and James MacAllister. *The Lost Tapes of Ian McHarg: Collaboration with Nature.* White River Junction, VT: Chelsea Green Publishing, 2006.

Margulis, Lynn, James Corner, and Brian Hawthorne. *Ian McHarg Conversations with Students: Dwelling in Nature.* New York: Princeton Architectural Press, 2007.

Marsh, Benjamin C. "Causes of Congestion," in *Proceedings of the Second National Conference on City Planning and the Problem of Congestion, Rochester, New York, May 2–4, 1910.* Boston, MA: National Conference on City Planning. http://urbanplanning.library.cornell.edu/DOCS/marshpop.htm, accessed March 30, 2019.

Marsh, George Perkins. *Man and Nature: Or, Physical Geography Modified by Human Action.* New York: Charles Scribner & Co., 1864.

Marx, Karl. *Capital: A Critique of Political Economy Volume I, Part I, The Process of Capitalist Production,* trans. Friedrich Engels. New York: Appleton & Co., 1889.

Matalene, H. W. *Romanticism and Culture: A Tribute to Morse Peckham ad a Bibliography of his Work.* Columbia, SC: Camden House, 1984.

Matthews, G.V.T. *The Ramsar Convention on Wetlands: Its History and Development.* Gland: Ramsar Convention Bureau, 1993. www.ramsar.org/sites/default/files/documents/pdf/lib/Matthews-history.pdf, accessed March 4, 2019.

McCormick, Jack. *The Pine Barrens: A Preliminary Ecological Inventory.* Trenton, NJ: State Museum, 1971.

———. "Succession," in *Via 1, Ecology and Design,* eds. Sauer, James Bryan, Thomas Gilmore. New York: Grossman Publishers, Inc., 1968, 22–35.

McHarg, Ian. "Architecture in the Netherlands," *Quarterly Journal of the Royal Incorporation of Architects in Scotland* 94 (1953): 41–46.

———. "Can We Afford Open Space? A Survey of Landscape Costs," *Architects' Journal* 123 (March 1956): 260–275.

———. "Cape Hatteras National Seashore Recreation Area," Landscape Architecture Department, School of Fine Arts, University of Pennsylvania, 1956.

———. "The Court House Concept," *Architectural Record* 122 (September 1957): 193–200.

———. "The Courthouse Concept," in *Architects' Year Book 8*, ed. Trevor Dannatt. London: Elek Books Limited, 1957, 74–102.

———. *Design with Nature*. Garden City, NY: The Natural History Press, 1969.

———. "Ecology for the Evolution of Planning and Design," in *VIA 1: Ecology in Design*, ed. Rolf Sauer, James Bryan, Thomas Gilmore. New York: Grossman Publishers, Inc., 1968: 47–49.

———. "The Ecology of the City," *Journal of Architectural Education* 17 no. 2. The Architect and the City. The 1962 AIA0ACSA Seminar Papers Presented at the Cranbrook Academy of Art. Part I (November 1962): 101–103.

———. "The Ecology of the City," in *The Architect and the City: Papers from the AIA-ACSA Teacher Seminar Cranbrook Academy of Art, June 11–22, 1962*, ed. Marcus Whiffen. Cambridge, MA: The M.I.T. Press, 1966, 53–65.

———. "Ecological Determinism," in *Future Environments of North America*, eds. F. Frazer Darling and John P. Milton. Garden City, NY: The Natural History Press, 1966, 526–538.

———. "An Ecological Method for Landscape Architecture," *Landscape Architecture Magazine* 57 no. 2 (1967): 105–107.

———. *The Essential Writing of Ian McHarg: Writings on Design and Nature*, ed. Frederick R. Steiner. Washington, DC: Island Press, 2006.

———. "The Humane City: Must the Man of Distinction Move to the Suburbs?" *Landscape Architecture* XLVIII no. 2 (January 1958): 103–107.

———. "Man and Environment," in *The Urban Condition: People and Policy in the Metropolis*, ed. Leonard Duhl. New York: Basic Books, Inc., 1963, 44–58.

———. "Open Space and Housing," in *Architects' Yearbook 6*, ed. Trevor Dannatt. London: Elek Books Limited, 1955, 75–82.

———. *A Quest for Life: An Autobiography*. New York: John Wiley & Sons, Inc., 1966.

McHarg, Ian L., and Frederick R. Steiner. *To Heal the Earth: Selected Writings of Ian L. McHarg*. Washington, DC: Island Press, 1998.

McNeil, R. E., and Ian McHarg. *"Report on the Damage Caused to the Working of the Apulian Aqueduct by the German Army and on the Work Done to Repair the Aqueduct,"* October 14, 1943.

Meller, Helen. *Patrick Geddes Social Evolutionist and City Planner*. New York: Routledge, 1990.

Milani, Abbas. *The Shah*. New York: Palgrave MacMillan, 2011.

Milner, Edward J. "Providence Tomorrow?" *Providence Sunday Journal* (June 11, 1950): 5.

Moradi, Maryam. "Cooperation and Conflict between Iran and America at the Iran-United States Claims Tribunal," Ph.D. diss., University of Exeter, 2010.

Morris, Clarence. "The Rights and Duties of Beasts and Trees: A Law Teacher's Essay for Landscape Architects," *Journal Legal Education* 17 no. 185 (1964): 185–192.

Mumford, Eric. *The CIAM Discourse on Urbanism 1928–1960*. Cambridge, MA: The MIT Press, 2000.

Mumford, Lewis. *Art and Technics*. New York: Columbia University Press, 1952, 2000.

———. *The City in History: Its Origins, Its Transformations, and Its Prospects*. New York: Harcourt Brace Jovanovich, Publishers, 1961.

———. *The Culture of Cities*. New York: Harcourt Brace & Company, 1938, 1970.

———. "The Fourth Migration," *The Survey Graphic* LIV no. 1 (May 1, 1925): 130–133.

———. *The Golden Day: A Study in American Experience and Culture*. New York: Boni and Liveright Publishers, 1926.

———. "The Natural History of Urbanization," in *Man's Role in Changing the Face of the Earth*, ed. William L. Thomas, Jr. Chicago, IL: University of Chicago Press, 1956, 387–398.

———. "Regions – to Live In," *The Survey Graphic* LIV no. 1 (May 1, 1925): 151–152.

———. *Sticks and Stones: A Study of American Architecture & Civilization*. New York: W. W. Norton & Company, 1924.

———. "Summary Remarks: Prospect," in *Man's Role in Changing the Face of the Earth*, ed. William L. Thomas, Jr. Chicago, IL: University of Chicago Press, 1956, 1143–1149.

Myers, J. "A Biographical Memoir of Bessel Kok 1818–1979," (Washington, DC: National Academy of Sciences, 1987), www.nasonline.org/publications/biographical-memoirs/memoir-pdfs/kok-bessel.pdf, accessed March 4, 2019.

Nairn, Ian. *The American Landscape: A Critical View*. New York: Random House, 1965.

NASA Spinoff Technology Transfer Program, Consumer/Home/Recreation. "Nutritional Products from Space Research," https://spinoff.nasa.gov/spinoff1996/42.html, accessed March 4, 2019.

Nash, Nancy Lee. "Conservation: Bringing Back Wildlife," *Nature* 472 (April 21 2011): 290–291.

Nash, Roderick Frazier. *Wilderness and the American Mind*. New Haven, CT: Yale University Press, 1967.

Nasr, Seyyed Hossein. *The Encounter of Man and Nature: The Spiritual Crisis of Modern Man*. London: George Allen and Unwin Ltd., 1968.

———. *Science and Civilization in Islam*. Cambridge, MA: Harvard University Press, 1968.

National Park Service. "Civil War Defenses of Washington," www.nps.gov/cwdw/planyourvisit/upload/Civial%20War%20defenses%20of%20Washington_final.pdf, accessed March 4, 2019.

New York Public Library Archives and Manuscripts, Century Foundation Records, "Biographical/Historical Records" http://archives.nypl.org/mss/18811, accessed March 4, 2019.

Ndubisi, Forster. *Ecological Planning: A Historical Approach*. Baltimore, MD: John Hopkins University Press, 2002.

Nelson, George. "The Modern House in America," *The Architectural Forum Magazine of Building* 71 no. 1 (July 1939).

Neuhart, John, Marilyn Neuhart, and Ray Eames. *Eames Design: The Work of the Office of Charles and Ray Eames*. New York: Harry H. Abrams, Inc. Publishers, 1989.

News Bureau Martin Company, A Division of the Martin Marietta Company. Press Release, Gravity Independent Photosynthetic Gas Exchanger, GIPSE Release

No. 3342-3163 (March 1, 1963), Courtesy of the Glen L. Martin Maryland Aviation Museum.

Newton, Norman T. *An Approach to Design*. Cambridge, MA: Addison-Wesley Press, Inc., 1951.

———. *Design on the Land: The Development of Landscape Architecture*. Cambridge, MA: The Belknap Press of Harvard University Press, 1971.

New York City Department of Parks and Recreation. "Freshkills Park," www.nycgovparks.org/park-features/freshkills-park, accessed March 4, 2019.

Odum, Eugene P. *Fundamentals of Ecology*. Philadelphia, PA: W. B. Saunders Company, 1953.

Olgyay, Victor. "The Temperate House," *The Architectural Forum Magazine of Building* 94 no. 3 (March 1951): 179–197.

Oliver, Robert W., and the Oral History Research Office, Columbia University, 1968, The World Bank, Oral History, "Crena de Iongh, Daniel," 1961 transcripts, http://oralhistory.worldbank.org/person/crena-de-iongh-daniel, accessed March 4, 2019.

Olmsted, Frederick Law. *Civilizing American Cities: Writings on Landscape*, ed. S. B. Sutton. New York: Da Capo Press, 1997.

Olmsted, Vaux, and Company. *Observations on the Progress of Improvements in Street Plans, with Special Reference to the Parkway Proposed to Be Laid Out in Brooklyn*. Brooklyn, NY: I. van Anden's Print, 1868.

Ortega y Gasset, José. *The Revolt of the Masses*. New York: W.W. Norton & Company, 1932.

Ostroff, Daniel. *An Eames Anthology*. New Haven, CT: Yale University Press, 2015.

Padian, Kevin. "From Geologizer to Theorist," *Bioscience* 68 no. 12 (December 2018): 1020–1021.

PBS. "Lady Bird Johnson, The Beautification Campaign," www.pbs.org/ladybird/shattereddreams/shattereddreams_report.html, accessed March 4, 2019.

Pearlman, Jill. "Joseph Hudnut's Other Modernism at the Harvard Bauhaus," *The Journal of the Society of Architectural Historians* 56 no. 4 (December 1997): 452–477.

Peckham, Morse. "Darwinism and Darwinisticism," *Victorian Studies* 3 no. 1 Darwin Anniversary Issue (September 1959): 19–40.

Pennsylvania Historical & Museum Commission. *Pennsylvania Architectural Field Guide*, "Second Empire/Mansard Style 1860–1900," www.phmc.state.pa.us/portal/communities/architecture/styles/second-empire.html, accessed March 4, 2019.

Persia: The Qajar Dynasty (Firouz, Farmanfarmaian, Farman-Farmaian, and Mossadeq) Genealogy, www.royalark.net/Persia/qajar9.htm, accessed March 4, 2019.

Peterson, Jon A. "The Birth of Organized City Planning in the United State, 1909–1910," *Journal of the American Planning Association* 75 no. 2 (Spring 2009): 123–133.

Potomac Planning Task Force. *The Potomac: A Report on Its Imperiled Future and a Guide for Orderly Development*. Washington, DC: United States Government Printing Office, 1997.

Quadir, Tark M. *Traditional Islamic Environmentalism*. New York: University Press of America, 2013.

Ramsden, Edmund, and Jon Adams. "Escaping the Laboratory: The Rodent Experiments of John B. Calhoun and their Cultural Influence," *The Journal of Social History* 42 no. 3 (March 1, 2009): 761–797.

Rees, Matthew C. "The Structure of Scientific Revolutions at Fifty," *The New Atlantis* 37 (Fall 2012): 71–86.

Reps, John W. "William Penn and the Planning of Philadelphia," *The Town Planning Review* 27 no. 1 (April 1956): 27–39.

Research Institute for Advanced Science. *1957 Annual Report.*

Rexroth, Kenneth. "The Visionary Paintings of Morris Graves," *Perspectives USA* 10 (Winter 1955): 58–66.

Rifkin, Jeremy. *Entropy: A New World Order.* New York: Bantam Books, 1981.

Riis, Jacob A. *How the Other Half Lives.* New York: Charles Scribner's Sons, 1890.

Rizvi, Kishwar. "Art History and the Nation: Arthur Upham Pope and the Discourse on 'Persian Art' in the Early Twentieth Century," *Muqarnas an Annual of the Visual Culture of the Islamic World* 24 no. 1 (2007): 45–67.

Rodenwaldt, Gerhardt. *The Acropolis.* Oxford: Blackwell Publishing, 1957.

Sabatino, Michelangelo. *Pride in Modesty: Modernist Architecture and the Vernacular Tradition in Italy.* Toronto: University of Toronto Press, 2011.

Salvini, R. "The Great Atlantic Storm of 1962," (March 6, 2012), www.njtvonline. org/news/uncategorized/the-great-atlantic-storm-of-1962/, accessed March 4, 2019.

Santayana, George. *The Philosophy of Santayana,* ed. Irwin Edman. New York: Random House, The Modern Library, 1936.

Sargent, Winthrop. "Mystic Painters of the Northwest," *Life Magazine* (September 28, 1953): 84–89.

Schayegh, Cyrus. "Iran's Karaj Dam Affair: Mass Consumerism, the Politics of Promise, and the Cold War in the Third World," *Comparative Studies in Society and History* 54 no. 3 (2012): 612–643.

Schneider, Peter. "Louis Kahn and the Little Book of Tea: Echoes of the Tao Te Ching in Louis Kahn's Thought," *International Journal of Humanities and Social Science* 14 no. 2 (Special Issue – July 2012): 22–27.

Schrader, Paul. "Poetry of Ideas: The Films of Charles Eames," *Film Quarterly* 23, no. 3 (Spring 1970): 2–19.

Schrödinger, E. *What Is Life?* Cambridge: Cambridge University Press, 1944.

Schwarzenbach, Alexis. *Saving the World's Wildlife: WWF – The First 50 Years.* London: Profile Books, 2011.

Schweitzer, Albert. *Out of My Life and Thought: An Autobiography,* trans. C.T. Campion. New York: Henry Holt, 1946.

Sears, Paul B. *Deserts on the March.* Norman: University of Oklahoma Press, 1935.

———. "The Future of the Naturalist," *American Naturalist* 78 no. 774 (January-February 1944): 43–53.

———. *The Living Landscape.* New York: Basic Books, 1962.

———. "The Processes of Environmental Change by Man," in *Man's Role in Changing the Face of the Earth,* ed. William L. Thomas, Jr., Chicago, IL: University of Chicago Press, 1956, 471–484.

Sedaghatfar, Jahangir. *Entopia.* Kent, OH: Kent State School of Architecture, 1971.

Sennott, R. Stephen, ed. *Encyclopedia of Twentieth Century Architecture: Volume 1 A-F.* London: Fitzroy Dearborn, 2004.

Selye, Hans. "The General Adaptation Syndrome and the Diseases of Adaptation," *The Journal of Allergy and Clinical Immunology* 17 no. 4 (July 1946): 358–398.

Sert, Josep Lluís. "Centres of Community Life," in *The Heart of the City: Towards the Humanization of Urban Life*, eds. J. Tyrwhitt, J. L. Sert, and E. N. Rogers. London: Lund Humphries & Co. Ltd., 1952, 3–16.

Shannon, C. E. "A Mathematical Theory of Communication," *The Bell System Technical Journal* 27 (July, October 1948): 379–423, 623–656.

Shaw, John Mackintosh. "Arnold Toynbee's Evaluation of the Place of Christianity in Religious History," *Time Inc.* 69 no. 9 (9 June 1958): 274–277.

Shepard, Paul. "The Place of Nature in Man's World," *Atlantic Naturalist* (April 1958): 85–89.

Shepard, Paul, and Daniel McKinley, eds. *Environmental Essays on the Planet as a Home*. Boston, MA: Houghton Mifflin Company, 1963.

Shoshkes, Ellen. "Jacqueline Tyrwhitt Translates Patrick Geddes for Post World War Two Planning," *Landscape and Urban Planning* 166 (2017): 15–24.

Silver, Noel. *Encyclopaedia Iranica*, "Pope, Arthur Upham," July 20, 2005, www.iranicaonline.org/articles/pope-arthur-upham, accessed March 4, 2019.

Smil, Vaclav. *Energy a Beginners Guide*. Oxford: One World Publications, 2006.

Smithsonian Institution, *South Mall Campus Cultural Landscape Report*. Washington, DC: Smithsonian Institution, June 2018.

Smoak, Harry. *Meaning as Response: Experience, Behavior, and Interactive Environment Design*, Ph.D. Diss., Concordia University (2015), 10.

Spencer, Herbert. *First Principles*. New York: A. L. Burt Company, Publishers, 1862.

———. *The Principles of Biology Vol. 1*. London: William and Norgate, 1864.

———. "The Proper Sphere of Government, Letter IX, Print of a Series of Letters," Originally Published in *The NonConformist* (London: John Green, 1843), 34–35, https://play.google.com/store/books/details?id=UKlXAAAAcAAJ&rdid=book-UKlXAAAAcAAJ&rdot=1, accessed March 4, 2019.

Spengler, Oswald. *The Decline of the West: Perspectives of World-History Vol. 2*. New York: Alfred A. Knopf, 1945.

Spirn, Whiston Spirn. "Ian McHarg, Landscape Architecture, and Environmentalism," in *Environmentalism in Landscape Architecture*, ed. Michel Conan. Washington, DC: Dumbarton Oaks Research and Library Collection, 2000, 97–114.

Spooner, Brian. "Cultural Anthropology in Iran: Beginnings and Prospects," *Expedition* 13 no. 3 (Spring–Summer 1971): 66–71.

———. "The Cultural Ecology of Pastoral Nomads," *Addison Wesley Module in Anthropology* 45 (1973): 1–53.

Srole, Leo, Thomas S. Langner, Stanley T. Michael, Marvin K. Opler, and Thomas A.C. Rennie. *Mental Health in the Metropolis: The Midtown Manhattan Study*. New York: McGraw-Hill Book Company, Inc., 1962.

Stein, Clarence S. "Dinosaur Cities," *The Survey Graphic* LIV no. 1 (May 1, 1925): 134–138.

Steinitz, Carl, Paul Parker, and Lawrie Jordan. "Hand-Drawn Overlays: Their History and Prospective Uses," *Landscape Architecture* 66 (September 1976): 444–445.

Strange, John H. "Citizen Participation in Community Action and Model Cities Programs," *Public Administration Review* 32 Special Issue Curriculum Essays on Citizens, Politics, and Administration in Urban Neighborhoods (October 1972): 655–669.

Strong, Ann L., and George Thomas, eds. *The Book of School: 100 Years*. Philadelphia: University of Pennsylvania Graduate School of Education, 1990.

The American Presidency Project. "Address of Senator John F. Kennedy Accepting the Democratic Party Nomination for the Presidency of the United States – Memorial Coliseum, Los Angeles, July 15, 1960," www.presidency.ucsb.edu/node/274679, accessed March 4, 2019.

The American Presidency Project. "Lyndon B. Johnson 576- Remarks at the Signing of the Highway Beautification Act, October 22, 1965," www.presidency.ucsb.edu/node/241177, accessed March 4, 2019.

The Architectural Forum Magazine of Building, "Measure" 89 no. 3 (November 1948).

The Aspen Institute. "A Brief History of the Aspen Institute," www.aspeninstitute.org/about/heritage/, accessed March 4, 2019.

The British Academy. "Jean Gottmann 1915–1994," *Proceeding of the British Academy, Biographical Memoirs of Fellows, II*, ed. P. J. Marshall. Oxford: Oxford University Press Scholarship Online, 2012: 201–218.

The Century Foundation, "About the Twentieth Century Fund," https://tcf.org/about/, accessed March 4, 2019.

The Comprehensive Plan for Tehran. "The Comprehensive Plan for Tehran First Stage – Concept Development Study and Evaluation – Volume II," and "The Comprehensive Plan for Tehran First Stage – Concept Development Study and Evaluation – Volume III," (Plan and Budget Organization Urban Development and Housing Department Tehran Development Council Secretariat, January 1976).

The Delaware River Basin Commission "The Role of the Delaware River Master in Interstate Flow Management," http://nj.gov/drbc/programs/flow/river_master.html, accessed March 4, 2019.

———. *Delaware River Basin Compact* (1961, Reprinted 2009), www.nj.gov/drbc/library/documents/compact.pdf, accessed March 4, 2019.

The Ford Foundation. Annual Report 1965, "Resources and Environment Grant Funding," New York: Hillison & Etten Company, 1965.

———. Annual Report 1967, "Resources and Environment Grant Funding," New York: Hillison & Etten Company, 1967.

The LBJ Presidential Library. "President Lyndon Baines Johnson, 'Special Message to the Congress on Conservation and Restoration of Natural Beauty, February 8, 1965,'" www.lbjlibrary.net/collections/selected-speeches/1965/02-08-1965.html, accessed March 4, 2019.

The Mandala Collaborative/Wallace, McHarg, Roberts and Todd, *Pardisan, Plan for an Environmental Park in Tehran, for the Imperial Government of Iran, Department of Environment, 1975*. Philadelphia, PA: WMRT, 1975.

———. *Pardisan Storylines*. Philadelphia, PA: WMRT, 1975.

———. *Plan for an Environmental Park in Tehran, Iran for the Imperial Government of Iran, Technical Report*. Philadelphia, PA: WMRT, 1975.

———. *World Wide Adaptive Strategies*. Philadelphia, PA: WMRT, 1975.

The Mandala Collaborative/Wallace McHarg Roberts Todd, and Glen Fleck. *Project Film Pardisan*. Philadelphia, PA: WMRT, 1975.

The National Academies of Sciences Engineering and Medicine, Evolution Resources. "Definitions of Evolutionary Terms," www.nas.edu/evolution/Definitions.html, accessed March 4, 2019.

The New York Times Archives. "Daniel Crena de Iongh, 82, Dies; Served World Development Bank, November 28, 1970," www.nytimes.com/1970/11/28/

archives/daniel-crena-de-iongh-82-dies-served-world-development-bank.html, accessed March 4, 2019.

The Seyyed Hossein Nasr Foundation, "About Seyyed Hossein Nasr," www. nasrfoundation.org/biography.html, accessed March 4, 2019.

The United Nations. "Declaration of the United Nations Conference on the Human Environment," in *Report of the United Nations Conference on the Human Environment, Stockholm, 5–16 June 1972*. New York: The United Nations, 1973.

Thomas, William L., ed. *Man's Role in Changing the Face of the Earth*. Chicago, IL: The University of Chicago Press, 1956.

Tichi, Cecelia. *Embodiment of a Nation: Human Form in American Places*. Cambridge, MA: Harvard University Press, 2001.

Teilhard de Chardin, Pierre. *Man's Place in Nature*. New York: Harper & Row, Publishers, 1956.

———. *The Phenomenon of Man*. New York: Harper & Row Publishers, 1959.

Toynbee, Arnold. *An Historian's Approach to Religion*. New York: Oxford University Press, 1956.

———. *A Study of History Abridgement of Volumes I–VI*. New York: Oxford University Press, 1974.

Trieb, Marc. "Nature Recalled," in *Rediscovering Landscapes: Essays in Contemporary Landscape Architecture*, ed. James Corner. New York: Princeton Architectural Press, 1999, 29–44.

Turner, Frederick Jackson. "Frederick Jackson Turner: The Significance of the Frontier in American History 1993, Excerpts," http://nationalhumanitiescenter. org/pds/gilded/empire/text1/turner.pdf, accessed March 4, 2019.

———. *The Frontier in American History*. New York: Henry Holt and Company, 1921: 321. Project Gutenberg eBook #22994, www.gutenberg.org/ files/22994/22994-h/22994-h.htm, accessed March 4, 2019.

Tyng, Anne Griswold. "Simultaneous Randomness and Order: The Fibonacci-Divine Proportion as a Universal Forming Principle." University of Pennsylvania, Ph.D., 1975, Architecture.

Tyrwhitt, Jacquelyn. "Town Planning," in *Architects Yearbook 1*, eds. Trevor Dannatt and Jane Drew. London: Elek Books Limited, 1945, 11–29.

Tyrwhitt, J., J. L. Sert, and E. N. Rogers, eds. *The Heart of the City: Towards the Humanization of Urban Life*. London: Lund Humphries & Co. Ltd., 1952.

Udall, Stewart L. *The Quiet Crisis*. New York: Holt, Rinehart and Winston, 1963.

University of Pennsylvania. "Dr. Muhlenberg, Penn Design," www.upenn.edu/ emeritus/memoriam/Muhlenberg.html, accessed March 4, 2019.

Van Eyck, Aldo, Paul Parin, and Fritz Morgenthaler. "A Miracle of Modernization," in *Via 1, Ecology in Design*, eds. Rolf Sauer, James Bryan, Thomas Gilmore. New York: Grossman Publishers, Inc., 1968, 96–125.

Van Roosmalen, Pauline K. M. "London 1944: Greater London Plan," in *Mastering the City: Northern European City Planning, 1900–2000*, eds. Koos Bosma and Helma Hellinga. Rotterdam: NAi Publishers, 1997, 266–273.

Van Wyhe, John. *The Complete Work of Charles Darwin Online*, "Catalogue of the *Beagle* Library," (2000), Darwin Online http://darwin-online.org.uk/ BeagleLibrary/Beagle_Library_Catalogue.htm, accessed March 4, 2019.

Waddington, C. H. "The Modular Principle and Biologic Form," in *Module, Proportion, Symmetry, Rhythm*, ed. Gyorgy Kepes. New York: George Braziller, 1966, 20–37.

————. Waddington, C. H. *The Strategy of the Gene*. London: Allen and Unwin, 1957.

Wallace, David. A. *Urban Planning My Way*. Chicago, IL: Planners Press, 2004.

Wallace, David, et al. *Metropolitan Open Space and Natural Processes*. Philadelphia: University of Pennsylvania Press, 1970.

Wallace-McHarg Associates. *Plan for the Valleys*. Philadelphia, PA: Wallace-McHarg, Associates, 1964.

Wallace McHarg Roberts, and Todd. *Pardisan, A Feasibility Study for an Environmental Park in Tehran, Iran, for the Imperial Government of Iran*. Philadelphia, PA: WMRT, 1973.

————. *Pardisan, A Feasibility Study for an Environmental Park in Tehran, Iran, for the Imperial Government of Iran, Appendix*. Philadelphia, PA: WMRT, 1973.

————. *A Comprehensive Highway Selection Method Applied to I-95 Between Delaware and Raritan Rivers*. Philadelphia, PA: WMRT, 1965.

————. "The Concept of Pardisan," Philadelphia: WMRT, 1973.

————. *The Least Social Cost Corridor for Richmond Parkway*. Philadelphia, PA: WMRT, 1968.

————. *Towards a Comprehensive Landscape Plan for Washington, D.C.* Philadelphia, PA: WMRT, 1967.

Ward, Barbara and René Dubois. *Only One Earth: The Care and Maintenance of a Small Planet*. New York: W.W. Norton & Company, 1972.

Warner, Sam Bass. *The Urban Wilderness: A History of the American City*. Berkeley: University of California Press, 1972.

Watts, Alan W. "Beat Zen, Square Zen, and Zen," *Chicago Review* 12, no. 2 (Summer 1958): 3–11.

Weindling P. "Julian Huxley and the Continuity of Eugenics in Twentieth-century Britain," *Journal of Modern European History* 10 no. 4 (November 1, 2012): 480–489.

Welter, Volker. *Biopolis*. Cambridge, MA: The M.I.T. Press, 2002.

————. "The Valley Region: From Figure of Thought to Figure on the Ground," in *New Geographies: Grounding Metabolism*, eds. Daniel Ibanez and Nikos Katsikis. Cambridge, MA: Harvard University Press, 2014, 78–87.

Wenner-Gren Foundation. "History, 'Man's Role in Changing the Face of the Earth,'" www.wennergren.org/history/mans-role-changing-face-earth, accessed March 4, 2019.

Wheaton, William L. C. "Jean Gottmann. Megalopolis: The Urbanized Northeastern Seaboard of the United States," *The Annals of the American Academy of Political and Social Science* 341 no. 1 (May, 1962): 166–167.

White House Conference on Natural Beauty. *Beauty for America: Proceedings of the White House Conference on Natural Beauty, Washington, D.C., May 24–25, 1965*. Washington, DC: US Government Printing Office, 1965.

Whitman, Walt, eds. William Everson, and James David Hart. *American Bard the Original Preface to Leaves of Grass*. New York: Viking, 1982.

Wikimapia. "Pardisan Forest Park (Tehran)," http://wikimapia.org/1538316/Pardisan-Forest-Park, accessed March 4, 2019.

Williams, Raymond. *Keywords: A Vocabulary of Culture and Society*. New York: Oxford University Press, 1983.

Williamson, G. Scott. "The Individual and the Community," in *The Heart of the City: Towards the Humanization of Urban Life*, eds. J. Tyrwhitt, J. L. Sert, and E. N. Rogers. London: Lund Humphries & Co. Ltd., 1952, 30–35.

Williamson, G. Scott, and I. H. Pearse. *Biologists in Search of Material: An Interim Report on the Work of The Pioneer Health Center.* London: Faber & Faber Limited, 1938.

Wilson, Edward O., and G. Evelyn Hutchinson. *"Robert Helmer MacArthur 1930–1972."* National Academy of Sciences, Washington, DC: National Academy of Sciences, 1989. www.nasonline.org/publications/biographical-memoirs/memoir-pdfs/mac-arthur-robert-h.pdf, accessed March 4, 2019.

Wolfe, Tom. *From Bauhaus to Our House.* New York: Picador, Farrar, Straus and Giroux, 1981.

World Wildlife Fund. "History," www.worldwildlife.org/about/history, accessed March 4, 2019.

Worster, Donald. *Nature's Economy: A History of Ecological Ideas.* New York: Cambridge University Press, 1997.

Woudstra, Jan. "The 'Sheffield Method and the First Department of Landscape Architecture in Great Britain," *Garden History* 38 no. 2 (Winter 2010): 242–266.

Wurster, William. "The Outdoors in Residential Design," *Architectural Forum* (September 1949): 68–69.

Zalduendo, Ines Maria. "Jacqueline Tyrwhitt's Correspondence Courses: Town Planning in the Trenches." Digital Access to scholarship, https://dash.harvard.edu/handle/1/13442987, accessed March 4, 2019.

Index

Note: *Italic* page numbers refer to figures and page numbers followed by "n" denote endnotes.